Communications
in Computer and Information Science 1345

Editorial Board Members

More information about this series at http://www.springer.com/series/7899

Ashish Awasthi · Sunil Jacob John ·
Satyananda Panda (Eds.)

Computational Sciences – Modelling, Computing and Soft Computing

First International Conference, CSMCS 2020
Kozhikode, Kerala, India, September 10–12, 2020
Revised Selected Papers

Springer

Editors
Ashish Awasthi
National Institute of Technology Calicut
Kozhikode, Kerala, India

Sunil Jacob John
National Institute of Technology Calicut
Kozhikode, Kerala, India

Satyananda Panda
National Institute of Technology Calicut
Kozhikode, Kerala, India

ISSN 1865-0929   ISSN 1865-0937 (electronic)
Communications in Computer and Information Science
ISBN 978-981-16-4771-0   ISBN 978-981-16-4772-7 (eBook)
https://doi.org/10.1007/978-981-16-4772-7

This Springer imprint is published by the registered company Springer Nature Singapore Pte Ltd.
The registered company address is: 152 Beach Road, #21-01/04 Gateway East, Singapore 189721, Singapore

Preface

It is indeed a pleasure for us to present the CCIS proceedings of The International Conference on Computational Sciences—Modelling, Computing, and Soft Computing (CSMCS 2020) to the authors, participants, speakers, faculty members, research scholars, and interested persons from industry and academia. We sincerely hope that everyone will find this proceedings useful, insightful, and motivating.

CSMCS 2020 aimed to bring together the leading academicians, scientists, and researchers from academia and industry to share their experiences and views in the field of computational sciences. CSMCS 2020 provided a platform for scientists and engineers to discuss their research findings including applications to industry in all areas under the theme of the conference. The conference was held virtually (online) during September 10–12, 2020, at the National Institute of Technology Calicut, India, due to the prevailing COVID-19 situation. There were three simultaneous related tracks, namely modelling, computing, and soft computing. All papers received in CSMCS 2020 have gone through a rigorous review process consisting of three stages. At the first stage review have been done by the Abstract Review Committee (ARC), program chairs have done at the second stage and at the final third stage papers were sent for peer reviews. The acceptance rate for the CCIS is less than 20%.

Mathematical modeling is one of the most ubiquitous branches of mathematics. This track's scope lay in the formulation, analysis, and simulation of mathematical models of real-world problems. Two papers are included in the proceedings under this track.

Computational sciences have given rise to the significant development in advanced techniques in modeling and simulations. This track was aimed at catering to the needs of modern and advanced industries, and it included computational partial differential equations (PDEs), computational finance, and computational complexity, amongst other topics. The proceedings include 12 papers presented in this conference track.

Soft computing refers to a collection of computational techniques that study, model, and analyze complex systems for which the conventional methods have not yielded the best solutions. This track focused on the theoretical basis of non-standard reasoning; research in fuzzy logic programming; design of calculi for uncertainty, imprecision, and vagueness; soft sets; rough sets; multisets; artificial neural networks, etc. Five papers are included from this track.

Also, two interdisciplinary papers related to general computing are included in the proceedings, in addition the above mentioned tracks.

We sincerely express our gratitude to our fellow members of the Program Committee and the members of the Scientific Advisory Committee for their consistent support and guidance throughout this conference. We are also thankful to our speakers, chairs, and participants for their valuable time and effort. We also acknowledge the support rendered by our reviewers to select the best possible papers for this proceedings. We put on record our thanks to all authors for contributing their research to CSMCS 2020. We also place on record our heartfelt thanks to the National Institute of Technology Calicut, India, for providing the institute's infrastructural facilities and the financial support for this conference under TEQIP Phase III. We are most appreciative of the help provided Springer's CCIS team for assembling the conference proceedings.

January 2021   Ashish Awasthi
                                                   Sunil Jacob John
                                                  Satyananda Panda

Organization

Patron

P. S. Sathidevi   National Institute of Technology Calicut, India

Conveners

Ashish Awasthi	National Institute of Technology Calicut, India
Satyananda Panda	National Institute of Technology Calicut, India
Sunil Jacob John	National Institute of Technology Calicut, India

Program Committee

A. K. Nandakumaran	IISc Bangalore, India
A. K. Pani	IIT Bombay, India
Aleksander Grm	University of Ljubljana, Slovenia
Alpesh Kumar	Rajiv Gandhi Institute of Petroleum Technology, India
Appadu Appanah Rao	Nelson Mandela University, South Africa
Ashish Awasthi	National Institute of Technology Calicut, India
Aswathy R. K.	National Institute of Technology Calicut, India
Aswin V. S.	Mahindra University École Centrale School of Engineering, India
B. V. Rathish	IIT Kanpur, India
Babitha K. V.	Government College Kasaragode, India
Baiju T.	Manipal Institute of Technology, India
Chithra A. V.	National Institute of Technology Calicut, India
Dirbude Sumer Bharat	National Institute of Technology Calicut, India
Dominic P. Clemence	North Carolina A&T State University, USA
Farzad Ismail	Universiti Sains Malaysia, Malaysia
Gayathri Varma	Indian Statistical Institute, Delhi, India
Govindan Rangarajan	IISc Bangalore, India
Harish Garg	Thapar Institute of Engineering & Technology, India
Jacob M. J.	National Institute of Technology Calicut, India
Jaya Paul	St. Peter's College, Kolenchery, India
Jessy John C.	National Institute of Technology Calicut, India
Kapil Kumar Sharma	South Asian University, Delhi, India
Kiran Kumar Patra	VIT-AP University, India
Krishnan Paramasivam	National Institute of Technology Calicut, India
Kuldeep Singh Patel	IIIT Naya Raipur, India
Lakshmi C.	Bharata Mata College, India
Lineesh M. C.	National Institute of Technology Calicut, India
Lokpati Tripathi	IIT Goa, India

Mahesh Kumar	National Institute of Technology Calicut, India
Mani Mehra	IIT Delhi, India
Mohan K. Kadalbajoo	IIT Kanpur, India
Naveen Jha	Government Engineering College Bharatpur, India
Neela Nataraj	IIT Bombay, India
Niladri Chatterjee	IIT Delhi, India
Olga Gil Medrano	University of Valencia, Spain
Pankaj Mishra	Deshbandhu College, University of Delhi, India
Peeyush Chandra	IIT Kanpur, India
Priyadarsan K. P.	CKG Memorial Government College Perambra, India
Ram Ajor Maurya	National Institute of Technology Calicut, India
Ratheesh K. P.	Government Arts and Science College, Malappuram, India
Ritesh Kumar Dubey	SRM Institute of Science and Technology, India
S. P. Tiwari	IIT Dhanbad, India
S. Sundar	IIT Madras, India
Samarjit Kar	NIT Durgapur, India
Sanjay P. K.	National Institute of Technology Calicut, India
Satyananda Panda	National Institute of Technology Calicut, India
Shijina V.	Co-operative Arts & Science College, Madayi, India
Shinoj T. K.	ISRO, Trivandrum, India
Shiv Prasad Yadav	IIT Roorkee, India
Shruti Dubey	IIT Madras, India
Shuvam Sen	Tezpur University, India
Siju K. C.	Vidya Academy of Science & Technology, India
Simon Peter	National Institute of Technology Calicut, India
Snehashish Chakraverty	NIT Rourkela, India
Stefano Bianchini	University of Bologna, Italy
Sunil Jacob John	National Institute of Technology Calicut, India
Sunil Mathew	National Institute of Technology Calicut, India
Sunitha M. S.	National Institute of Technology Calicut, India
Suresh Kumar N.	National Institute of Technology Calicut, India
Sushama C. M.	National Institute of Technology Calicut, India
Tamal Pramanick	National Institute of Technology Calicut, India
Tanmoy Som	IIT Varanasi, India
Vijitha Mukundan	Sacred Heart College, Chalakudy, India
Vikas Gupta	The LNM Institute of Information Technology, India
Vivek Kumar Aggarwal	Delhi Technological University, India
Wil Schilders	Eindhoven University of Technology, The Netherlands
Zhonghua Qiao	The Hong Kong Polytechnic University, Hong Kong

Scientific Advisory Committee

Olga Gil Medrano	University of Valencia, Spain
Dominic P. Clemence	North Carolina A & T State University
John R. Ockendon	University of Oxford

Florentin Smarandache	University of New Mexico
Wil Schilders	T U Eindhoven
Siddartha Mishra	ETH Zurich
Saied Jafari	College of Vestsjaelland
Reza Langari	Texas A&M University
J. N. Singh	Barry University
Bilge Inan	kilis 7 aralik university
Mohan K. Kadalbajoo	IIT Kanpur
Niladri Chatterjee	IIT Delhi
Govindan Rangarajan	IISc, Bangalore
A. K. Nandakumaran	IISc Bangalore
A. K. Pani	IIT Bombay
Neela Nataraj	IIT Bombay
S. Sundar	IIT Madras
Peeyush Chandra	IIT Kanpur
Shiv Prasad Yadav	IIT Roorkee
Tanmoy Som	IIT Varanasi
S. P. Tiwari	IIT Dhanbad
Snehashish Chakraverty	NIT Rourkela
Samarjit Kar	NIT Durgapur
Harish Garg	Thapar Institute of Engineering & Technology, Patiala
Kapil Kumar Sharma	South Asian University, Delhi
B. V. Rathish	IIT Kanpur

Reviewers

Sreedharan R.

Madhusudanan Pillai

Waquar Ahamad

Jay Prakash

Sudev Das

T. J. Sarvoththama Jothi

Ranjith Maniyeri

Jidesh P.

Arumuga Perumal D.

Aradhana Dutt Jauhari

Yedida V. S. S. Sanyasiraju

Dhirendra Bahuguna

Bini A .A.

Rajkumar

N. Kamatchi

Sharmistha Ghosh

Piyush Tiwari

K. M. Kathiresan

N. Kishore Kumar

Ramis M. K.

Favas T. K.

Rajesh Thumbakara

Vinodkumar P. B.

Aswathy R. K.

A. Balu

Brinda R. K.

Parbati Sahoo

Mini Rani

Dhanya Mol

Minu K. K.

Bajeel P. N.

Kavitha Raj

Chithralekha K.

Lincy George

Rajesh Kumar

R. Sivaraj

Praveen Nagarajan

Contents

Computing

Soft Computing

General Computing

Modelling

Computing

An Auxiliary Inequality Based Method for Stabilization and Mesh-Adaptation of Steady and Time-Dependent Differential Equations

Vivek Kumar[1($\boxtimes$)] ⓘ and Balaji Srinivasan[2] ⓘ

[1] Department of Applied Mathematics, Delhi Technological University,
Bawana Road, Delhi 110042, India
[2] Department of Mechanical Engineering, Indian Institute of Technology,
Chennai 600036, India

Abstract. We develop a general discrete inequality based on the entropy idea in hyperbolic conservation laws, and demonstrate that enforcing this auxiliary inequality can be utilized for a number of steady and time-dependent problems. We demonstrate that, for any existing central difference based method, addition of this auxiliary inequality at the discrete level, enables one to achieve several desired purposes. Firstly, the violation of the inequality allows us to determine unphysical regions in the numerical solution without any a-priori knowledge of the solution. Secondly, the sign of the discrete production also functions as an excellent indicator for mesh adaptation in several problems in general and singular perturbation problems in particular. Thirdly, the operator can be used to derive robust schemes for convection dominated problems. Most importantly, all these are achieved without any ad-hoc, user introduced, parameters. We provide a range of numerical results demonstrating the efficacy of the method and its applicability to both steady and time dependent problems.

Keywords: Central finite difference schemes · Entropy ·
Layer-adaptive meshes

1 Introduction

Computation of convection dominated flows is made difficult even in the laminar case due to several numerical artefacts. These numerical artefacts may be either cosmetic, (such as, oscillatory numerical solutions in physically monotonic regions [1]) or they may be incipient serious instabilities such as the sonic glitch [2] or the carbuncle, etc. [3]. The typical approach to handling such artefacts is by a judicious combination of stabilization and mesh adaptation [4].

While there is abundant literature on both these approaches, there is rarely any universality in the specifics of the approach. There are also situation dependent ad-hoc parameters in the problem and a strong dependence on geometrical

© Springer Nature Singapore Pte Ltd. 2021
A. Awasthi et al. (Eds.): CSMCS 2020, CCIS 1345, pp. 3–24, 2021.
https://doi.org/10.1007/978-981-16-4772-7_1

parameters such as the value of the local gradient. Overall, a unified, non-ad-hoc approach to improving diagnosis of unphysical behavior, stabilization and meshing would help in increasing robustness of solution to a variety of problems. In the case of hyperbolic conservation laws, [5] demonstrated that a local entropy measure could be employed to derive numerical schemes. Guermond et al. [6] have recently extended this to derive stabilization schemes for higher order finite element schemes. Both these attempt to obtain the appropriate weak solution in non-smooth regions. We demonstrate here that it is possible to derive an auxiliary discrete inequality even in the case of smooth regions. This auxiliary inequality can be utilized to (a) detect unphysical behavior, (b) derive stabilization schemes and (c) derive adaptive mesh schemes for steady and unsteady differential equations. The organization of the rest of the paper is as follows. In the Sect. 2, we outline our methodology for deriving and utilizing the auxiliary inequality. In the Sect. 3, we present numerical results on a variety of test cases. We offer our conclusions and suggestions for future possibilities in the Sect. 4.

2   The Auxiliary Entropy Inequality

2.1   Physical Motivation

As mentioned in the previous section, the computation of convection dominated is complicated by several factors. The clearest demonstration of this occurs in the case of the computation of hyperbolic conservation laws where the numerical solution may converge to the incorrect weak solution [7, 8]. Theoretically, this is handled by ensuring that that the numerical solution converges to the correct entropy solution. We extend this idea to design an auxiliary entropy inequality for all the purposes outlined above – diagnosis of unphysical solutions, mesh adaptation, derivation of stabilization schemes. In essence, our idea is as follows – *Just as an entropy inequality separates the correct weak solution from the incorrect ones, it is possible to identify incorrect numerical behavior by the violation of an auxiliary, entropy like inequality.*

We may physically motivate the inequality by observing the function that the entropy inequality plays in the case of hyperbolic conservation laws. In the latter case, the entropy inequality is often equivalent to the presence of a vanishingly small, positive, viscosity [7]. Similarly, any general, discrete, auxiliary inequality would determine if the numerical scheme is adding sufficient numerical dissipation locally. It is important to note here that even though the governing differential equation may be dissipative, in the case of small viscosity coefficients or large mesh sizes, the numerical discretization may not have sufficient dissipation to mask the destabilizing effect of the numerical scheme. As will become clear in the derivations below, the auxiliary inequality measures precisely this lack of dissipation.

2.2   A Simple Prototype

We now show how such an auxiliary entropy may be derived using a simple prototype equation. We choose the naive, central discretization of the convection diffusion problem as it has often been used as a departure point for deriving numerical schemes for the Navier-Stokes equations [1].

The steady state convection-diffusion equation is given as

$$\frac{du}{dx} = \varepsilon \frac{d^2u}{dx^2} \tag{1}$$

with some specified boundary conditions.

For insufficiently resolved boundary layers, the central discretization of this equation results in highly oscillatory solutions though the exact solution is monotonic [1]. We may determine the dissipation inherent in the physical solution by multiplying (1) with $2u$ and obtaining

$$\frac{d(u^2)}{dx} - \varepsilon \frac{d^2(u^2)}{dx^2} = -2\varepsilon \left(\frac{du}{dx} \right)^2 \tag{2}$$

By invoking an analogy to the Navier-Stokes equations [9], the term on the right hand side of the (2) may be physically interpreted as the dissipation of the energy u^2. Therefore, all smooth solutions to the original equation satisfy the following auxiliary inequality

$$\frac{dS}{dx} - \varepsilon \frac{d^2S}{dx^2} \le 0 \tag{3}$$

where $S(u) = u^2$ is defined as the entropy variable with analogy to hyperbolic conservation laws.

As mentioned above, while continuous solutions of discretizations of (1) will necessarily the auxiliary inequality, not all discretizations may satisfy a discrete version of (3). So, in order to ensure that there is sufficient dissipation locally, at the discrete level, we discretize each of the derivative operations in the auxiliary inequality identically to the corresponding term in the original (1).

For instance, corresponding to the 2nd order central difference discretization of (1) on a uniform grid

$$\frac{u_{i+1} - u_{i-1}}{2h} = \varepsilon \left(\frac{u_{i+1} - 2u_i + u_{i-1}}{h^2} \right) \tag{4}$$

we would have the natural discretization of (3) as

$$P_i = \frac{S_{i+1} - S_{i-1}}{2h} - \varepsilon \left(\frac{S_{i+1} - 2S_i + S_{i-1}}{h^2} \right) \le 0 \tag{5}$$

Inequality (5) is the discrete auxiliary inequality corresponding to the central discretization of the steady state convection diffusion (1). Despite its simplicity, the discrete auxiliary inequality has several remarkable properties. The most important among these is that the central discretization shows unphysical oscillations if and only if the discrete inequality (5) is violated by the solution. This can be seen as follows. Multiplying (4) by $(u_{i+1} + u_{i-1})$ we get

$$\frac{S_{i+1} - S_{i-1}}{2h} = \varepsilon (u_{i+1} + u_{i-1}) \left(\frac{u_{i+1} - 2u_i + u_{i-1}}{h^2} \right) \tag{6}$$

Subtracting the above (6) from (5) and rearranging gives

$$P_i = -2\varepsilon \left(\frac{u_{i+1} - u_i}{h}\right)\left(\frac{u_i - u_{i-1}}{h}\right). \tag{7}$$

Equation (7) demonstrates immediately that the auxiliary inequality is satisfied/violated whenever the original solution is monotonic/oscillatory. We refer the reader to [10–12] for a detailed discussion of this property as well as other properties of this inequality.

The above property of being able to discern unphysical numerical behavior purely by *a posteriori* determination of the satisfaction of an auxiliary inequality, without *a priori* knowledge of solution behavior can provide a unified, non-ad-hoc way to address the various issues raised above.

1. **Diagnosis:** Once the discretization as well as the governing equation are known, the corresponding discrete entropy production operator P_i can be computed a *posteriori* and evaluated. Any local violation of the inequality may be utilized to determine both the location and magnitude of unphysical numerical behavior. We would immediately know that the numerical solution in the corresponding region is physically suspect. We emphasize that we require no *a priori* knowledge of solution behavior.
2. **Mesh adaptation:** A simple, non-ad-hoc mesh adaptation scheme may be derived by adapting the mesh at all regions where the inequality is violated. We note that this adaptation scheme differs from others in the literature in that it depends on a criterion that depends both on the governing equation being solved as well as the specific discretization scheme being chosen (since both these are woven into the auxiliary inequality). As we will see in the next section this allows us perform mesh adaptation with remarkably sparse initial meshes.
3. **Stabilization:** We may add stabilization terms until the corresponding inequality is satisfied locally everywhere. We demonstrate in the next section that this allows us to deal with persistent problems such as the sonic glitch in the Burgers equation.

2.3   Extensions to General Differential Equations

The above procedure may be easily applied to derive auxiliary inequalities for singular perturbation problems, elliptic equations and parabolic equations in one and higher dimensions. We show a couple of illustrative extensions below and the rest in Sect. 3.

Second Order ODEs: The above principle can be extended to any general scalar singular perturbation problem of the form

$$-\varepsilon u''(x) - a(x)u'(x) + b(x)u(x) = f \tag{8}$$

Here, $a(x)$ represents the (possibly spatially varying) background velocity and ϵ is the diffusion coefficient as before. Once again, as ϵ becomes smaller, convective processes dominate over diffusive processes. ϵ also represents the singular perturbation parameter. Without loss of generality, we consider cases where $\epsilon > 0$ by allowing the sign of $a(x)$, $b(x)$ to vary arbitrarily. Proceeding as before, we multiply (8) by $2u$ and rearrange to obtain

$$- \varepsilon S''(x) - a(x)S'(x) - 2uf + 2b(x)u(x) = -2\varepsilon(u')^2 \tag{9}$$

The left hand side of the above (9) represents the entropy production P for (8). Our numerical results [11] show that in certain cases (especially for mesh adaptation), tighter results are obtained for the stabilizing $b(x) > 0$ case when we use

$$P = -\varepsilon S''(x) - a(x)S'(x) - 2uf = -2b(x)u(x) - 2\varepsilon(u')^2 \leq 0 \tag{10}$$

Unsteady Advection Diffusion: For the unsteady advection-diffusion equation

$$u_t + au_x = \varepsilon u_{xx} \tag{11}$$

the corresponding entropy inequality is obtained as

$$P = S_t + aS_x - \varepsilon S_{xx} = -\varepsilon(u_x)^2 \leq 0 \tag{12}$$

The discrete version of this inequality is obtained by mimicking the discretization of (11) term by term. For instance for the FTCS scheme one obtains

$$P_i = \frac{S_i^{n+1} - S_i^n}{\Delta t} + a\frac{S_{i+1}^n - S_{i-1}^n}{2h} - \varepsilon\frac{S_{i+1}^n - 2S_i^n + S_{i-1}^n}{h^2} \tag{13}$$

Viscous Burgers Equation: The viscous Burgers equation is given by

$$u_t + \left(\frac{u^2}{2}\right)_x = \varepsilon u_{xx} \tag{14}$$

This non-linear equation is popularly used as a benchmark case for compressible flows. The numerical problems caused here are of a slightly different nature than those in the convection diffusion equation; In the Burgers shock, oscillations are found around the nearly discontinuous shock, whereas in the convection-diffusion case they are found in high-gradient smooth regions. The corresponding entropy inequality is derived as before (by multiplying the original Eq. (14) by $2u$) as

$$P = S_t + \left(\frac{2u^3}{3}\right)_x - \varepsilon S_{xx} = -2\varepsilon(u_x)^2 \leq 0 \tag{15}$$

The flux term is chosen so as to retain the conservative nature of the entropy flux in the vanishing viscosity case. As before, the terms in (15) are discretized exactly like the corresponding terms in (14).

3   Results

We now demonstrate that the auxiliary inequality may be utilized to successfully diagnose unphysical numerical solutions, stabilize them and also to adapt meshes for a wide range of steady and time dependent differential equations.

3.1   Diagnosis

As discussed earlier, we determine the region and magnitude of unphysical numerical artifact by determining the local inequality violation of the solution. We demonstrate this for a steady case, an unsteady case and a nonlinear case.

Steady Convection Diffusion Equation: Figure 1 shows how the local violation of the auxiliary inequality directly correlates with unphysical numerical behavior. It may also be seen that the extent of violation of the inequality matches extremely well with the magnitude of unphysical oscillations.

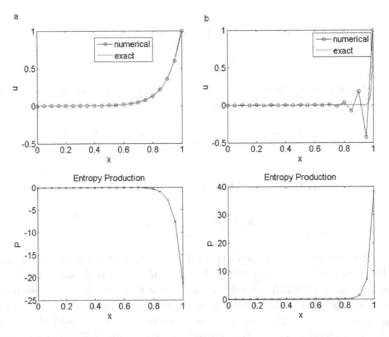

Fig. 1. Convection diffusion equation for $N = 20$. (a) Resolved solution, $\epsilon = 0.1$ (b) Under-resolved solution, $\epsilon = 0.01$

Unsteady Advection Diffusion Equation: We use the formulation described above to demonstrate how the inequality correctly diagnoses problematic discretizations for this unsteady case. Figure 2 shows the time evolution of a "square wave" initially non-zero in $(-0.1, 0.1)$ for a FTCS discretization. We use two different sets of coefficients. The convection-dominated case can be seen to have oscillations. These are directly correlated with the violation of the entropy inequality. Further, the inequality indicator correctly identifies the smooth sub-regions as entropy satisfying regions [14,15].

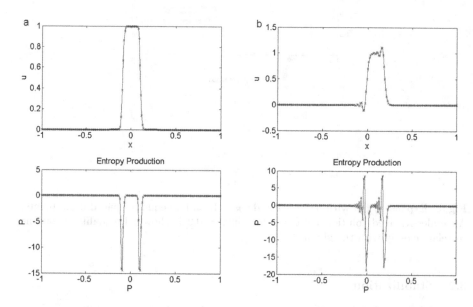

Fig. 2. Unsteady advection-Diffusion equation with 200 equally spaced mesh points of a square wave. (a) Non-oscillatory solution for $a = 1$, $\epsilon = 0.01$ (b) Stable, oscillatory solution for $a = 10, \epsilon = 0.01$.

Viscous Burgers Equation: We consider an insufficiently resolved expansion fan, which famously leads to a numerical artifact known as the sonic glitch [2]. This phenomenon is known to be "entropy satisfying". This is true in the sense that at infinitesimal resolution the solution does converge to the correct weak solution. However, at finite resolution, this may still lead to unphysical behavior – a situation that the auxiliary inequality is designed to detect.

Figure 3 shows an under-resolved expansion for a Riemann problem with initial conditions being $u_L = -1$ and $u_R = 1$. We choose $\epsilon = 0.5$ for this problem to clearly demonstrate the presence of the sonic glitch. It can be seen that the auxiliary inequality correctly classifies the physical and unphysical portions of the solution. A comparison with the inviscid (i.e. $\epsilon = 0$) solution also demonstrates that the viscous numerical solution is, paradoxically, less diffusive than the inviscid solution. This clearly explains the origin of the unphysical behavior

– incorrect numerical dissipation. We would also like to point out that this disproves the claim in [2] that central methods do not show the sonic glitch and that the glitch is not an entropy violation. It is clear from our results that the glitch is a *local* entropy violation and that *under-resolved* central schemes may exhibit the glitch.

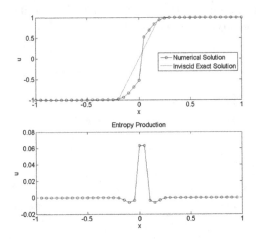

Fig. 3. Expansion fan with $u_L = -1$ and $u_R = 1$, and 40 equally spaced mesh points. An under-resolved solution for $\epsilon = 0.05$, time $= 0.2$ is shown. Inequality violations correlate closely with the glitch.

3.2 Stabilization

The violation of the auxiliary inequality may be corrected by adding compensatory diffusive terms with a free coefficient to the governing equation. This coefficient may then be adjusted in order to satisfy the inequality correctly.

Upwinding the Convection Diffusion Equation: For instance, for the steady state convection diffusion equation (1) the physical diffusion ϵ may be augmented by a numerical diffusion α. The corresponding inequality is also modified by this simple augmentation. One can very quickly demonstrate that a constant $\alpha = h/2$ is the least amount of diffusion that ensures unconditional satisfaction of the inequality. The resultant numerical discretization is given by

$$\frac{u_i - u_{i-1}}{h} = \varepsilon \left(\frac{u_{i+1} - 2u_i + u_{i-1}}{h^2} \right), \tag{16}$$

which is the well known upwind discretization. This has the corresponding inequality

$$P_i = \frac{S_i - S_{i-1}}{h} - \varepsilon \left(\frac{S_{i+1} - 2S_i + S_{i-1}}{h^2} \right) \tag{17}$$

Multiplying (16) by $u_i + u_{i-1}$ and subtracting from (17) we get $P_i = -2\varepsilon \left(\frac{u_i - u_{i-1}}{h} \right)^2$. Hence, the discrete entropy inequality is satisfied for all positive values of ϵ.

Stabilizing the Sonic Glitch: We may also adapt the above method to allow for locally varying diffusion coefficients as done in the entropy viscosity method of Guermond et al. [6]. We adapt this method to fix the sonic glitch by additionally enforcing Galilean invariance for the additional viscosity (else the solution may become too diffusive). For every grid point, the entropy production is calculated using the local velocities $[u_{i-1} - u_i, 0, u_{i+1} - u_i]$. The corresponding local entropy function is $S = (u - u_i)^2$ and the entropy flux is modified to $F = \frac{2}{3} \left(u - \frac{3u_i}{2} \right) u^2$. The corresponding inequality is $P = S_t + F_x - \varepsilon S_{xx} = -2\varepsilon(u_x)^2 \leq 0$. We show results for the extreme case of $\epsilon = 0$ (finite ϵ work better but this choice allows direct comparison with the Godunov method). As can be seen in the Fig. 4, the stabilization works remarkably well in comparison to the solution obtained by using the Godunov method [7] for the convective terms.

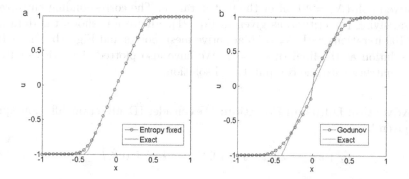

Fig. 4. Solution at t = 0.2 for an expansion fan. (a) Stabilized solution which removes the sonic glitch (b) Godunov solution with clear sonic glitch around the sonic point.

3.3   Numerical Results

One Dimensional (1D) Heat Equation: We consider 1D heat equation given as

$$u_t = \epsilon u_{xx}, \quad x \in (0,1), 0 < \epsilon \ll 1, 0 \leq t \leq T, \tag{18}$$

with initial conditions

$$u(x, t = 0) = \exp(-100(4x - 1)^2). \tag{19}$$

The series solution for zero boundary conditions is given as

$$u(x,t) = \sum_{n=1}^{\infty} B_n \sin(\frac{n\pi x}{L}) \exp(-\epsilon(\frac{n\pi}{L})^2 t), \tag{20}$$

where

$$B_n = \frac{2}{L} \int_0^L u(x, t=0) \sin(\frac{n\pi x}{L}) dx \; ; \quad L = 1 \text{ in this case.}$$

The entropy production equation for the above heat Eq. (18) can be obtained by multiplying with a factor $2u$, which becomes

$$2uu_t = 2u\epsilon u_{xx}$$

which can be further written as (putting $S(t,x) = u^2$)

$$S_t - \epsilon S_{xx} = -2\epsilon u_x^2 \le 0. \tag{21}$$

The parameter ϵ is known as diffusion coefficient. Figure 5(a) shows the initial solution at the initial uniform mesh for $N = 16$ and $\epsilon = .01$ and the corresponding entropy production. Ideally, the entropy should be non positive throughout the domain but it violates this criteria only in the region where solution needs more resolution. The points starts accumulating automatically in the exact location where the entropy is being violated. Figure 5(b) shows the solution and the adaptive mesh ($N_a = 40$) after the first iteration. The corresponding entropy is also satisfying the criteria as given in Eq. (21). Figure 6(a) shows the diffusion after 100 iterations with $N_a = 76$ adaptive mesh points and Fig. 6(b) shows the final solution at the final time $t = 1$. We have also plotted the series solution and it matches with the computational solution.

1D Advection Diffusion Equation: We consider 1D advection diffusion equation given as

$$u_t + au_x = \epsilon u_{xx}, \quad x \in (0,1), 0 < \epsilon \ll 1, 0 \le t \le T, \tag{22}$$

with initial conditions

$$u(x, t=0) = \exp(-100(4x-1)^2). \tag{23}$$

Its exact solution, using operator splitting, is given as

$$u(x,t) = h(x - at, t), \quad \text{where } h \text{ is the solution of the equation } h_t = \epsilon h_{xx}. \tag{24}$$

Its corresponding entropy production equation can be written as

$$S_t + aS_x - \epsilon S_{xx} = -2\epsilon u_x^2 \le 0. \tag{25}$$

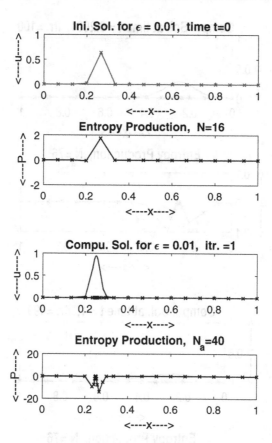

Fig. 5. Results for the test case of heat equation with black as computed solution and blue as series solution (Color figure online)

Figure 7(a) represents the initial solution and the corresponding entropy production for the initial starting uniform mesh for diffusion coefficient $\epsilon = .01$. Again, the entropy is being violated in the exact location where the solution is under resolved. Figure 7(b) shows the adaptive mesh with mesh point $N_a = 40$ and the solution after the first iteration is over. Figure 8 shows the results for the reduced diffusion coefficient $\epsilon = .0001$ during the first iteration. Figure 9(a) demonstrate the solution at the 26th iteration and Fig. 9(b) shows the solution after 50 iterations and the corresponding entropy and the adaptive mesh points.

Steady State 1D Burgers Equation: We also consider the nonlinear steady state Burgers equation

$$uu_x = \epsilon u_{xx}, \quad a \leq x \leq b \tag{26}$$

with boundary conditions as

$$u(a) = \alpha, u(b) = \beta \tag{27}$$

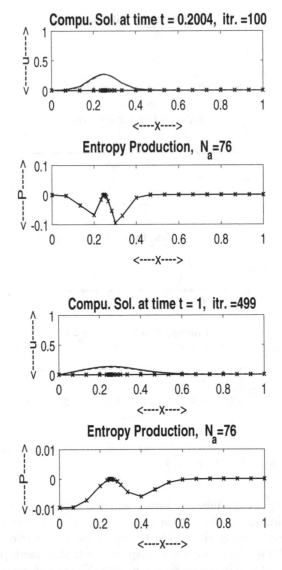

Fig. 6. Results for the test case of heat equation with black as computed solution and blue as series solution (Color figure online)

The above Eq. (26) can be written in the conserved form as

$$\left(\frac{u^2}{2}\right)_x = \epsilon u_{xx}. \tag{28}$$

In this problem whenever α and β have opposite sign, the solution must cross the x-axis (at least once). The location of this crossing may be quite susceptible to the values of α and β, therefore it is a very interesting example to demonstrate

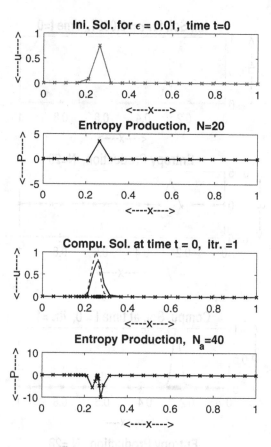

Fig. 7. Results for the advection equation. Black line shows computed sol. and the blue line shows the series sol. (Color figure online)

the robustness of the proposed method for finding the location and then producing the adaptive mesh using the concept of entropy production. The entropy production equation for the conserved form (28) (on multiplying by $2u$) is given as

$$- \epsilon S_{xx} + \left(\frac{2u^2}{3}\right)_x = -2\epsilon(u_x)^2, \quad \text{where } S = u^2. \tag{29}$$

We solve the conserved form (28) using central finite difference scheme for initial $N = 11$ uniform mesh points and the nonlinearity is handled using the Newton's iterations. Figure 10(a) shows the results for $\epsilon = .01$ using proposed adaptive mesh technique corresponding to the entropy production as given in the Eq. (29) with resolution $N_a = 19$. Figure 10(b) gives the results for very small diffusion parameter $\epsilon = .0001$ with adaptive mesh having resolution $N_a = 39$. These

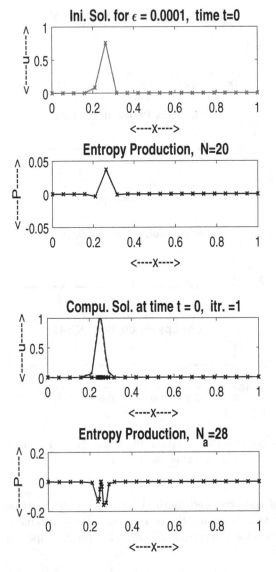

Fig. 8. Results for the advection equation. Black line shows computed sol. and the blue line shows the series sol. (Color figure online)

results has been calculated for boundary conditions $\alpha = 1$ and $\beta = -1$ and the solution cross the x-axis at the location close to $x = 0.12$. Figure 11(a) shows the results for the parameters $\alpha = .995, \beta = -1$ and $\epsilon = .001$ for adaptive mesh with $N_a = 55$ mesh points. We have observed that the location of the crossing of the solution to x-axis ($x = -.06$) has changed to the left of the zero.

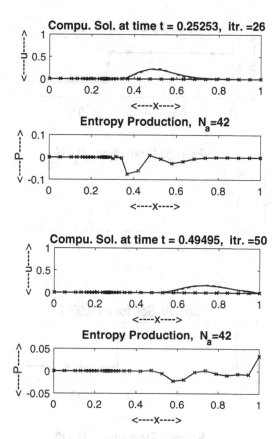

Fig. 9. Results for the advection equation using adaptive mesh points $N_a = 100$ for $\epsilon = .0001$

Figure 11(b) also demonstrate the results for $\alpha = 1.005$ and for this value the crossing happens at $x = .07$. As in this test problem, the location of the crossing of the solution is sensitive to the values of α and β, no piecewise uniform mesh based on pre-knowledge of location of the crossing will be able to produce correct solutions.

2D Elliptic Equation: We also consider a 2D elliptic test problem given as

$$- \epsilon(u_{xx} + u_{yy}) - (x - \frac{1}{2})u_x = 0, \quad \text{with domain } \Omega = (0,1)^2. \tag{30}$$

with Dirichlet boundary conditions as

$$u(0,y) = 1, \quad u(1,y) = 2, \quad u(x,0) = u(x,1) = 0.$$

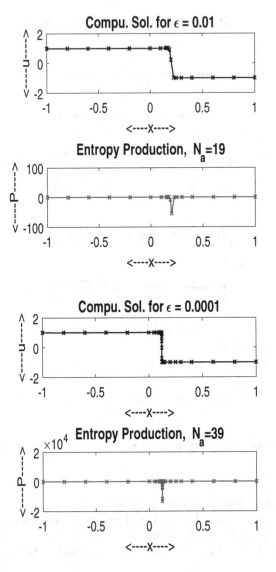

Fig. 10. Results for the steady state Burgers equation with $\alpha = 1$ and $\beta = -1$.

This equation has interior layer at the location $x = 1/2$ [13]. This problem serves as a good model for complex fluid flow with relatively small diffusion (ϵ) and dominating convection to demonstrate the extension of the proposed method of adaptivity to higher dimensional problems. The problem is discretized using central finite difference on nonuniform cartesian grid with initial uniform grid points $(x_i, y_j), i = 1, 2, 3, \ldots, m$; $j = 1, 2, 3, \ldots, n$. Let $u_{i,j}$ represents an

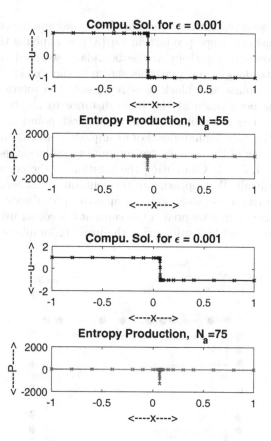

Fig. 11. Results for the steady state Burgers equation (a) For $\alpha = .995$ and $\beta = -1$ (b) For $\alpha = 1.005$ and $\beta = -1$

approximation to $u(x_i, y_j)$, using 5 points stencil, the discretized equation can be written as

$$-\epsilon\{\frac{2u_{i-1,j}}{(x_{i-1} - x_i)(x_{i-1} - x_{i+1})} + u_{i,j}(\frac{2}{(x_i - x_{i-1})(x_i - x_{i+1})} + \frac{2}{(y_j - y_{j-1})(y_j - y_{j+1})})$$
$$+\frac{2u_{i+1,j}}{(x_{i+1} - x_{i-1})(x_{i+1} - x_i)} + \frac{2u_{i,j-1}}{(y_{j-1} - y_j)(y_{j-1} - y_{j+1})} + \frac{2u_{i,j+1}}{(y_{j+1} - y_{j-1})(y_{j+1} - y_j)}\}$$
$$+(x_i - 1/2)\frac{u_{i+1,j} - u_{i-1,j}}{(x_{i+1} - x_{i-1})} = 0.$$

On multiplying Eq. (30) by a factor of $2u$ and considering entropy as $S(x,y) = u^2$ with $S_x = 2uu_x$, $S_y = 2uu_y$, $S_{xx} = 2(uu_{xx} + u_x^2)$, $S_{yy} = 2(uu_{yy} + u_y^2)$, the entropy production equation can be written as

$$- \epsilon(S_{xx} + S_{yy}) - (x - \frac{1}{2})S_x = -\epsilon(u_x + u_y)^2 \leq 0. \tag{31}$$

Once we find the solution on the uniform mesh, we compute the entropy through-out the domain using entropy production Eq. (31) by applying the central finite difference. For computing entropy at the boundary, we need to introduce the ghost mesh points along x and y axes as shown in the Fig. 12. Red dots shows the corner ghost points and black dots represents the interior ghost points. Ghost points has been taken at the same distance to the boundaries to the next grid points. The solution at the interior ghost points can be computed using the Eq. (30) for the boundaries. For example to compute the interior ghost points at the bottom of the domain, we compute the solution using Eq. (30) for $i = 2, 3, \ldots, m - 1, j = 1$. Computing the solution at the corner ghost points is rather more difficult. We approximate the solution at the corner points using predictor and corrector methods. For example to approximate the solution at the bottom left corner, first we predict the value at left corner using double mesh method and then correct this value using the usual finite difference operator.

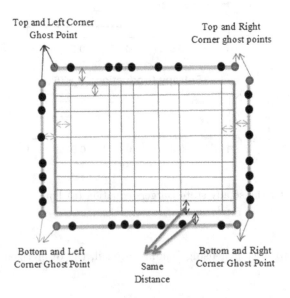

Fig. 12. Ghost points to calculate entropy at the boundaries (Color figure online)

Once we are able to compute the entropy at all the grid points in the domain, we locate the location where the entropy production is positive, violating the Eq. (31), and then add one point to the left and one point to the right of the x, y coordinates. Figure 13(a) shows the solution and the corresponding entropy for the initial course grid with 9×8 points for $\epsilon = .01$ and Fig. 13(b) gives the solution for $\epsilon = .00001$. It can been seen that entropy is not satisfied as expected

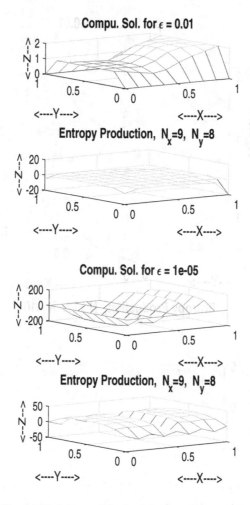

Fig. 13. Solution with the initial uniform mesh.

in the Eq. (31) in the some part of the domain. Then, as discussed earlier, points being added in the neighbourhood of the (x, y) location of the maximum entropy violation. Figure 14 represents the well resolved solution corresponding to the adaptive mesh having 13×10 mesh points. Figure 15 shows the solution for $\epsilon = .00001$ for the corresponding adaptive mesh. As we reduce the ϵ, interior layer becomes quite sharp and more mesh points are needed for well resolved solution as evident in the Fig. 15.

Compu. Sol. for $\epsilon = 0.01$

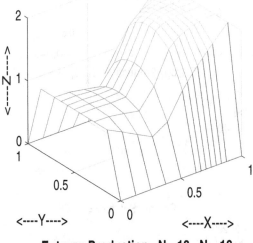

Entropy Production, N$_x$=13, N$_y$=10

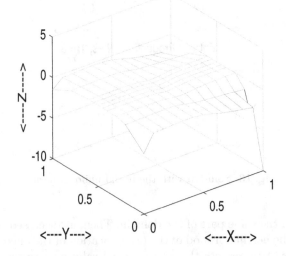

Fig. 14. Results for the adaptive mesh for $\epsilon = .01$

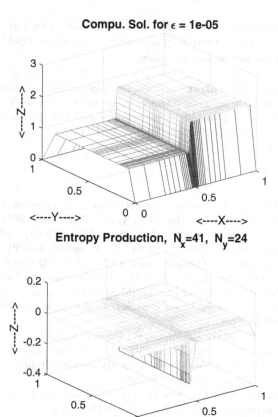

Fig. 15. Results for the adaptive mesh for $\epsilon = .00001$ using adaptive mesh and corresponding entropy production.

4   Conclusions

We have presented in this paper a unified framework by which one may address several problems that occur in computations of convection dominated problems. Through the enforcement of a discrete auxiliary inequality that depends both on the governing equation and the discretization, we were able to (a) diagnose unphysical solutions without a priori knowledge of the solution (b) stabilize solutions (c) derive mesh adaptation schemes. All these could be done with a single operator with no ad-hoc parameters. The method performed well on a wide range of test cases including steady and time-dependent problems. The method still needs to be adapted so that it may be used in practical flow problems. Firstly, the method needs to be extended to be capable of handling unstructured meshes and associated discretization approaches such as finite element and finite volume. This would allow for complex geometries. Secondly, theory has to be developed to handle multiple variables instead of just the scalar cases demonstrated here. In particular, it would be interesting to investigate if there would be more than one inequality in operation in such cases.

Acknowledgement. The authors thank the anonymous referees for their constructive comments that lead to a significant improvement of this article. VK would like to acknowledge the National board of higher mathematics (NBHM) for research Grant no. - Ref. No. 2/48(6)/2016/NBHM(R.P.)/R & D II /15455.

References

1. Oñate, E., Manzan, M.: Stabilization techniques for finite element analysis of convection-diffusion problems. Dev. Heat Transf. **7**, 71–118 (2001)
2. Tang, H.: On the sonic point glitch. J. Comput. Phys. **202**(2), 507–132 (2005)
3. Dumbser, M., Moschetta, J.M., Gressier, J.: A matrix stability analysis of the carbuncle phenomenon. J. Comput. Phys. **197**(2), 647–670 (2004)
4. Miller, J.J.H., O'Riordan, E., Shishkin, I.G.: Fitted Numerical Methods for Singular Perturbation Problems, 2nd edn. World Scientific, Singapore (2012)
5. Merriam, M.L.: An entropy-based approach to nonlinear stability. NASA Tech. Memo. **101086**(64), 1–154 (1989)
6. Guermond, J.L., Pasquetti, R., Popov, B.: Entropy viscosity method for nonlinear conservation laws. J. Comput. Phys. **230**(11), 4248–4267 (2011)
7. Leveque, R.J.: Finite Volume Methods for Hyperbolic Problems, 1st edn. Cambridge University Press, Cambridge (2002)
8. Huang, W., Russell, R.D.: Adaptive Moving Mesh Methods, 1st edn. Springer, New York (2010). https://doi.org/10.1007/978-1-4419-7916-2
9. Kundu, P., Cohen, I.: Fluid Mechanics, 5th edn. Academic Press, Cambridge (2014)
10. Srinivasan, B., Kumar, V.: The versatility of an entropy inequality for the robust computation of convection dominated problems. Procedia Comput. Sci. **108C**, 887–896 (2017)
11. Kumar, V., Srinivasan, B.: An adaptive mesh strategy for singularly perturbed convection diffusion problems. Appl. Math. Model. **39**(7), 2081–2091 (2015)
12. Kumar, V., Srinivasan, B.: A novel adaptive mesh strategy for singularly perturbed parabolic convection diffusion problems. Differ. Equ. Dyn. Syst. **27**(1–3), 203–220 (2019)
13. Kumar, V.: High-order compact finite-difference scheme for singularly-perturbed reaction-diffusion problems on a new mesh of Shishkin type. J. Optimiz. Theory App. **143**(1), 123–147 (2009)
14. Kumar, V., Rao, S.V.R.: Composite scheme using localized relaxation with nonstandard finite difference method for hyperbolic conservation laws. J. Sound Vib. **311**(3–5), 786–801 (2008)
15. Lochab, R., Kumar, V.: An improved flux limiter using fuzzy modifiers for Hyperbolic Conservation Laws. Math. Comput. Simulat. **181**, 16–37 (2021)

Existence and Uniqueness of Time-Fractional Diffusion Equation on a Metric Star Graph

Vaibhav Mehandiratta[1], Mani Mehra[1(✉)], and Günter Leugering[2]

[1] Department of Mathematics, Indian Institute of Technology, Delhi, India
mmehra@maths.iitd.ac.in
[2] Lehrstuhl Angewandte Mathematik II, Friedrich-Alexander-Universität
Erlangen-Nürnberg (FAU), Cauerstr. 11, 91058 Erlangen, Germany
guenter.leugering@fau.de

Abstract. In this paper, we study the time-fractional diffusion equation on a metric star graph. The existence and uniqueness of the weak solution are investigated and the proof is based on eigenfunction expansions. Some priori estimates and regularity results of the solution are proved.

Keywords: Time-fractional diffusion equation · Caputo fractional derivative · Weak solution · Star graph

1 Introduction

We consider a graph $\mathcal{G} = (V, E)$ consisting of a finite set of vertices (nodes) $V = \{v_0, v_1, \ldots, v_k\}$ and a finite set of edges E (such as heat conducting elements) connecting these nodes. The graph considered in this work is a metric graph. A metric graph is a graph in which each edge is endowed with an implicit metric structure. More precisely, each edge e_i, $i = 1, 2, \ldots, k$ is parametrised by an interval $(0, l_i)$ such that $0 < l_i < \infty$. The study of partial differential equations (PDEs) on networks or metric graphs is not just the analysis of known mathematical objects on special domains, since in our context, graphs or networked domains are not manifolds. Thus, we investigate PDEs on single edges of graph (interpreted as continuous curves or manifolds) [37] along with certain transmission conditions such as continuity and Kirchoff condition at junction node. Hence, we define a coordinate system on each edge e_i by taking v_0 as the origin and $x \in (0, l_i)$ as the coordinate. We consider a time-fractional diffusion equation (TFDE) on a metric star graph $\mathcal{G}$, which is a graph consisting of k edges incident to a common vertex v_0 (see Fig. 1):

$$_cD_{0,t}^\alpha y(x,t) = \frac{\partial^2 y(x,t)}{\partial x^2} + f(x,t), \quad x \in \mathcal{G}, \ t \in (0,T), \ 0 < \alpha < 1. \quad (1.1)$$

$$y(x,0) = y^0(x), \quad x \in \mathcal{G}. \quad (1.2)$$

© Springer Nature Singapore Pte Ltd. 2021
A. Awasthi et al. (Eds.): CSMCS 2020, CCIS 1345, pp. 25–41, 2021.
https://doi.org/10.1007/978-981-16-4772-7_2

More precisely, at each edge we have the following fractional diffusion equation

$$_{c}D_{0,t}^{\alpha}y_i(x,t) = \frac{\partial^2 y_i(x,t)}{\partial x^2} + f_i(x,t), \quad x \in (0,l_i), \ t \in (0,T), \ 0 < \alpha < 1, \quad (1.3)$$

$$y_i(x,0) = y_i^0(x), \quad x \in (0,l_i), \quad i = 1,2\ldots,k, \tag{1.4}$$

along with the continuity and Kirchoff conditions at junction node v_0 as

$$y_i(0,t) = y_j(0,t), \ i \neq j, \ i,j = 1,2,\ldots,k, \ t \in (0,T), \tag{1.5}$$

$$\sum_{i=1}^{k} \frac{\partial y_i(0,t)}{\partial x} = 0, \tag{1.6}$$

and Dirichlet boundary conditions at boundary nodes v_i

$$y_i(l_i,t) = 0, \quad t \in (0,T), \ i = 1,2,\ldots,k. \tag{1.7}$$

Here $_{c}D_{0,t}^{\alpha}$ denotes the Caputo fractional derivative of order α with respect to t which is defined as

$$_{c}D_{0,t}^{\alpha}y(x,t) = \frac{1}{\Gamma(1-\alpha)} \left(\int_0^t (t-\xi)^{-\alpha} \frac{\partial y(x,\xi)}{\partial \xi} d\xi \right), \quad 0 < \alpha < 1, \quad t \in (0,T),$$

where $\Gamma(.)$ denotes the Euler gamma function. In this paper, we prove the existence and uniqueness of the weak solution of initial-value problem (IVP) (1.1)–(1.2) whose restriction to the edge e_i gives the weak solution of initial-boundary value problem (IBVP) (1.3)–(1.7). When α approaches 1, the Caputo fractional derivative $_{c}D_{0,t}^{\alpha}y$ approaches the ordinary derivative $\frac{\partial y}{\partial t}$ and, thus, IBVP (1.3)–(1.7) represents the standard diffusion equation on graphs for which existence and uniqueness was proved in [40]. In [24], authors established the existence and uniqueness of solution for nonlinear fractional boundary value problem on a star graph, see also [23,27,42]. Hence, this work could be seen as the extension of [24] for time dependent problem.

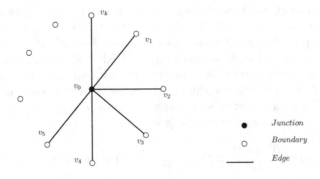

Fig. 1. A star graph consisting of k edges

The origin of the study of differential equation on graphs can be traced back to 1980s with Lumer's work [20] on ramification spaces. Considerable work related to eigenvalue problems (Sturm-Liouville type problems) on networks, i.e. metric graphs has been done, for instance see the article by von Below [2] and [32]. Partial differential equations on graphs or multi-link structures play an important role in the field of science and engineering. For instance, controlled vibrations of networks of strings (hyperbolic wave equations) [7], water wave propagation in open channel networks (Burgers type equation) [41], naturally lead to partial differential equation on graphs. Evolutionary problems (such as linear parabolic equations) on metric graphs were considered in [3]. The dynamic networks of strings and beams along with their control properties were studied by Lagnese et al. [14]. The progress of problems defined on metric graphs until 2006 has been presented in an excellent survey by Dager and Zuazua [5]. Since then, modeling, analysis and optimal control problems for linear and nonlinear partial differential equations on metric graphs have become an active area of research. In [41], Yoshioka et al. considered the Burger type equation models on connected graph and discussed the existence and uniqueness of the model along with the energy estimates. For more problems on metric graphs, we refer [1,8,34] and references therein.

On the other hand, fractional calculus find its importance in different fields of science and engineering [4,6,9,12,13,22,25,38,39]. Moreover, there are some classical models, describing the biological process, such as [26] that can also be modeled using fractional differential/difference equations. A strong motivation for the study and analysis of fractional diffusion equations comes from the fact that they efficiently describe the phenomenon of anomalous diffusion [19]. Fractional diffusion equations on bounded domains have been studied by various authors. For instance in [17], Luchko gave the maximum principle for the time-fractional diffusion equation, while in [18] he established the existence and uniqueness results for time fractional diffusion equation using eigenfunction expansion by taking source term $f = 0$. In [36], the existence results for fractional diffusion-wave equations were established by Sakamoto and Yamamoto, while IBVP for a coupled fractional diffusion system was discussed in [15]. For more results on fractional time dependent problems, we refer [16,21,31] and references therein.

To the best of author's knowledge, there has not been any published work related to the existence and uniqueness results for the time-fractional diffusion equation on metric graphs so far. In this paper, we focus on proving the existence and uniqueness of IBVP (1.3)–(1.7) and study the regularity of solution given by the eigenfunction expansions.

2 Preliminaries

First of all, we define the following function spaces on a star graph $\mathcal{G}$:

$$L_2(\mathcal{G}) = \prod_{i=1}^{k} L_2(0, l_i) \quad \text{and} \quad H_m(\mathcal{G}) = \prod_{i=1}^{k} H_m(0, l_i)$$

with the corresponding inner products

$$\langle y, w \rangle_{L_2(\mathcal{G})} := \sum_{i=1}^{k} \langle y_i, w_i \rangle_{L_2(0,l_i)} \quad \text{and} \quad \langle y, w \rangle_{H_m(\mathcal{G})} := \sum_{i=1}^{k} \langle y_i, w_i \rangle_{H_m(0,l_i)},$$

where $L_2(0, l_i)$ and $H_m(0, l_i)$ are standard Sobolev spaces. The spaces $L_2(\mathcal{G})$ and $H_m(\mathcal{G})$ are Hilbert spaces with the inner products $\langle \cdot, \cdot \rangle_{L_2(\mathcal{G})}$ and $\langle \cdot, \cdot \rangle_{H_m(\mathcal{G})}$, respectively (see [28]). We define the operator $\mathcal{L}$ on the Hilbert space $L_2(\mathcal{G})$ as follows:

$$D(-\mathcal{L}) = \Big\{ y \in L_2(\mathcal{G}) : y_i \in H_2(0, l_i),$$

$$y_i(l_i) = 0, y_i(0) = y_j(0), \;\; i \neq j, \; i, j = 1, 2, \dots, k \text{ and } \sum_{i=1}^{k} y_i'(0) = 0 \Big\},$$

$$\forall y \in D(-\mathcal{L}): \; \mathcal{L}y = \left(\frac{\partial^2 y_i}{\partial x^2} \right)_{i=1}^{k}.$$

Remark 1. The operator $-\mathcal{L}$ is a non-negative self-adjoint operator since it is the Friedrichs extension of the triple $(L_2(\mathcal{G}); V; a)$ defined by [10]

$$V = \Big\{ y \in \prod_{i=1}^{k} H_1(0, l_i) : y_i(l_i) = 0, y_i(0) = y_j(0), \; i \neq j, \; i, j = 1, 2, \dots, k \Big\},$$

which is a Hilbert space with the inner product $\langle y, w \rangle_V := \sum_{i=1}^{k} \int_0^{l_i} y_i' w_i' dx$ and a is the corresponding bilinear form defined as $a(y, w) = \sum_{i=1}^{k} \int_0^{l_i} y_i'(x) w_i'(x) dx$. The spectrum of operator $-\mathcal{L}$ consist of eigenvalues, having the form

$$0 < \mu_1(\mathcal{G}) \leq \mu_2(\mathcal{G}) \leq \dots \to \infty;$$

and the eigenfunction $\Psi_n = (\psi_{n,1}, \psi_{n,2}, \dots, \psi_{n_k})$ corresponding to eigenvalue μ_n: $-\mathcal{L}\Psi_n = \mu_n \Psi_n$, $n \in \mathbb{N}$. Then the sequence $\{\Psi_n\}_{n \in \mathbb{N}}$ forms an orthonormal basis of $L_2(\mathcal{G})$ (see [29,35]). Hence $\{\mu_n, \Psi_n\}_{n \in \mathbb{N}}$ is the eigensystem of following problem:

$$\psi_{n,i}''(x) = -\mu_n \psi_{n,i}(x), \quad 0 < x < l_i, \tag{2.1}$$

$$\psi_{n,i}(l_i) = 0, \; i = 1, 2, \dots, k, \tag{2.2}$$

$$\psi_{n,i}(0) = \psi_{n,j}(0), \; i, j = 1, 2, \dots, k, \; i \neq j, \tag{2.3}$$

$$\sum_{i=1}^{k} \psi_{n,i}'(0) = 0. \tag{2.4}$$

Now, using the spectral decomposition of operator $\mathcal{L}$, one can define the fractional power $(-\mathcal{L})^{\gamma}$, $\gamma \in \mathbb{R}$, as $(-\mathcal{L})^{\gamma} y = ((-\mathcal{M})^{\gamma} y_i)_{i=1}^{k}$, where $(\mathcal{M})^{\gamma} y_i = \sum_{n=1}^{\infty} \mu_n^{\gamma} \langle y, \Psi_n \rangle \psi_{n,i}$, with the norm

$$\|(-\mathcal{L})^\gamma y\|_{L_2(\mathcal{G})}^2 = \sum_{i=1}^{k} \|(-\mathcal{M})^\gamma y_i\|_{L_2(0,l_i)}^2 = \sum_{n=1}^{\infty} \mu_n^{2\gamma} |\langle y, \Psi_n \rangle|^2.$$

Then we define

$$D\left((-\mathcal{L})^\gamma\right) = \left\{ y \in L_2(\mathcal{G}) : \sum_{n=1}^{\infty} \mu_n^{2\gamma} |\langle y, \Psi_n \rangle|^2 < \infty \right\}.$$

It follows that $D\left((-\mathcal{L})^\gamma\right)$ forms a Hilbert space equipped with the norm

$$\|y\|_{D((-\mathcal{L})^\gamma)} = \|(-\mathcal{L})^\gamma y\|_{L_2(\mathcal{G})} = \left(\sum_{n=1}^{\infty} \mu_n^{2\gamma} |\langle y, \Psi_n \rangle|^2 \right)^{\frac{1}{2}}. \qquad (2.5)$$

Remark 2. Using Parseval's identity, we have [30]

$$\|y\|_V^2 \sim \|y\|_{D(-\mathcal{L}^{1/2})}^2,$$

while in general $D\left((-\mathcal{L})^\gamma\right) \subset H_{2\gamma}(\mathcal{G})$ holds for $\gamma > 0$. Hence, in view of (2.5), the spaces V, $L_2(\mathcal{G})$ and $H_2(\mathcal{G})$ can be characterised as follows:

$$V = \left\{ y = \sum_{n=1}^{\infty} \langle y, \Psi_n \rangle \Psi_n : \quad \|y\|_V^2 = \sum_{n=1}^{\infty} \mu_n |\langle y, \Psi_n \rangle|^2 < \infty \right\}$$

$$L_2(\mathcal{G}) = \left\{ y = \sum_{n=1}^{\infty} \langle y, \Psi_n \rangle \Psi_n : \quad \|y\|_{L_2(\mathcal{G})}^2 = \sum_{n=1}^{\infty} |\langle y, \Psi_n \rangle|^2 < \infty \right\}$$

and

$$H_2(\mathcal{G}) = \left\{ y = \sum_{n=1}^{\infty} \langle y, \Psi_n \rangle \Psi_n : \quad \|y\|_{H_2(\mathcal{G})}^2 = \sum_{n=1}^{\infty} \mu_n^2 |\langle y, \Psi_n \rangle|^2 < \infty \right\}.$$

Now, we give the following definition and propositions (for proofs see [33]) regarding Mittag-Leffler function which will be used further.

Definition 1. *The Mittag-Leffler function is defined as follows*

$$E_{\alpha,\beta}(z) = \sum_{j=0}^{\infty} \frac{z^j}{\Gamma(\alpha j + \beta)}, \quad z \in \mathbb{C},$$

where $\alpha > 0$ and $\beta \in \mathbb{R}$ are arbitrary constants.

Proposition 1. *Let $0 < \alpha < 2$, $\beta \in \mathbb{R}$ be arbitrary and μ be such that $\pi\alpha/2 < \mu < \min\{\pi, \pi\alpha\}$, then there exists a positive constant $C = C(\alpha, \beta, \mu)$ such that $|E_{\alpha,\beta}(z)| \leq \frac{C}{1+|z|}, \quad \mu \leq |\arg(z)| \leq \pi.$*

Proposition 2. *Let $0 < \alpha < 1$ and $\eta > 0$, then $0 < E_{\alpha,\alpha}(-\eta) < \frac{1}{\Gamma(\alpha)}$. Furthermore, $E_{\alpha,\alpha}(-\eta)$ is a monotonic decreasing function with $\eta > 0$.*

Proposition 3. *Let $0 < \alpha < 1$ and $t > 0$, then $0 < E_{\alpha,1}(-t) < 1$. Furthermore, $E_{\alpha,1}(-t)$ is completely monotonic that is $(-1^n)\frac{d^n}{dt^n}E_{\alpha,1}(-t) \geq 0, \quad n \in \mathbb{N}$.*

Proposition 4. *Let $\alpha > 0$, $\lambda > 0$ and $m \in \mathbb{N}$, then*

$$\frac{d^m}{dt^m}E_{\alpha,1}(-\lambda t^\alpha) = -\lambda t^{\alpha-m}E_{\alpha,\alpha-m+1}(-\lambda t^\alpha), \quad t > 0.$$

Proposition 5. *Let $\alpha > 0$ and $\lambda > 0$, then we have*

$$_cD_{0,t}^\alpha E_{\alpha,1}(-\lambda t^\alpha) = -\lambda E_{\alpha,1}(-\lambda t^\alpha), \quad t > 0.$$

Lemma 1. *Let $f(\cdot,t) \in L_2(\mathcal{G})$, $y^0 \in L_2(\mathcal{G})$, then the solution $y_i(x,t)$ of IBVP (1.3)–(1.7) has the form*

$$y_i(x,t) = \sum_{n=1}^{\infty} \langle y^0, \Psi_n \rangle E_{\alpha,1}(-\mu_n t^\alpha)\psi_{n,i}(x)$$

$$+ \sum_{n=1}^{\infty} \left(\int_0^t \langle f(x,\xi), \Psi_n \rangle (t-\xi)^{\alpha-1}E_{\alpha,\alpha}\left(-\mu_n(t-\xi)^\alpha\right) d\xi \right) \psi_{n,i}(x),$$

$$(2.6)$$

where $\{\mu_n, \Psi_n\}_{n\in\mathbb{N}}$ is the eigensystem of (2.1)–(2.4) and $\langle \cdot, \cdot \rangle$ denotes the inner product in $L_2(\mathcal{G})$.

Proof. We will use the method of eigenfunction expansions for the solution of (1.3). Hence, we write the solution in the form

$$y_i(x,t) = \sum_{n=1}^{\infty} T_n(t)\psi_{n,i}(x), \quad i = 1, 2, \ldots, k. \qquad (2.7)$$

Thus, we obtain

$$\frac{\partial^2 y_i(x,t)}{\partial x^2} = \sum_{n=1}^{\infty} T_n(t)\psi_{n,i}''(x) \quad \text{and} \quad _cD_{0,t}^\alpha y_i(x,t) = \sum_{n=1}^{\infty} \left(_cD_{0,t}^\alpha T_n(t)\right)\psi_{n,i}(x).$$

After substituting the value of above expressions in Eq. (1.3), we get

$$\sum_{n=1}^{\infty} \left[\left(_cD_{0,t}^\alpha T_n(t)\right)\psi_{n,i}(x) - T_n(t)\psi_{n,i}''(x)\right] = f_i(x,t)$$

and

$$\sum_{n=1}^{\infty} \left[_cD_{0,t}^\alpha T_n(t) + \mu_n T_n(t)\right]\psi_{n,i}(x) = f_i(x,t), \qquad (2.8)$$

where we used the fact that $\psi_{n,i}''(x) = -\mu_n \psi_{n,i}(x)$, $x \in (0, l_i)$. Now, we expand the functions $y^0(x)$ and $f(x,t)$ in terms of Fourier series, given by,

$$f(x,t) = \sum_{n=1}^{\infty} f_n(t)\Psi_n(x) \quad \text{and} \quad y^0(x) = \sum_{n=1}^{\infty} a_n\Psi_n(x), \tag{2.9}$$

which gives

$$f_i(x,t) = \sum_{n=1}^{\infty} f_n(t)\psi_{n,i}(x) \quad \text{and} \quad y_i^0(x) = \sum_{n=1}^{\infty} a_n\psi_{n,i}(x), \tag{2.10}$$

where

$$f_n(t) = \langle f(x,t), \Psi_n(x) \rangle \quad \text{and} \quad a_n = \langle y^0(x), \Psi_n(x) \rangle.$$

Hence, from Eqs. (2.8) and (2.9), we obtain

$$\sum_{n=1}^{\infty} \left[{}_cD_{0,t}^{\alpha}T_n(t) + \mu_n T_n(t) \right] \Psi_n(x) = \sum_{n=1}^{\infty} f_n(t)\Psi_n(x).$$

Using the uniqueness of Fourier series we get the family of fractional ODE's

$$_cD_{0,t}^{\alpha}T_n(t) + \mu_n T_n(t) = f_n(t) \tag{2.11}$$

and

$$y_i(x,0) = \sum_{n=1}^{\infty} T_n(0)\psi_{n,i}(x) = y_i^0(x) = \sum_{n=1}^{\infty} a_n\psi_{n,i}(x),$$

so that

$$T_n(0) = a_n \quad n \geq 1. \tag{2.12}$$

The solution of fractional differential equation (2.11) subject to initial condition (2.12) is given by [11]

$$T_n(t) = a_n E_{\alpha,1}(-\mu_n t^{\alpha}) + \int_0^t (t-\xi)^{\alpha-1} E_{\alpha,\alpha}\left(-\mu_n(t-\xi)^{\alpha}\right) f_n(\xi)d\xi.$$

Hence, from Eq. (2.7), we have

$$y_i(x,t) = \sum_{n=1}^{\infty} a_n E_{\alpha,1}(-\mu_n t^{\alpha})\psi_{n,i}(x)$$

$$+ \sum_{n=1}^{\infty} \left(\int_0^t (t-\xi)^{\alpha-1} E_{\alpha,\alpha}\left(-\mu_n(t-\xi)^{\alpha}\right) f_n(\xi)d\xi \right) \psi_{n,i}(x). \tag{2.13}$$

After substituting the value of a_n and $f_n(t)$ in Eq. (2.13), we obtain the desired result.

3 Existence and Uniqueness Results of a Weak Solution

In this section, the existence and uniqueness of weak solutions will be proved. Therefore, let us first define the weak solution as follows.

Definition 2. *We define y as a weak solution of (1.1)–(1.2) if (1.1) holds in $L_2(\mathcal{G})$ and $y(\cdot, t) \in V$ for almost all $t \in (0, T)$ and satisfy*

$$\lim_{t \to 0} \|y(\cdot, t) - y^0\|_{L_2(\mathcal{G})} = 0.$$

Now we state our first main result as follows.

Theorem 1. *Let $y^0 \in L_2(\mathcal{G})$ and $f(x, t) \in L_\infty(0, T; L_2(\mathcal{G}))$. Then there exists a unique weak solution $y \in C([0, T]; L_2(\mathcal{G})) \cap C((0, T]; D(-\mathcal{L}))$ such that $_C D_{0,t}^\alpha y \in L_\infty(0, T; L_2(\mathcal{G}))$. Furthermore, there exists a positive constant C_1 such that*

$$\|y\|_{C([0,T];L_2(\mathcal{G}))} \leq C_1 \left(\|y^0\|_{L_2(\mathcal{G})} + \|f\|_{L_\infty(0,T;L_2(\mathcal{G}))} \right), \tag{3.1}$$

$$\|y(\cdot, t)\|_{\prod_{i=1}^k H_2(0,l_i)} \leq C_1 \left(\|y^0\|_{L_2(\mathcal{G})} t^{-\alpha} + \|f\|_{L_\infty(0,T;L_2(\mathcal{G}))} \right). \tag{3.2}$$

Proof. We will show that $y(x, t) = (y_i(x, t))_{i=1}^k$, where $y_i(x, t)$ is given by Eq. (2.6), is certainly the weak solution of (1.1)–(1.2). We assume $C > 0$ to be a generic constant in the following proof. Hence, using Eq. (2.6) and the fact that $\sum_{i=1}^k \langle \psi_{n,i}, \psi_{m,i} \rangle_{L_2(0,l_i)} = \begin{cases} 1 & \text{if } m = n \\ 0 & \text{for } m \neq n \end{cases}$, we have

$$\sum_{i=1}^k \|y_i(\cdot, t)\|_{L_2(0,l_i)}^2 = \sum_{n=1}^\infty \left| \langle y^0, \Psi_n \rangle E_{\alpha,1}(-\mu_n t^\alpha) \right|^2$$
$$+ \sum_{n=1}^\infty \left| \int_0^t \langle f(\cdot, \xi), \Psi_n \rangle (t - \xi)^{\alpha-1} E_{\alpha,\alpha} \left(-\mu_n (t - \xi)^\alpha \right) d\xi \right|^2.$$

Using Propositions 2 and 3, we get

$$\sum_{i=1}^k \|y_i(\cdot, t)\|_{L_2(0,l_i)}^2 \leq \sum_{n=1}^\infty \left| \langle y^0, \Psi_n \rangle \right|^2 + \sum_{n=1}^\infty \left| \int_0^t \langle f(\cdot, \xi), \Psi_n \rangle \frac{(t - \xi)^{\alpha-1}}{\Gamma(\alpha)} d\xi \right|^2$$

$$\leq \|y^0\|_{L_2(\mathcal{G})}^2 + \sum_{n=1}^\infty \sup_{0 \leq t \leq T} \left| \langle f(\cdot, t), \Psi_n \rangle \right|^2 \left(\frac{t^\alpha}{\Gamma(\alpha + 1)} \right)^2$$

$$\leq \|y^0\|_{L_2(\mathcal{G})}^2 + \|f\|_{L_\infty(0,T;L_2(\mathcal{G}))}^2 \frac{T^{2\alpha}}{(\Gamma(\alpha + 1))^2}.$$

Hence,

$$\|y(\cdot, t)\|_{L_2(\mathcal{G})} \leq C_1 \left(\|y^0\|_{L_2(\mathcal{G})} + \|f\|_{L_\infty(0,T;L_2(\mathcal{G}))} \right), \quad t \in [0, T].$$

Therefore, $y \in C([0,T]; L_2(\mathcal{G}))$. Now, it will be shown that $y \in C((0,T]; D(-\mathcal{L}))$ and $_cD_{0,t}^\alpha y \in L_\infty(0,T; L_2(\mathcal{G}))$. We have

$$(-\mathcal{M})y_i(x,t) = \sum_{n=1}^\infty \mu_n \langle y^0, \Psi_n \rangle E_{\alpha,1}(-\mu_n t^\alpha) \psi_{n,i}(x)$$

$$+ \sum_{n=1}^\infty \mu_n \left(\int_0^t \langle f(\cdot,\xi), \Psi_n \rangle (t-\xi)^{\alpha-1} E_{\alpha,\alpha}\left(-\mu_n(t-\xi)^\alpha\right) d\xi \right) \psi_{n,i}(x).$$

Now,

$$\|(-\mathcal{L})y(\cdot,t)\|_{L_2(\mathcal{G})}^2 = \sum_{i=1}^k \|(-\mathcal{M})y_i(\cdot,t)\|_{L_2(0,l_i)}^2$$

$$= \sum_{n=1}^\infty \mu_n^2 \left| \langle y^0, \Psi_n \rangle E_{\alpha,1}(-\mu_n t^\alpha) \right|^2$$

$$+ \sum_{n=1}^\infty \mu_n^2 \left| \int_0^t \langle f(\cdot,\xi), \Psi_n \rangle (t-\xi)^{\alpha-1} E_{\alpha,\alpha}\left(-\mu_n(t-\xi)^\alpha\right) d\xi \right|^2.$$

Also from Propositions 3 and 4

$$\int_0^t \left| \xi^{\alpha-1} E_{\alpha,\alpha}(-\mu_n \xi^\alpha) \right| d\xi = \int_0^t \xi^{\alpha-1} E_{\alpha,\alpha}(-\mu_n \xi^\alpha) d\xi$$

$$= -\frac{1}{\mu_n} \int_0^t \frac{d}{d\xi} E_{\alpha,1}(-\mu_n \xi^\alpha) d\xi = \frac{1}{\mu_n}(1 - E_{\alpha,1}(-\mu_n t^\alpha)) \le \frac{1}{\mu_n}. \tag{3.3}$$

Now, using Eq. (3.3), Proposition 1 and Young inequality for the convolution, we get

$$\|(-\mathcal{L})u\|_{L_2(\mathcal{G})}^2 \le \sum_{n=1}^\infty \mu_n^2 \left| \langle y^0, \Psi_n \rangle \right|^2 \left(\frac{C_1}{1+\mu_n t^\alpha} \right)^2$$

$$+ \sum_{n=1}^\infty \mu_n^2 \sup_{0 \le t \le T} \left| \langle f(\cdot,t), \Psi_n \rangle \right|^2 \left| \int_0^T t^{\alpha-1} E_{\alpha,\alpha}(-\mu_n t^\alpha) dt \right|^2.$$

Hence, we obtain

$$\|(-\mathcal{L})y\|_{L_2(\mathcal{G})}^2 \le \|y^0\|_{L_2(\mathcal{G})}^2 t^{-2\alpha} + \|f\|_{L_\infty(0,T;L_2(\mathcal{G}))}^2. \tag{3.4}$$

Since $-\mathcal{L}y$ is convergent in $L_2(\mathcal{G})$ uniformly on $t \in (t_0, T]$ for any given $t_0 > 0$, we deduce that $-\mathcal{L}y \in C((0,T]; L_2(\mathcal{G}))$, that is $-\mathcal{M}y_i \in C((0,T]; L_2(0,l_i))$, $i = 1, 2 \ldots, k$ and hence $y \in C((0,T]; D(-\mathcal{L}))$. Furthermore, we obtain the following estimate from Eq. (3.4)

$$\|y(\cdot,t)\|_{\prod_{i=1}^k H_2(0,l_i)} = \sum_{i=1}^k \|y_i(\cdot,t)\|_{H_2(0,l_i)}$$

$$= C' \|(-\mathcal{L})y(\cdot,t)\|_{L_2(\mathcal{G})}$$

$$\le C \left(\|y^0\|_{L_2(\mathcal{G})} t^{-\alpha} + \|f\|_{L_\infty(0,T;L_2(\mathcal{G}))} \right). \tag{3.5}$$

By (1.1), we get $_cD_{0,t}^\alpha y \in L_\infty(0,T; L_2(\mathcal{G}))$ and (1.1) holds in $L_2(\mathcal{G})$ for $t \in (0,T]$.

Next, we will show that $\lim_{t\to 0} \|y(\cdot,t) - y^0\|_{L_2(\mathcal{G})} = 0$. From Eqs. (2.6) and (2.10), we have

$$y_i(x,t) - y_i^0(x) = \sum_{n=1}^{\infty} \langle y^0, \Psi_n\rangle \left(E_{\alpha,1}(-\mu_n t^\alpha) - 1\right) \psi_{n,i}(x)$$

$$+ \sum_{n=1}^{\infty} \left(\int_0^t \langle f(x,\xi), \Psi_n\rangle (t-\xi)^{\alpha-1} E_{\alpha,\alpha}\left(-\mu_n(t-\xi)^\alpha\right) d\xi \right) \psi_{n,i}(x).$$

Hence,

$$\sum_{i=1}^{k} \|y_i(\cdot,t) - y_i^0(\cdot)\|_{L_2(0,l_i)}^2 \leq \sum_{n=1}^{\infty} \left| \langle y^0, \Psi_n\rangle \left(E_{\alpha,1}(-\mu_n t^\alpha) - 1\right) \right|^2$$

$$+ \sum_{n=1}^{\infty} \left| \int_0^t \langle f(x,\xi), \Psi_n\rangle (t-\xi)^{\alpha-1} E_{\alpha,\alpha}\left(-\mu_n(t-\xi)^\alpha\right) d\xi \right|^2$$

$$=: V_1(t) + V_2(t).$$

Clearly, $\lim_{t\to 0} V_2(t) = 0$, using Proposition 3

$$V_1(t) = \sum_{n=1}^{\infty} \left| \langle y^0, \Psi_n\rangle \left(E_{\alpha,1}(-\mu_n t^\alpha) - 1\right) \right|^2 \leq C \|y^0\|_{L_2(\mathcal{G})}^2$$

and $\lim_{t\to 0} (E_{\alpha,1}(-\mu_n t^\alpha) - 1) = 0$. Hence, by using Lebesgue dominated convergence theorem, we have $\lim_{t\to 0} V_1(t) = 0$. Hence $\lim_{t\to 0} \sum_{i=1}^{k} \|y_i(\cdot,t) - y_i^0(\cdot)\|_{L_2(0,l_i)} = 0$, which shows that

$$\lim_{t\to 0} \|y(\cdot,t) - y^0\|_{L_2(\mathcal{G})} = 0.$$

Finally, we show the uniqueness of the weak solution to IVP (1.1)–(1.2).

Uniqueness: Under the conditions $y^0 = 0$ and $f = 0$, we need to show that system (1.3)–(1.7) has only the trivial solution. On taking the inner product of (1.1) with $\Psi_n(x)$, applying Green's formula and setting $y^n(t) = (y(\cdot,t), \Psi_n)$, we obtain

$$_cD_{0,t}^\alpha y^n(t) = \int_{\mathcal{G}} \frac{\partial^2 y(x,t)}{\partial x^2} \Psi_n(x) dx$$

$$= \sum_{i=1}^{k} \int_0^{l_i} \frac{\partial^2 y_i(x,t)}{\partial x^2} \psi_{n,i}(x) dx$$

$$= -\sum_{i=1}^{k} \int_0^{l_i} \frac{\partial y_i(x,t)}{\partial x} \psi'_{n,i}(x) dx + \sum_{i=1}^{k} \frac{\partial y_i(x,t)}{\partial x} \psi_{n,i}(x) \Big|_0^{l_i}.$$

Using Eqs. (1.6) and (2.3), we get

$$\sum_{i=1}^{k} \frac{\partial y_i(x,t)}{\partial x} \psi_{n,i}(x) \Big|_0^{l_i} = \sum_{i=1}^{k} \frac{\partial y_i(l_i,t)}{\partial x} \psi_{n,i}(l_i) - \sum_{i=1}^{k} \frac{\partial y_i(0,t)}{\partial x} \psi_{n,i}(0)$$

$$= -\sum_{i=1}^{k} \frac{\partial y_i(0,t)}{\partial x} \psi_{n,i}(0) = -\phi_n(0) \sum_{i=1}^{k} \frac{\partial y_i(0,t)}{\partial x} = 0,$$

where $\psi_{n,i}(0) = \psi_{n,j}(0) = \phi_n(0)$, $i \neq j$, $i,j = 1,2,\ldots,k$. Hence, we get

$$_CD_{0,t}^{\alpha} y^n(t) = -\sum_{i=1}^{k} \int_0^{l_i} \frac{\partial y_i(x,t)}{\partial x} \psi'_{n,i}(x) dx$$

$$= \sum_{i=1}^{k} \int_0^{l_i} y_i(x,t) \psi''_{n,i}(x) dx - \sum_{i=1}^{k} y_i(x,t) \psi'_{n,i}(x) \Big|_0^{l_i}.$$

Again using Eqs. (1.5) and (2.4) and a similar approach as above, we get

$$\sum_{i=1}^{k} y_i(x,t) \psi'_{n,i}(x) \Big|_0^{l_i} = 0.$$

Therefore,

$$_CD_{0,t}^{\alpha} y^n(t) = \sum_{i=1}^{k} \int_0^{l_i} y_i(x,t) \psi''_{n,i}(x) dx$$

$$= -\mu_n \sum_{i=1}^{k} \int_0^{l_i} y_i(x,t) \psi_{n,i}(x) dx = -\mu_n \langle y(\cdot,t), \Psi_n \rangle.$$

Hence, we get the following initial value fractional differential equation

$$\begin{cases} _CD_{0,t}^{\alpha} y^n(t) = -\mu_n y^n(t), & t \in (0,T), \\ y^n(0) = 0. \end{cases}$$

Due to the existence and uniqueness of the above fractional differential equation, we get that $y^n(t) = 0$, $n = 1,2,\cdots$. Since Ψ_n forms a complete orthonormal basis of $L_2(\mathcal{G})$, we get $y = 0$ in $\mathcal{G} \times (0,T)$.

Theorem 2. *Let $y^0 \in V$, $f(x,t) \in L_\infty(0,T;L_2(\mathcal{G}))$. Then there exists a unique weak solution $y \in L_2((0,T];D(-\mathcal{L}))$ such that $_CD_{0,t}^{\alpha} y \in L_2(\mathcal{G} \times (0,T))$ and the following inequality holds:*

$$\|y\|_{L_2\left((0,T];\prod_{i=1}^{k} H_2(0,l_i)\right)} + \|_CD_{0,t}^{\alpha} y\|_{L_2(\mathcal{G} \times (0,T))} \leq C \left(\|y^0\|_V + \|f\|_{L_\infty(0,T;L_2(\mathcal{G}))} \right).$$

$$(3.6)$$

Proof. We have,

$$\|(-\mathcal{L})y(\cdot,t)\|_{L_2(\mathcal{G})}^2 = \sum_{n=1}^{\infty} \left|\mu_n \langle y^0, \Psi_n \rangle E_{\alpha,1}(-\mu_n t^\alpha)\right|^2$$
$$+ \sum_{n=1}^{\infty} \mu_n^2 \left|\int_0^t \langle f(\cdot,\xi), \Psi_n \rangle (t-\xi)^{\alpha-1} E_{\alpha,\alpha}\left(-\mu_n (t-\xi)^\alpha\right) d\xi\right|^2.$$

Now, using Proposition 1 and Young inequality for the convolution, we get

$$\|(-\mathcal{L})y(\cdot,t)\|_{L_2(\mathcal{G})}^2 \le \sum_{n=1}^{\infty} \mu_n \left|\langle y^0, \Psi_n \rangle\right|^2 \left(\frac{C_1\sqrt{\mu_n}}{1+\mu_n t^\alpha}\right)^2$$
$$+ \sum_{n=1}^{\infty} \mu_n^2 \sup_{0 \le t \le T} |\langle f(\cdot,t), \Psi_n \rangle|^2 \left|\int_0^T t^{\alpha-1} E_{\alpha,\alpha}(-\mu_n t^\alpha)dt\right|^2$$
$$= \sum_{n=1}^{\infty} \mu_n \left|\langle y^0, \Psi_n \rangle\right|^2 \left(\frac{C_1\sqrt{\mu_n t^\alpha}}{1+\mu_n t^\alpha}\right)^2 t^{-\alpha}$$
$$+ \sum_{n=1}^{\infty} \mu_n^2 \sup_{0 \le t \le T} |\langle f(\cdot,t), \Psi_n \rangle|^2 \left|\int_0^T t^{\alpha-1} E_{\alpha,\alpha}(-\mu_n \xi^\alpha)dt\right|^2$$
$$\le C\|y^0\|_V^2 t^{-\alpha} + \|f\|_{L_\infty(0,T;L_2(\mathcal{G}))}^2,$$

where we have used Eq. (3.3). Hence,

$$\|y\|_{L_2((0,T);\prod_{i=1}^k H_2(0,l_i))}^2 = \int_0^T \|y(\cdot,t)\|_{\prod_{i=1}^k H_2(0,l_i)}^2 dt$$
$$\le \int_0^T \left(C\|y^0\|_V^2 t^{-\alpha} + \|f\|_{L_\infty(0,T;L_2(\mathcal{G}))}^2\right) dt$$
$$= \frac{CT^{1-\alpha}}{1-\alpha}\|y^0\|_V^2 + T\|f\|_{L_\infty(0,T;L_2(\mathcal{G}))}^2$$
$$\le C_1 \left(\|y^0\|_V^2 + \|f\|_{L_\infty(0,T;L_2(\mathcal{G}))}^2\right).$$

Therefore, we have $y \in L_2((0,T); D(-\mathcal{L}))$. Now, using Proposition 5 and Lemma 2.8 in [16], we have

$$_cD_{0,t}^\alpha y_i(x,t) = -\sum_{n=1}^{\infty} \mu_n \langle y^0, \Psi_n \rangle E_{\alpha,1}(-\mu_n t^\alpha)\psi_{n,i}(x) + \sum_{n=1}^{\infty} \langle f(x,t), \Psi_n \rangle \psi_{n,i}(x)$$
$$- \sum_{n=1}^{\infty} \mu_n \left(\int_0^t \langle f(x,\xi), \Psi_n \rangle (t-\xi)^{\alpha-1} E_{\alpha,\alpha}\left(-\mu_n (t-\xi)^\alpha\right) d\xi\right) \psi_{n,i}(x).$$

Hence,

$$\sum_{i=1}^{k}\|_C D_{0,t}^{\alpha}y_i(\cdot,t)\|_{L_2(0,l_i)}^2 \leq \sum_{n=1}^{\infty}\left|\mu_n\langle y^0,\Psi_n\rangle E_{\alpha,1}(-\mu_n t^{\alpha})\right|^2 + \sum_{n=1}^{\infty}|\langle f(x,t),\Psi_n\rangle|^2$$

$$+\sum_{n=1}^{\infty}\mu_n^2\left|\int_0^t\langle f(x,\xi),\Psi_n\rangle(t-\xi)^{\alpha-1}E_{\alpha,\alpha}\left(-\mu_n(t-\xi)^{\alpha}\right)d\xi\right|^2$$

$$\leq \sum_{n=1}^{\infty}\mu_n\left|\langle y^0,\Psi_n\rangle\right|^2\left(\frac{C_1\sqrt{\mu_n t^{\alpha}}}{1+\mu_n t^{\alpha}}\right)^2 t^{-\alpha} + \sum_{n=1}^{\infty}|\langle f(x,t),\Psi_n\rangle|^2$$

$$+\sum_{n=1}^{\infty}\mu_n^2\sup_{0\leq t\leq T}|\langle f(\cdot,t),\Psi_n\rangle|^2\left|\int_0^T t^{\alpha-1}E_{\alpha,\alpha}(-\mu_n t^{\alpha})dt\right|^2.$$

Again, using Eq. (3.3), we obtain

$$\|_C D_{0,t}^{\alpha}y(\cdot,t)\|_{L_2(\mathcal{G})}^2 = \sum_{i=1}^{k}\|_C D_{0,t}^{\alpha}y_i(\cdot,t)\|_{L_2(0,l_i)}^2$$

$$\leq C\|y^0\|_V^2 t^{-\alpha} + \|f(\cdot,t)\|_{L_2(\mathcal{G})}^2 + \|f\|_{L_\infty(0,T;L_2(\mathcal{G}))}^2$$

$$\leq C_1\left(\|y^0\|_V^2 t^{-\alpha} + \|f\|_{L_\infty(0,T;L_2(\mathcal{G}))}^2\right).$$

Since $0<\alpha<1$, we see that $\|_C D_{0,t}^{\alpha}y\|_{L_2(\mathcal{G}\times(0,T))}\leq C\left(\|y^0\|_V + \|f\|_{L_\infty(0,T;L_2(\mathcal{G}))}\right)$. Therefore, we have $_C D_{0,t}^{\alpha}y \in L_2(\mathcal{G}\times(0,T))$.

The proof of $\lim_{t\to 0}\|y(\cdot,t)-y^0\|_{L_2(\mathcal{G})} = 0$ and uniqueness of weak solution is similar to the one derived in the proof of Theorem 1. Thus, the proof of Theorem 2 is complete.

Theorem 3. *Let $y^0 \in D(-\mathcal{L})$, $f(x,t) \in L_\infty(0,T;L_2(\mathcal{G}))$. Then there exists a unique weak solution $y \in C([0,T];L_2(\mathcal{G}))\cap C((0,T];D(-\mathcal{L}))$ such that $_C D_{0,t}^{\alpha}y \in L_2(\mathcal{G}\times(0,T))$. Furthermore, there exists a positive constant C_1 such that*

$$\|y\|_{C([0,T];\prod_{i=1}^{k}H_2(0,l_i))} + \|_C D_{0,t}^{\alpha}y\|_{L_2(\mathcal{G}\times(0,T))}$$

$$\leq C_1\left(\|y^0\|_{\prod_{i=1}^{k}H_2(0,l_i)} + \|f\|_{L_\infty(0,T;L_2(\mathcal{G}))}\right). \tag{3.7}$$

Proof. Under the assumption $y^0 \in D(-\mathcal{L})$, using Proposition 4, Proposition 3 and Young inequality for the convolution, we get

$$\|(-\mathcal{L})y(\cdot,t)\|_{L_2(\mathcal{G})}^2 = \sum_{n=1}^{\infty}\mu_n^2\left|\langle y^0,\Psi_n\rangle E_{\alpha,1}(-\mu_n t^{\alpha})\right|^2$$

$$+\sum_{n=1}^{\infty}\mu_n^2\left|\int_0^t\langle f(\cdot,\xi),\Psi_n\rangle(t-\xi)^{\alpha-1}E_{\alpha,\alpha}\left(-\mu_n(t-\xi)^{\alpha}\right)d\xi\right|^2$$

$$\leq \sum_{n=1}^{\infty} \mu_n^2 \left| \langle y^0, \Psi_n \rangle \right|^2$$

$$+ \sum_{n=1}^{\infty} \mu_n^2 \sup_{0 \leq t \leq T} \left| \langle f(\cdot, t), \Psi_n \rangle \right|^2 \left| \int_0^T t^{\alpha-1} E_{\alpha,\alpha}(-\mu_n t^\alpha) dt \right|^2$$

$$\leq \| y^0 \|^2_{\prod_{i=1}^{k} H_2(0,l_i)} + \| f \|^2_{L_\infty(0,T;L_2(\mathcal{G}))}.$$

Hence, $y(\cdot,t)\|_{\prod_{i=1}^{k} H_2(0,l_i)} \leq C \left(\| y^0 \|_{\prod_{i=1}^{k} H_2(0,l_i)} + \| f \|_{L_\infty(0,T;L_2(\mathcal{G}))} \right)$. Now,

$$\sum_{i=1}^{k} \|_C D_{0,t}^\alpha y_i(\cdot,t) \|^2_{L_2(0,l_i)} \leq \sum_{n=1}^{\infty} \mu_n^2 \left| \langle y^0, \Psi_n \rangle E_{\alpha,1}(-\mu_n t^\alpha) \right|^2 + \sum_{n=1}^{\infty} \left| \langle f(x,t), \Psi_n \rangle \right|^2$$

$$+ \sum_{n=1}^{\infty} \mu_n^2 \left| \int_0^t \langle f(x,\xi), \Psi_n \rangle (t-\xi)^{\alpha-1} E_{\alpha,\alpha} \left(-\mu_n (t-\xi)^\alpha \right) d\xi \right|^2$$

$$\leq \sum_{n=1}^{\infty} \mu_n^2 \left| \langle y^0, \Psi_n \rangle \right|^2 + \sum_{n=1}^{\infty} \left| \langle f(x,t), \Psi_n \rangle \right|^2$$

$$+ \sum_{n=1}^{\infty} \mu_n^2 \sup_{0 \leq t \leq T} \left| \langle f(\cdot,t), \Psi_n \rangle \right|^2 \left| \int_0^T t^{\alpha-1} E_{\alpha,\alpha}(-\mu_n t^\alpha) dt \right|^2.$$

Using Eq. (3.3), we obtain

$$\|_C D_{0,t}^\alpha y(\cdot,t) \|^2_{L_2(\mathcal{G})} = \sum_{i=1}^{k} \|_C D_{0,t}^\alpha y_i(\cdot,t) \|^2_{L_2(0,l_i)}$$

$$\leq \| y^0 \|^2_{\prod_{i=1}^{k} H_2(0,l_i)} + \| f(\cdot,t) \|^2_{L_2(\mathcal{G})} + \| f \|^2_{L_\infty(0,T;L_2(\mathcal{G}))}$$

$$\leq \| y^0 \|^2_{\prod_{i=1}^{k} H_2(0,l_i)} + C \| f \|^2_{L_\infty(0,T;L_2(\mathcal{G}))}.$$

Hence,

$$\|_C D_{0,t}^\alpha y \|^2_{L_2(\mathcal{G} \times (0,T))} = \int_0^T \|_C D_{0,t}^\alpha y(\cdot,t) \|^2_{L_2(\mathcal{G})} dt$$

$$\leq T \left(\| y^0 \|^2_{\prod_{i=1}^{k} H_2(0,l_i)} + C \| f \|^2_{L_\infty(0,T;L_2(\mathcal{G}))} \right)$$

$$\leq C_1 \left(\| y^0 \|^2_{\prod_{i=1}^{k} H_2(0,l_i)} + \| f \|^2_{L_\infty(0,T;L_2(\mathcal{G}))} \right).$$

In view of above theorems the following results are immediate.

Corollary 1. *Let $y^0 \in L_2(\mathcal{G})$ and $f = 0$. Then we obtain the following estimate for the unique weak solution $y \in C([0,T]; L_2(\mathcal{G})) \cap C((0,T]; D(-\mathcal{L}))$:*

$$\| y(\cdot,t) \|_{L_2(\mathcal{G})} \leq \frac{C}{1 + \mu_1 t^\alpha} \| y^0 \|_{L_2(\mathcal{G})}, \quad t \in (0,T). \tag{3.8}$$

Corollary 2. *Let* $y^0 \in D(-\mathcal{L})$ *and* $f = 0$. *Then there exists a positive constant* C_1 *such that*

$$\|y(\cdot,t)\|_{\prod_{i=1}^{k} H_2(0,l_i)} + \|_C D_{0,t}^{\alpha} y(\cdot,t)\|_{L_2(\mathcal{G})} \leq \frac{C_1}{1 + \mu_1 t^\alpha} \|y^0\|_{\prod_{i=1}^{k} H_2(0,l_i)}, \quad t \in (0,T).$$

(3.9)

4 Conclusion and Future Work

In this paper, the existence and uniqueness of time-fractional diffusion equation on a star graph has been investigated. By using the method of eigenfunction expansion the existence and uniqueness of the weak solution and the regularity of the solution are derived. In future, we investigate the numerical solution of time-fractional diffusion equation on a metric star graph using finite difference approximation.

Acknowledgements. The author would like to thank the Indo-German exchange program "Multiscale Modelling, Simulation and Optimization for Energy, Advanced Materials and Manufacturing". The program (grant number 1-3/2016 (IC)) is funded by University Grants Commission (India) and DAAD (Germany). The coordination of the program through the "Central Institute for Scientific Computing" at FAU, Erlangen is acknowledged.

References

1. Adami, R., Caaciapuoti, C., Finco, D., Noja, D.: Variational properties and orbital stability of standing waves for NLS equation on a star graph. J. Differ. Equ. **257**, 3738–3777 (2014)
2. Below, J.V.: A characteristic equation associated to an eigenvalue problem on c^∞-net. Linear Algebra Appl. **71**, 309–325 (1985)
3. Below, J.V.: Classical solvability of linear parabolic equations on networks. J. Differ. Equ. **72**(2), 316–337 (1988)
4. Bohannan, G.W.: Analog fractional order controller in temperature and motor control applications. J. Vib. Control **14**, 1487–1498 (2008)
5. Dáger, R., Zuazua, E.: Wave Propagation, Observation and Control in 1-d Flexible Multi-structures. Mathématiques & Applications [Mathematics & Applications], vol. 50. Springer, Berlin (2006). https://doi.org/10.1007/3-540-37726-3
6. Friedrich, C.: Rheological material functions for associating comb-shaped or H-shaped polymers a fractional calculus approach. Philos. Mag. Lett. **66**, 287–292 (1992)
7. Leugering, G.: On the semi-discretization of optimal control problems for networks of elastic strings: global optimality systems and domain decomposition. J. Comput. Appl. Math. **120**, 133–157 (2000)
8. Grigor'yan, A., Lin, Y., Yang, Y.: Yamabe type equations on graph. J. Differ. Equ. **261**, 4924–4943 (2016)
9. Hilfer, R.: Applications of Fractional Calculus in Physics. World Scientific, Singapore (2000)

10. Kato, T.: Perturbation Theory for Linear Operators. Springer, Berlin (1980). https://doi.org/10.1007/978-3-642-66282-9
11. Kilbas, A.A., Srivastava, H.M., Trujillo, J.J.: Theory and Applications of Fractional Differential Equations. Elsevier, Amsterdam (2006)
12. Kumar, N., Mehra, M.: Collocation method for solving nonlinear fractional optimal control problems by using Hermite scaling function with error estimates. Optim. Control Appl. Methods **42**, 417–444 (2021)
13. Kumar, N., Mehra, M.: Legendre wavelet collocation method for fractional optimal control problems with fractional Bolza cost. Numer. Methods Partial Differ. Equ. **37**, 1693–1724 (2021)
14. Lagnese, J.E., Leugering, G., Schmidt, E.J.P.G.: Modeling, Analysis and Control of Dynamic Elastic Multi-link Structures. Systems & Control: Foundations & Applications, Birkhäuser Boston Inc., Boston (1994). https://doi.org/10.1007/978-1-4612-0273-8
15. Li, L., Jin, L., Fang, S.: Existence and uniqueness of the solution to a coupled fractional diffusion system. Adv. Differ. Equ. **2015**(1), 1–14 (2015). https://doi.org/10.1186/s13662-015-0707-0
16. Li, Y.S., Wei, T.: An inverse time-dependent source problem for a time-space fractional diffusion equation. Appl. Math. Comput. **336**, 257–271 (2018)
17. Luchko, Y.: Maximum principle for the generalized time-fractional diffusion equation. J. Math. Anal. Appl. **351**, 218–223 (2009)
18. Luchko, Y.: Some uniqueness and existence results for the initial-boundary value problems for the generalized time-fractional diffusion equations. Comput. Math. Appl. **59**, 1766–1772 (2010)
19. Luchko, Y.: Anomalous diffusion: models, their analysis, and interpretation. In: Rogosin, S., Koroleva, A. (eds.) Advances in Applied Analysis, pp. 115–145. Springer, Basel (2012). https://doi.org/10.1007/978-3-0348-0417-2_3
20. Lumer, G.: Connecting of local operators and evolution equations on a network. Lect. Notes Math. **787**, 219–234 (1980)
21. Mainardi, F.: The fundamental solutions for the fractional diffusion-wave equation. Appl. Math. Lett. **9**, 23–28 (1996)
22. Mehandiratta, V., Mehra, M., Leugering, G.: An approach based on Haar wavelet for the approximation of fractional calculus with application to initial and boundary value problems. Math. Methods Appl. Sci. **44**, 3195–3213 (2020)
23. Mehandiratta, V., Mehra, M., Leugering, G.: Fractional optimal control problems on a star graph: optimality system and numerical solution. Math. Control Related Fields **11**, 189–209 (2021)
24. Mehandiratta, V., Mehra, M., Leugering, G.: Existence and uniqueness results for a nonlinear Caputo fractional boundary value problem on a star graph. J. Math. Anal. Appl. **477**, 1243–1264 (2019)
25. Mehandiratta, V., Mehra, M., Leugering, G.: Existence results and stability analysis for a nonlinear fractional boundary value problem on a circular ring with an attached edge: A study of fractional calculus on metric graph. Netw. Heterog. Media. **16**, 155–185 (2021)
26. Mehra, M., Mallik, R.K.: Solutions of differential-difference equations arising from mathematical models of granulocytopoiesis. Differ. Equ. Dyn. Syst. **22**(1), 33–49 (2014)
27. Mophou, G., Leugering, G., Fotsing, P.S.: Optimal control of a fractional Sturm-Liouville problem on a star graph. Optimization **70**, 659–687 (2020)
28. Mugnolo, D.: Gaussian estimates for a heat equation on a network. Netw. Heterogen. Media **2**, 55–79 (2007)

29. Nicaise, S.: Some results on spectral theory over networks, applied to nerve impulse transmission. In: Brezinski, C., Draux, A., Magnus, A.P., Maroni, P., Ronveaux, A. (eds.) Polynômes Orthogonaux et Applications. LNM, vol. 1171, pp. 532–541. Springer, Heidelberg (1985). https://doi.org/10.1007/BFb0076584

30. Nicaise, S., Zair, O.: Identifiability, stability and reconstruction results of point sources by boundary measurements in heteregeneous trees. Revista Matematica Complutense **16**, 151–178 (2003)

31. Patel, K.S., Mehra, M.: Fourth order compact scheme for space fractional advection-diffusion reaction equations with variable coefficients. J. Comput. Appl. Math. **380**, 112963 (2020)

32. Penkin, O.M., Pokornyi, Y.V., Provotorova, E.N.: On one vector boundary-value problem. Bound. Value Probl. **171**, 64–70 (1983)

33. Podlubny, I.: Fractional Differential Equations. Academic Press, San Diego (1999)

34. Pokornyi, Y.V., Borovskikh, A.V.: Differential equations on networks (geometric graphs). J. Math. Sci. **119**, 691–718 (2004)

35. Provotorov, V.V.: Eigenfunctions of the Sturm-Liouville problem on a star graph. Sbornik Math. **199**, 1523–1545 (2008)

36. Sakamoto, K., Yamamoto, M.: Initial value/boundary value problems for fractional diffusion-wave equations and applications to some inverse problems. J. Math. Anal. Appl. **382**, 426–447 (2011)

37. Shukla, A., Mehra, M., Leugering, G.: A fast adaptive spectral graph wavelet method for the viscous burgers' equation on a star-shaped connected graph. Math. Methods Appl. Sci. **43**(13), 7595–7614 (2020)

38. Singh, A.K., Mehra, M.: Uncertainty quantification in fractional stochastic integro-differential equations using Legendre wavelet collocation method. Lect. Notes Comput. Sci. **12138**, 58–71 (2020)

39. Singh, A.K., Mehra, M.: Wavelet collocation method based on Legendre polynomials and its application in solving the stochastic fractional-integro differential equations. J. Comput. Sci. **51**, 101342 (2021)

40. Walther, M.: Simulation-based model reduction for partial differential equations on networks. Ph.D. thesis, FAU Studies Mathematics and Physics, Erlangen (2018)

41. Yoshioka, H., Unami, K., Fujihara., M.: Burgers type equation models on connected graphs and their application to open channel hydraulics (2014). http://hdl.handle.net/2433/195771

42. Zhang, W., Liu, W.: Existence and Ulam's type stability results for a class of fractional boundary value problems on a star graph. Math. Methods Appl. Sci. **43**, 8568–8594 (2020)

Learning Numerical Viscosity Using Artificial Neural Regression Network

Ritesh Kumar Dubey[1]([✉]), Anupam Gupta[2], Vikas Kumar Jayswal[3],
and Prashant Kumar Pandey[3]

[1] Research Institute and Department of Mathematics, SRM Institute of Science
and Technology, Chennai, India
riteshkd@srmist.edu.in
[2] Blockapps AI, Bangalore, India
[3] Department of Mathematics, SRM Institute of Science and Technology,
Chennai, India
{vikaskus,prasanth}@srmist.edu.in

Abstract. Numerical diffusion plays an important role in deciding the
characteristics of numerical schemes for flow problems containing discontinuities. In this work, we attempt to learn the numerical viscosity of underlying three-point shock-capturing schemes using non-linear
regression neural network in a supervised learning paradigm. Details on
network architecture, used data type and training are elaborated. Computed results by underlying schemes using exact numerical diffusion and
predicted diffusion by trained network are given and compared. These
results show that the network gives a good approximation of numerical
diffusion and computed solutions are indistinguishable from the solution
using exact numerical diffusion.

1 Introduction

Hyperbolic conservation laws are an important class of problems in engineering
and sciences. They are a natural model for many physical phenomena for example
fluid flow problem, nonlinear acoustic, atmospheric modeling. Let $\Omega \subset \mathbb{R}$ then
in 1D setting conservation laws can be written in the form

$$u_t + f(u)_x = 0, \quad (x,t) \in \Omega \times \mathbb{R}^+ \tag{1.1}$$

along with the initial condition $u(x,0) = u_0(x)$. Variable $u : \mathbb{R} \to \mathbb{R}$ is said
to be the vector of conserved quantity and $f : \mathbb{R} \to \mathbb{R}$ is known as associated
non-linear flux function of u. It is well known that finding a closed-form solution
of (1.1) is not always possible due to the nonlinearity of the flux function. In
fact, even with a smooth initial condition, solution of (1.1) develops discontinuities like shocks and rarefaction in finite time which makes even the numerical approximation of its solution very tedious. In particular, many well known
classical schemes fail to crisply resolve these discontinuities without spurious
oscillations and often converge to nonphysical weak solution [Smo12, DDD+05].

© Springer Nature Singapore Pte Ltd. 2021
A. Awasthi et al. (Eds.): CSMCS 2020, CCIS 1345, pp. 42–55, 2021.
https://doi.org/10.1007/978-981-16-4772-7_3

A few theoretical attempts are made to define a framework for the physically correct solution for systems of conservation laws e.g., vanishing viscosity solution [BB05] and entropy stable solutions [Lax73] etc. These challenges lead to the development of several shock capturing methods for solving conservation laws in the past few years. Some of the classical shock capturing methods are Lax–Wendroff, Lax–Friedrich, and MacCormack method [Lan98]. Examples of modern shock-capturing schemes which formally high order, non-oscillatory, and give high resolution for discontinuities are limiters based second-order total variation diminishing schemes [Har83, Swe84, TB00], essentially non-oscillatory(ENO) scheme proposed by Harten et al. [HEOC87], WENO scheme [LOC+94, Shu12]. These methods although are interesting in their approach but the core underlying feature is their adaptive nature. In other words, to avoid Gibbs oscillations near discontinuities limiters based high-resolution schemes use adaptive artificial numerical viscosity whereas schemes like ENO or WENO use an adaptive stencil or adaptive weights respectively. In this work, we focus on the classical and modern three points finite-difference shock-capturing schemes.

2   Finite Difference Discretization

In order to get finite difference (volume) scheme we discretize the spatial domain Ω in the form of intervals $I_i = [x_{i-\frac{1}{2}}, x_{i+\frac{1}{2}}]$ with uniform spacing $\triangle x = x_{i+\frac{1}{2}} - x_{i-\frac{1}{2}}$ so that x_i is the center of I_i, time domain $[0, T]$ is discretize with uniform spacing $\triangle t$ such that $T = n\triangle t$,where n is the number of time steps. A scheme to solve (1.1) in conservative form is given by following equation.

$$u_i^{n+1} = u_i^n - \frac{\triangle t}{\triangle x}(F_{i+\frac{1}{2}}^n - F_{i-\frac{1}{2}}^n), \qquad (2.1)$$

where $u_i^n = u(x_i, t^n)$ is grid function, $F_{i+\frac{1}{2}}^n$ is a consistent numerical flux such that $F_{i+\frac{1}{2}}^n(u, u...u) = f(u)$. On dropping the superscript n for time level, for scalar conservation laws numerical flux of any centred three points scheme can be written as

$$F_{i+\frac{1}{2}} = \bar{f}_{i+\frac{1}{2}} - \frac{1}{2}\mathcal{D}_{i+\frac{1}{2}}(u_{i+1} - u_i) \qquad (2.2)$$

where $\bar{f}_{i+\frac{1}{2}} = \frac{f_i + f_{i+1}}{2}$, $f_i = f(u_i)$ and $\mathcal{D}$ is the coefficient of numerical viscosity which determines the shock-capturing nature of the numerical approximation by the resulting scheme. In particular, the presence of excessive numerical diffusion makes the underlying scheme too diffusive to capture the discontinuities whereas insufficient diffusion causes oscillatory approximation of the discontinuities. Classical shock capturing methods like first-order Lax Friedrichs and second-order Lax-Wendroff use the fixed amount of diffusion (at least in linear case) whilst high-resolution shock-capturing schemes use nonlinear diffusion depending on the regularity of the solution. In fact, a key point in the shock-capturing ability of the resulting scheme is the choice of suitable numerical diffusion.

Recently, artificial neural networks (ANNs) are applied for the computation of solutions of partial differential equations. These neural networks of computational models are capable of learning any input-output function containing a high degree of non-linearity and complexity. These networks are although simple, yet powerful function approximator [Hor91]. Once trained on a suitable data-set they can be used to make predictions for unknown data generally said to be test data set. Due to this capability of generalization they are used as universal black boxes in many applications such as speech recognition, image segmentation, medical imaging, etc. Few recent applications are neural networks as trouble cell detector [RH18] in RKDG solver, in a recent work Discacciati et al. [DHR20] used a regression type ANN to learn artificial viscosity in DG scheme. Recently a novel ANN called PINN (Physics informed neural network) is introduced in [RPK17] for learning PDEs from data.

In this work, our goal is to study and implement ANN for learning numerical viscosity of three points shock-capturing schemes. For this, we rely on a specific architecture of ANNs called Multi-layer perception(MLP). In the MLP network, neurons are arranged in a fashion called layers the data fed to the network process from the input layer to the output layer through hidden layers. The costly training of the network is done offline using a suitably designed data-set. The trained network is used in the numerical solver.

3   Shock Capturing Schemes

In this section, numerical flux of a few representative three point schemes are considered in diffusion form (2.2). Let $\lambda = \frac{\Delta t}{\Delta x}$ where the time step Δt is determined by the relationship $\Delta t = \frac{\Delta x}{max_u(|f'(u)|)}$.

3.1   Local Lax-Friedrich Scheme

The finite difference Local Lax -Friedrich numerical flux [Hes17] is given by

$$F_{i+\frac{1}{2}}^n = \frac{1}{2}(f(u_{i+1}^n) + f(u_i^n)) - \frac{\alpha}{2}(u_{i+1} - u_i)$$ (3.1)

where $\alpha = \max_{j=i,i+1}(|f'(u_j)|)$. On comparison the Eq. (3.1) with the dissipation form (2.2) we get dissipation coefficient to be $\mathcal{D}^{LxF} = \alpha$.

3.2   Lax-Wendroff Scheme

Lax -wendroff numerical flux is given by

$$F_{i+\frac{1}{2}}^n = \frac{1}{2}(f(u_{i+1}^n) + f(u_i^n)) - \frac{\lambda}{2}f'(\bar{u}_{i+\frac{1}{2}})(f(u_{i+1}^n) - f(u_i^n)),$$ (3.2)

where $\lambda = \frac{\Delta t}{\Delta x}$ and $\bar{u}_{i+\frac{1}{2}} = \frac{(u_{i+1}+u_i)}{2}$. By using above flux in Eq. (2.1) we get the desired scheme. Upon comparing the above flux with form (2.2) we get the non-linear dissipation coefficient $\mathcal{D}$.

3.3   Entropy Stable TVD Scheme

It is well known that the solution of conservation law (1.1) develops discontinuity in finite time due to non-linearity that is why we need to extend the notion of solution to the more general class of solution called a weak solution. Weak solutions are generally non-unique and a physically relevant weak solution needs to satisfy an additional entropy inequality given by,

$$\eta_t + q(u)_x \leq 0, \tag{3.3}$$

where $\eta \equiv \eta(u)$ is convex function, said to be entropy function and $q(u)$ is known as entropy flux. They satisfy the following compatibility condition

$$v^T f_u = q_u^T \tag{3.4}$$

$v = \eta_u$ called entropy variable. Scheme (2.1) is said to be entropy conservative if solution satisfies following semi-discrete entropy equation

$$\frac{d\eta(u_i(t))}{dt} + \frac{1}{\triangle x}(\hat{q}_{i+\frac{1}{2}} - \hat{q}_{i-\frac{1}{2}}) = 0 \tag{3.5}$$

where $u_i(t)$ is grid value of u at (x_i, t), $\hat{q}_{i+\frac{1}{2}}$ is numerical entropy flux consistent with q that is,

$$\hat{q}_{i+\frac{1}{2}} = \hat{q}(u, u, ..., u) = q(u) \tag{3.6}$$

One method to make entropy conservative scheme is to find numerical flux $F^*_{i+\frac{1}{2}}$ which satisfies following relation refer [Tad87]

$$[v]_{i+\frac{1}{2}} \cdot F^*_{i+\frac{1}{2}} = [\psi]_{i+\frac{1}{2}}, \quad [\psi]_{i+\frac{1}{2}} = \psi_{i+1} - \psi_i \tag{3.7}$$

where ψ given as

$$\psi(v) = v \cdot g(v) - q(u(v)), \quad g(v) \equiv f(u(v)). \tag{3.8}$$

However, a scheme using entropy conservative flux F^* may give oscillation near shock which can be overcome by adding positive diffusion term in the flux. The resulting entropy stable flux can be written in the form

$$F_{i+\frac{1}{2}} = F^*_{i+\frac{1}{2}} - \frac{1}{2}\tilde{D}_{i+\frac{1}{2}}[v]_{i+\frac{1}{2}}, \quad \tilde{D}_{i+\frac{1}{2}} > 0 \tag{3.9}$$

where $\tilde{D}_{i+\frac{1}{2}}$ is diffusion matrix. On writing it in viscosity form (2.2) we get

$$F_{i+\frac{1}{2}} = \bar{f}_{i+\frac{1}{2}} - \frac{1}{2}\mathcal{D}_{i+\frac{1}{2}}[v]_{i+\frac{1}{2}}, \tag{3.10}$$

The above flux (3.9) result in to entropy stable and TVD scheme for following choices of diffusion coefficients $\mathcal{D}$ refer [DB18].

$$\mathcal{D}^1_{i+\frac{1}{2}} = |b_{i+\frac{1}{2}}| + |Q^*_{i+\frac{1}{2}}| + Q^*_{i+\frac{1}{2}} \tag{3.11a}$$

$$\mathcal{D}^2_{i+\frac{1}{2}} = |b_{i+\frac{1}{2}}| + \left(-Q^*_{i+\frac{1}{2}}\right)^+ + Q^*_{i+\frac{1}{2}} \tag{3.11b}$$

$$\mathcal{D}^3_{i+\frac{1}{2}} = \left(|b_{i+\frac{1}{2}}| - Q^*_{i+\frac{1}{2}}\right)^+ + Q^*_{i+\frac{1}{2}} \tag{3.11c}$$

where $z^+ = \frac{(z+|z|)}{2}$ and $Q^*_{i+\frac{1}{2}}$ is said to be numerical viscosity which is given by

$$Q^*_{i+\frac{1}{2}} = \int_{\xi=-\frac{1}{2}}^{\xi=\frac{1}{2}} \left(\frac{1}{4} - \xi^2\right) g''(v_{i+\frac{1}{2}}(\xi))d\xi \cdot [v], \quad v_{i+\frac{1}{2}}(\xi) = \bar{v}_{i+\frac{1}{2}} + \xi[v]_{i+\frac{1}{2}} \quad (3.12)$$

where g is defined in (3.8) and $b_{i+\frac{1}{2}} = \frac{[f]_{i+\frac{1}{2}}}{[v]_{i+\frac{1}{2}}}$. For scalar conservation laws choice of entropy function $\eta = u^2/2$ leads to easily computable expression for b and Q^* in the diffusion coefficients (3.11). In particular for inviscid Burgers equation they are defined as

$$b_{i+\frac{1}{2}} = \bar{u}_{i+\frac{1}{2}} \qquad (3.13a)$$

$$Q^*_{i+\frac{1}{2}} = \frac{1}{12}(u_{i+1} - u_i) \qquad (3.13b)$$

4   Artificial Neural Networks (ANNs)

Artificial neural networks are computational models mimicking the behavior of the human brain. They process the information with the help of simple units called neurons in a similar way the biological neurons do in the human brain. Each neuron is connected by edges called connections in this way they form a parallel processing network. The arrangement of neurons defines the type of network architecture. We describe here simple yet the powerful architecture of neural networks called multi-layer perceptions (MLPs). In MLPs, neurons are grouped in layers. The first layer is said to be a source (input) layer made of n_I number of neurons and the last layer is called the output layer containing n_O number of neurons. All the $M \geq 1$ layers between input and output layers are called hidden layers made of n_H^m number of neurons ($m = 1, 2...., M$). In fully connected MLP, each neuron in a layer is connected to the neurons in the immediate next layer as shown in Fig. 1.

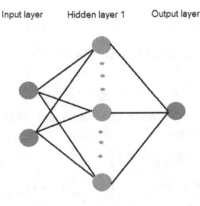

Input layer   Hidden layer 1   Output layer

Fig. 1. An MLP with one hidden layer. Number of neurons $n_I = 2$ $n_H = 20$ and $n_O = 1$.

4.1 Network Training

There are several paradigm of learning we focus here on supervised one. In the supervised learning the input for each forward pass through network and target values are given a priory. In this setting, approximation of a nonlinear function $\Phi : \mathbb{R}^{n_I} \to \mathbb{R}^{n_o}$ can be considered as a neural network map $\mathcal{N}$

$$\mathcal{N} : \mathbb{R}^{n_I} \to \mathbb{R}^{n_o} \text{ s.t. } \hat{y}_i = \mathcal{N}(x_i; W, b) \tag{4.1}$$

where the input data is $X = \{x_1, x_2, \ldots, x_N\}$ and corresponding target values $Y = \{y_1, y_2, \ldots y_N\}$ are such that $y_i = \Phi(x_i)$. Note that, each sample $(x_i, y_i) \in \mathbb{R}^{n_I} \times \mathbb{R}^{n_o}, i = 1, 2, \ldots N$. The learnable/trainable parameters W and b are called weight and bias of the network. In order to get a good approximation of the non-linear function Φ, the weights and biases of neural network $\mathcal{N}$ are trained so that $\mathcal{N}(X; W, b) \approx \Phi(X))$.

The parameter training aims to minimize the difference between the predicted value $\hat{Y} = \mathcal{N}(X; W, b)$ and the target truth value $Y = \Phi(X)$. This is done by minimizing a suitable cost function $C(Y, \hat{Y})$ with respect to the network parameters W and b. The idea is to train the network by optimal values of weights and biases of the network connections so that it minimizes C for the training data-set $\mathbb{T}$ fed into it.

The minimization of the cost function C is done by an iterative process called gradient descent based back-propagation algorithm. In simple word, the elements of weights and bias matrices are updated as

$$w_{i,j}^k = w_{i,j}^k - \eta \frac{\partial C}{\partial w_{i,j}} \tag{4.2a}$$

$$b_i^k = b_i^k - \eta \frac{\partial C}{\partial b_i} \tag{4.2b}$$

where $w_{i,j}^k$ is weight for node j in layer l_k for incoming node i, b_i^k is the associated bias and η is the learning rate. The trained network is tested on validation and test data which is outside the train data to determine its generalization performance. This helps to avoid over-fitting, this property of the network is called generalization for further details of ANNs see [GBC16, Kri05].

5 Viscosity Network

5.1 Network Architecture

We took a MLP network with one hidden layer as shown in Fig. 1 consist of $n_H^1 = 20$ neurons, input layer with $n_I = 2$ neuron ,output layer has $n_O = 1$ neuron. For calculating discrepancy in the prediction of the network with respect to actual output we chose Mean square error (MSE) loss as cost function $C(y, \hat{y})$.

$$C(y, \hat{y}) = \frac{1}{n} \sum_{1}^{n} (y_i - \hat{y}_i)^2 \tag{5.1}$$

where n is sample size. In order to introduce non-linearity in the network Relu as non-linear activation function [NH10, HSM+00] is used and shown in Fig. 2. The Adam optimization algorithm for training the MLP network is used. Note that Adam is an extension of stochastic gradient descent optimization [KB14]. In it, the entire training and validation data is divided into small batches and then pass into the training loop. Once the whole training set is exhausted, the training is said to have completed one epoch.

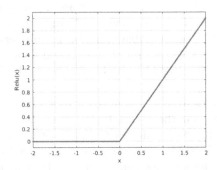

Fig. 2. Relu activation function, $Relu(x) = \max(x, 0)$.

5.2 Training and Validation Data Creation

The viscosity (diffusion) coefficient in the considered numerical fluxes can be viewed as a mapping $\alpha : \mathbb{R}^2 \to \mathbb{R}$. The input to network is given pairwise data $\mathbf{u} = (u_i, u_{i+1}), i = 1, 2, ...N$, where N is the number of samples. The input data X is generated through

- Smooth functions $S_i(x)$, jump functions $J_i(x)$ and their combinations.
- Random variants of smooth jump function i.e., $m_i S_i(x) \, k_i J_i(x)$, where k_i, m_i are random numbers from a Gaussian distribution and $i = 1, 2...N$ for some fixed N.

The target data Y is generated using the diffusion coefficients of respective fluxes defined in Eqs. (3.1), (3.2), (3.11). In order to train and validate the neural network, the generated data set $\mathbb{D} = (X, Y)$ with approximately $62,000$ samples is divided in to the train data $\mathbb{T} = (X_t, Y_t)$ and validation data $\mathbb{V} = (X_v, Y_v)$ in 70% and 30% ratios respectively. The trained model is saved for best validation accuracy i.e., for which the cost $C(y_v, \hat{y}_v)$ is minimum. This is one of the proven and tested approach to train the network without over-fitting with respect to training data $\mathbb{T}$. It is needed to be emphasized that in the field of machine learning the standard prevailing practice is to use normalized/scaled data a priory. However, since the considered network in this work has two input neurons, therefore underlying networks for each diffusion could be trained efficiently with raw data without any scaling. This makes the application of a trained network more efficient when it comes to numerical computation as there also scaling and the re-scaling step is omitted. The entire code for network training is written using PyTorch API [PGC+17] https://pytorch.org/. The learning rate in Adams optimizer is set to $\eta = 10^{-3}$ with all other arguments are set to default. The train batch size is 1024 and validation batch size 256 is used.

6   Numerical Results

We have given various numerical tests for scalar conservation laws to see the performance of the network in predicting the numerical diffusion. We consider the transport equation with constant speed 1 and Burger's equation with periodic boundary conditions. The training and validation loss plots of all the trained diffusion networks are given in Fig. 3 and Fig. 4. The trained network is used in Python codes of numerical schemes We followed few conventions to represent solutions by numerical schemes using exact diffusion coefficient and predicted one by the network, which are given by following

- For Entropy stable TVD scheme with diffusion $\mathcal{D}^1_{i+\frac{1}{2}}$, $\mathcal{D}^3_{i+\frac{1}{2}}$ labels ESTVD1 and ESTVD3 respectively are used whereas Lax-Wendroff and local Lax-Friedrichs scheme are represented by LxW and LLxF respectively.
- For the solutions calculated by using the network we use NN_LxW and NN_LLxF for Lax-Wendroff and Lax-Friedrich scheme for ESTVD1 and ESTVD3 we used NN_ESTVD1 and NN_ESTVD3.
- To represent the training and validation loss of network we use notations train_loss and val_loss respectively.

6.1   Network Loss Plots

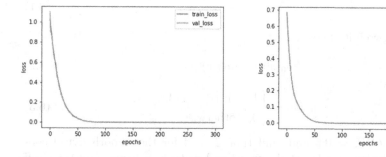

Fig. 3. Loss plot of the viscosity network trained for the Advection equation with LxF scheme (left), LxW sxcheme (right)

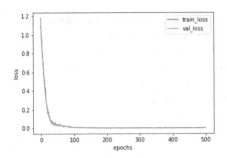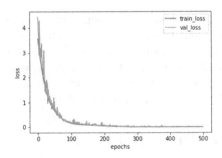

Fig. 4. Loss plots of the viscosity networks trained for Burger's equation with LLxF scheme (left), LxW scheme (right)

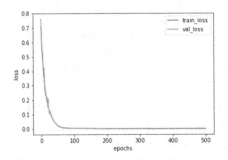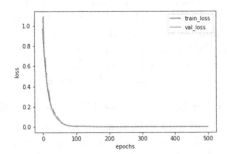

Fig. 5. Loss plots of the viscosity networks trained for Burger's equation with ESTVD1 scheme (left), ESTVD2 scheme (right)

6.2   Solution of Scalar Conservation Law

6.2.1   Advection Equation

The first set of test problems considered is following scalar Advection equation

$$u_t + u_x = 0 \tag{6.1}$$

with following two initial conditions

$$u_0(x) = sin(\pi x) \tag{6.2}$$

$$u_0(x) = \begin{cases} 1, & \text{if } |x| < 1/3. \\ 0, & \text{otherwise.} \end{cases} \tag{6.3}$$

we have taken $CFL = 0.5$ and final time $T = 4$ for the smooth initial condition (6.2). Solution is calculated using Lax-Friedrichs scheme, Lax-Wendroff scheme with exact and predicted diffusion by neural network(NN). Computed and exact solutions are shown in left side of the Fig. 6 and 7.

Similarly the solution at time $T = 2$ and $CFL = 0.5$, with respect to the (6.3) is shown in right side of Fig. 6 and 7.

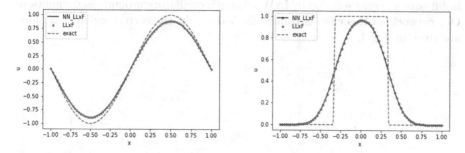

Fig. 6. Solution of Advection equation using Lax-Friedrichs scheme and trained network with initial conditions (6.2) (left), (6.3) (right).

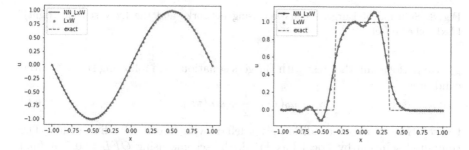

Fig. 7. Solution of Advection equation using Lax-Wendroff scheme and trained network with initial conditions (6.2) (left), (6.3) (right).

Note that, the solutions of the transport equation is simply a translation of initial data by fixed speed here $f'(u) = 1$. The computed solutions obtained by LxF is diffusive whereas the solution obtained by LxW is oscillatory near discontinuity as expected.

6.2.2 Burger's Equation
In the second test we take Burger's equation

$$u_t + \left(\frac{u^2}{2} \right)_x = 0 \tag{6.4}$$

with following initial condition

$$u_0(x) = \begin{cases} 1, & \text{if } |x| < 1/3. \\ 0, & \text{otherwise.} \end{cases} \tag{6.5}$$

The initial condition defined above introduces a steady shock and a rarefaction in its solution. Numerical results are given in Fig. 8 calculation are performed with CFL = 0.5 and final time $T = 0.3$. The solution by Local Lax-Friedrichs scheme is diffusive whereas solution by LxW scheme is oscillatory around discontinuities. Our network, in this case, is trained for 500 epochs the corresponding loss plots are given in Fig. 4.

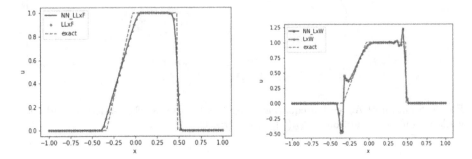

Fig. 8. Solution of Burger's equation using networks and the LxW scheme (right), LLxF scheme (left)

We perform another test with Burgers equation with following smooth initial condition

$$u_0(x) = \frac{1}{2} + sin(\pi x) \qquad (6.6)$$

the solution by above condition gives left moving shock after finite time. The computed solution by Local Lax Friedrichs scheme using $CFL = 0.5$ at final time $T = 0.5$ are shown in Fig. 9.

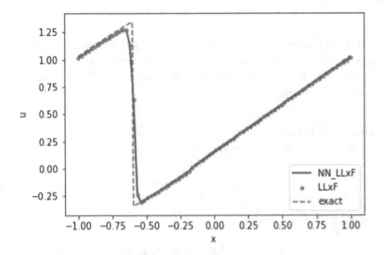

Fig. 9. Solution of Burger's equation using network and LLxF scheme

The networks to learn numerical viscosity of ESTVD schemes are trained for 500 epochs and the loss plots are shown in Fig. 5. Solution using ESTVD1 and ESTVD3 for Burgers equation with initial condition (6.5) are given in Fig. 10 using $CFL = 0.3$ at final time $T = 0.3$. Similarly the solution by ESTVD1 and ESTVD3 corresponding to initial condition (6.6) are given in Fig. 11 using $CFL = 0.5$ at final time $T = 0.5$.

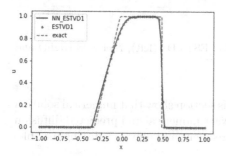 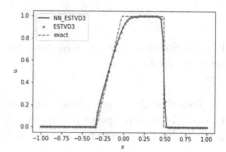

Fig. 10. The solution of Burger's equation with ESTVD1, ESTVD3 and corresponding networks for 6.3

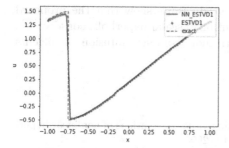 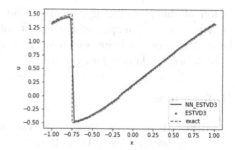

Fig. 11. Solution of Burger's equation by using ESTVD1 (left), ESTVD3 (right) and networks corresponding to initial condition (6.6)

In another test, consider the Burgers equation with initial condition

$$u_0(x) = -sin(\pi x), \ x \in [-1, 1] \tag{6.7}$$

The solution of Burgers equation with above initial condition develops a stationary shock centerd at $x = 0$. The computed solutions by ESTVD1 and ESTVD3 for Burgers equation with initial condition (6.7) are given in Fig. 12 using $CFL = 0.5$ at final time $T = 0.5$.

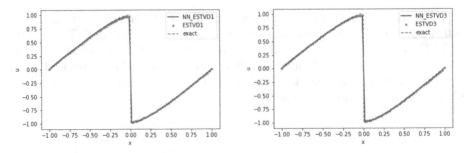

Fig. 12. Solution of Burger's equation by using ESTVD1 (left), ESTVD3 (right) and networks, for 6.7

Note that all the presented results in this section show that numerical solution by underlying three-point schemes using exact numerical and predicted diffusion using trained networks has a very good agreement and almost indistinguishable.

7   Concluding Remarks

In this work, a numerical diffusion network of multi-layer perception type is designed and implemented for non-linear regression to learn the numerical diffusion of shock-capturing schemes. Computed results show that the proposed model works well in approximating the solution of scalar hyperbolic conservation laws. This work is being extended to learn more complex diffusion coefficients in high order schemes for systems.

References

[BB05]   Bianchini, S., Bressan, A.: Vanishing viscosity solutions of nonlinear hyperbolic systems. Ann. Math. 223–342 (2005)

[DB18]   Dubey, R.K., Biswas, B.: Suitable diffusion for constructing non-oscillatory entropy stable schemes. J. Comput. Phys. **372**, 912–930 (2018)

[DDD+05] Dafermos, C.M., Dafermos, C.M.: Hyperbolic conservation laws in continuum physics, vol. 3. Springer, Berlin (2005). https://doi.org/10.1007/978-3-662-49451-6

[DHR20]   Discacciati, N., Hesthaven, J.S., Ray, D.: Controlling oscillations in high-order discontinuous galerkin schemes using artificial viscosity tuned by neural networks. J. Comput. Phys. **409**, 109304 (2020)

[GBC16]   Goodfellow, I., Bengio, Y., Courville, A.: Deep Learning. MIT press, Cambridge (2016)

[Har83]   Harten, A.: High resolution schemes for hyperbolic conservation laws. J. Comput. Phys. **49**(3), 357–393 (1983)

[HEOC87] Harten, A., Engquist, B., Osher, S., Chakravarthy, S.R.: Uniformly high order accurate essentially non-oscillatory schemes, III. In: Hussaini, M.Y., van Leer, B., Van Rosendale, J. (eds.) Upwind and High-Resolution Schemes. Springer, Berlin, Heidelberg (1987). https://doi.org/10.1007/978-3-642-60543-7_12

[Hes17]   Hesthaven, J.S.: Numerical methods for conservation laws: from analysis to algorithms. SIAM (2017)

[Hor91]   Hornik, K.: Approximation capabilities of multilayer feedforward networks. Neural Netw. **4**(2), 251–257 (1991)

[HSM+00] Hahnloser, R.H.R., Sarpeshkar, R., Mahowald, M.A., Douglas, R.J., Seung, H.S.: Digital selection and analogue amplification coexist in a cortex-inspired silicon circuit. Nature **405**(6789), 947–951 (2000)

[KB14]   Kingma, D.P., Adam, B.J.: A method for stochastic optimization. arXiv preprint arXiv:1412.6980 (2014)

[Kri05]   Kriesel, D.: A brief introduction to neural networks (2005). http://www.dkriesel.com/en/science/neural_networks. Accessed 30 Oct 2015

[Lan98]   Culbert, B.: Laney. Cambridge University Press, Computational gasdynamics (1998)

[Lax73]   Lax, P.D.: Hyperbolic systems of conservation laws and the mathematical theory of shock waves. SIAM (1973)

[LOC+94] Liu, X.-D., Osher, S., Chan, T., et al.: Weighted essentially non-oscillatory schemes. J. Comput. Phys. **115**(1), 200–212 (1994)

[NH10]   Nair, V., Hinton, G.E.: Rectified linear units improve restricted boltzmann machines. In: Fürnkranz, J., Joachims, T. (eds) ICML, pp. 807–814. Omnipress (2010)

[PGC+17] Paszke, A., et al.: Automatic differentiation in pytorch. In: NIPS-W (2017)

[RH18]   Ray, D., Hesthaven, J.S.: An artificial neural network as a troubled-cell indicator. J. Comput. Phys. **367**, 166–191 (2018)

[RPK17]   Raissi, M., Perdikaris, P., Karniadakis, G.E.: Physics informed deep learning (part i): Data-driven solutions of nonlinear partial differential equations. arXiv preprint arXiv:1711.10561 (2017)

[Shu12]   Shu, C.-W.: Efficient algorithms for solving partial differential equations with discontinuous solutions. Not. AMS **59**(5) (2012)

[Smo12]   Smoller, J.: Shock waves and reaction–diffusion equations, vol. 258. Springer Science & Business Media (2012)

[Swe84]   Sweby, P.K.: High resolution schemes using flux limiters for hyperbolic conservation laws. SIAM J. Numer. Anal. **21**(5), 995–1011 (1984)

[Tad87]   Tadmor, E.: The numerical viscosity of entropy stable schemes for systems of conservation laws. I. Math. Comp. **49**, 91–103 (1987)

[TB00]   Toro, E.F., Billett, S.J.: Centred tvd schemes for hyperbolic conservation laws. IMA J. Numer. Anal. **20**(1), 47–79 (2000)

Computational Study of Some Numerical Methods for the Generalized Burgers-Huxley Equation

Appanah Rao Appadu[1]($\boxtimes$), Yusuf Olatunji Tijani[1], and Justin Munyakazi[2]

[1] Department of Mathematics and Applied Mathematics, Nelson Mandela University, Port Elizabeth 6031, South Africa
Rao.Appadu@mandela.ac.za
[2] Department of Mathematics and Applied Mathematics,
University of the Western Cape, Private Bag X17, Bellville 7535, South Africa

Abstract. In this work, we initially construct three finite difference methods in order to solve the 1D Generalized Burgers-Huxley (GBH) equation which consists of advective, dissipative and reactive terms. We considered a numerical experiment where the coefficients of advective and reactive parameters are equal. These methods are two versions of non-standard finite difference and an explicit exponential finite difference method. Satisfactory results are obtained. To improve the results, each scheme is modified by using the technique of remainder effect. The modified schemes proved to be very efficient. By computing the L_1 and L_∞ errors as well as CPU time and rate of convergence, the performance of all the schemes are analysed.

Keywords: Generalized Burgers-Huxley equation · Nonstandard finite difference method · Explicit exponential finite difference method · Rate of convergence · L_1 and L_∞ errors

1 Introduction

The Generalized Burgers-Huxley equation which can be seen as the prototype for describing the relations between convection, reaction and diffusion is of great scientific and engineering importance. Some of the methods used for obtaining approximate and numerical solutions to the GBH equation are variational iteration method [1], adomian decomposition method [2] to list but a few. İnan [3] constructed an explicit exponential finite difference scheme (EEFDM) to solve the generalized Huxley and Burgers-Huxley equations. Zibaei *et al.* [4] constructed nonstandard finite difference method and a finite difference method using the exact solution to solve Burgers-Huxley equation. Recently, Appadu *et al.* [5] constructed four numerical methods to solve Burgers-Huxley equation and this was the first time that a comparison was made between nonstandard and exponential finite difference methods. The use of both nonstandard and exponential finite difference methods are very new for the Burgers-Huxley equation.

© Springer Nature Singapore Pte Ltd. 2021
A. Awasthi et al. (Eds.): CSMCS 2020, CCIS 1345, pp. 56–67, 2021.
https://doi.org/10.1007/978-981-16-4772-7_4

In this work, we solve the generalized Burgers-Huxley equation which is given by

$$u_t = u_{xx} - \alpha u^\delta u_x + \beta u (1 - u^\delta)(u^\delta - \gamma), \tag{1}$$

for $x \in [0,1]$ and $t \in [0, 1.0]$ subject to the initial condition $u(x,0) = (\frac{\gamma}{2} + \frac{\gamma}{2}\tanh\{\sigma\gamma x\})^{\frac{1}{\delta}}$, where $\alpha > 0$, $\beta > 0$, $0 < \gamma < 1$ and $\delta > 0$ is a positive constant, $\sigma = \dfrac{\delta(\rho - \alpha)}{4(1+\delta)}$ and $\rho = \sqrt{\alpha^2 + 4\beta(1+\delta)}$. α, β are the coefficients of advection and reaction respectively and γ is the species carrying capacity. The exact solution to Eq. (1) is obtained from Wang et al. [6] and lies in the interval $(0, \gamma^{\frac{1}{\delta}})$. In this study, we work with $\delta = 4$, $\alpha = 1.0$, $\beta = 1.0$ and $\gamma = 0.01$. The time of the experiment is $T_{max} = 1.0$. The performance of the methods was measured by computing L_1 and L_∞ errors as well as the rate of convergence (using L_∞ norm) in time as follows:

$$L_1 = h \sum_{j=1}^{N} |u(x_j, t_n) - U(x_j, t_n)|, \tag{2}$$

$$L_\infty = \max |u(x_j, t_n) - U(x_j, t_n)|, \tag{3}$$

and

$$R_t = \frac{\log(e_d/e_{2d})}{\log 2}, \tag{4}$$

where e_{2d} and e_d are the L_∞ errors corresponding to the number of grid points $2d$ and d respectively.

Tables 1, 2, 3, 4, 5 and 6 show the L_1, L_∞ errors and rate of convergence in time. Figures 1, 2, 3, 4, 5 and 6 shed light on the performance of the schemes.

2   Nonstandard Finite Difference Scheme (NSFD)

The derivations are mostly based on the idea of dynamical consistency which are positivity, boundedness and monotonicity of the solutions, see Mickens [7].

2.1   NSFD1

Two versions of NSFD schemes were constructed by Appadu et al. [5] for discretising the Burgers-Huxley equation. We modify these schemes to discretise the Generalized Burgers-Huxley equation.

We propose the following scheme for Eq. (1):

$$\frac{U_j^{n+1} - U_j^n}{\phi(k)} = \left[\frac{U_{j+1}^n - 2U_j^n + U_{j-1}^n}{[\psi(h)]^2}\right] - \alpha U_j^{n+1}(U_j^n)^3 \frac{U_j^n - U_{j-1}^n}{\psi(h)} + \beta(1+\gamma)\left[2(U_j^n)^5 - (U_j^n)^4 U_j^{n+1}\right]$$
$$- \beta\gamma U_j^{n+1} - \beta U_j^{n+1}(U_j^n)^8. \tag{5}$$

The denominator functions are $\psi(h) = \dfrac{e^{\beta h} - 1}{\beta}$ and $\phi(k) = \dfrac{e^{\beta k} - 1}{\beta}$. We obtain

$$U_j^{n+1} = \frac{(1 - 2R)U_j^n + R(U_{j+1}^n + U_{j-1}^n) + 2\phi(k)\beta(1 + \gamma)(U_j^n)^5}{1 + \alpha r(U_j^n)^3(U_j^n - U_{j-1}^n) + \phi(k)\beta\gamma + \phi(k)\beta(1 + \gamma)(U_j^n)^4 + \phi(k)\beta(U_j^n)^8}, \quad (6)$$

where $R = \dfrac{\phi(k)}{[\psi(h)]^2}$ and $r = \dfrac{\phi(k)}{\psi(h)}$.

For positivity, we require $1 - 2R \geq 0$ and $1 - \alpha r\gamma \geq 0$. Using $h = 0.1$, we require $k \leq 5.515 \times 10^{-3}$ or $k \leq T_{max}/182$.

Boundedness

We assume $U_j^n \in \left[0, \gamma^{\frac{1}{4}}\right]$ for all considered values of n and j. Therefore,

$$\left(U_j^{n+1} - \gamma^{\frac{1}{4}}\right)\left(1 + \alpha r(U_j^n)^3(U_j^n - U_{j-1}^n) + \phi(k)\beta\gamma + \phi(k)\beta(1 + \gamma)(U_j^n)^4 + \phi(k)\beta(U_j^n)^8\right)$$

$$= (1 - 2R)U_j^n + R(U_{j+1}^n + U_{j-1}^n) + 2\phi(k)\beta(1 + \gamma)(U_j^n)^5 - \gamma^{\frac{1}{4}} - \alpha r\gamma^{\frac{1}{4}}(U_j^n)^3(U_j^n - U_{j-1}^n) - \phi(k)\beta\gamma^{\frac{5}{4}}$$

$$-\phi(k)\beta\gamma^{\frac{1}{4}}(1 + \gamma)(U_j^n)^4 - \phi(k)\beta\gamma^{\frac{1}{4}}(U_j^n)^8 \leq -\alpha r\gamma^{\frac{1}{4}}(U_j^n)^3(U_j^n - U_{j-1}^n) \leq 0. \quad (7)$$

Table 1. L_1 and L_∞ errors, CPU times and rate of convergence (in time) for $\alpha = 1.0, \beta = 1.0, \gamma = 0.01$ at some different time-step size k with spatial mesh size $h = 0.1$ using NSFD1 at $T_{max} = 1.0$.

Time step (k)	L_1 Error	L_∞ Error	R_t	CPU (s)
$T_{max}/2000$	1.967743×10^{-4}	2.981740×10^{-4}	-	0.3044
$T_{max}/4000$	1.967834×10^{-4}	2.981878×10^{-4}	-6.667×10^{-5}	0.5403
$T_{max}/8000$	1.967880×10^{-4}	2.981947×10^{-4}	-3.338×10^{-5}	1.1334
$T_{max}/16000$	1.967902×10^{-4}	2.981982×10^{-4}	-1.693×10^{-5}	2.5852

2.2   NSFD1-ϵ

Ruxun *et al.* [8] designed a stable, high resolution scheme by controlling the numerical effects of dispersion and dissipation. They used a simple approach by reforming the Lax-Wendroff (LW) scheme when used to discretise the 1-D linear advection equation. LW-ϵ was derived which is monotonic, positive and second-order accurate in space. Some more work on use of remainder effect technique can be found in Appadu *et al.* [9], Agbavon *et al.* [10]. We construct NSFD1-ϵ by adding the expression $\epsilon_1 U_{j+1}^n + \epsilon_2 U_j^n + \epsilon_3 U_{j-1}^n$ to the right hand side of Eq. (6). We have

$$U_j^{n+1} = \frac{(1 - 2R)U_j^n + R(U_{j+1}^n + U_{j-1}^n) + 2\phi(k)\beta(1 + \gamma)(U_j^n)^5}{1 + \alpha r(U_j^n)^3(U_j^n - U_{j-1}^n) + \phi(k)\beta\gamma + \phi(k)\beta(1 + \gamma)(U_j^n)^4 + \phi(k)\beta(U_j^n)^8} + \epsilon_1 U_{j+1}^n + \epsilon_2 U_j^n + \epsilon_3 U_{j-1}^n.$$

$$(8)$$

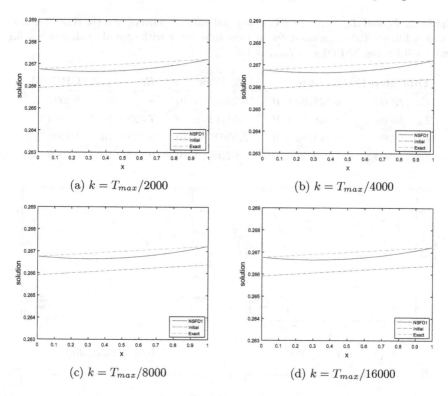

Fig. 1. Plot of initial and numerical, exact profiles at time $T_{max} = 1.0$ with $h = 0.1$ and some different values of k. The numerical method used is NSFD1.

For consistency, we have $\epsilon_1 = \epsilon_3 = \epsilon$ and $\epsilon_2 = -2\epsilon$. The NSFD1-ϵ scheme is given by

$$U_j^{n+1} = \frac{(1 - 2R)U_j^n + R(U_{j+1}^n + U_{j-1}^n) + 2\phi(k)\beta(1 + \gamma)(U_j^n)^5 + \epsilon\Omega(U_{j+1}^n - 2U_j^n + U_{j-1}^n)}{\Omega}, \quad (9)$$

where $\Omega = 1 + \alpha r(U_j^n)^3(U_j^n - U_{j-1}^n) + \phi(k)\beta\gamma + \phi(k)\beta(1 + \gamma)(U_j^n)^4 + \phi(k)\beta(U_j^n)^8$. We require $0 < \epsilon << 1$, see [8]. It is noted that the scheme has first order accuracy in time and first order accuracy in space on performing truncation analysis. For positivity, $1 - 2R - 2\epsilon\Omega \geq 0$ and $1 - \alpha r\gamma \geq 0$. Using $h = 0.1$, we obtain $k \leq 4.414 \times 10^{-3}$ or $k \leq T_{max}/227$.

Boundedness

We assume $0 \leq U_j^n \leq \gamma^{\frac{1}{4}}$ for all considered values of n and j, thus

$$(U_j^{n+1} - \gamma^{\frac{1}{4}})\Omega = (1 - 2R)U_j^n + R(U_{j+1}^n + U_{j-1}^n) + 2\phi(k)\beta(1 + \gamma)(U_j^n)^5 + \epsilon\Omega(U_{j+1}^n - 2U_j^n + U_{j-1}^n)$$

$$- \gamma^{\frac{1}{4}} - \alpha r\gamma^{\frac{1}{4}}(U_j^n)^3(U_j^n - U_{j-1}^n) - \phi(k)\beta\gamma^{\frac{5}{4}} - \phi(k)\beta\gamma^{\frac{1}{4}}(1 + \gamma)(U_j^n)^4 - \phi(k)\beta\gamma^{\frac{1}{4}}(U_j^n)^8$$

$$\leq -\alpha r\gamma^{\frac{1}{4}}(U_j^n)^3(U_j^n - U_{j-1}^n) \leq 0. \quad (10)$$

Table 2. L_1 and L_∞ errors, CPU times and rate of convergence (in time) for $\alpha = 1.0, \beta = 1.0, \gamma = 0.01$ at some different time-step size k with spatial mesh size $h = 0.1$ and $\epsilon = 0.1$ using NSFD1-ϵ at $T_{max} = 1.0$.

Time step (k)	L_1 Error	L_∞ Error	R_t	CPU (s)
$T_{max}/2000$	6.131780×10^{-5}	9.290899×10^{-5}	–	0.2692
$T_{max}/4000$	3.638588×10^{-5}	5.513125×10^{-5}	7.529×10^{-1}	0.5284
$T_{max}/8000$	2.013180×10^{-5}	3.050307×10^{-5}	8.539×10^{-1}	1.1998
$T_{max}/16000$	1.070422×10^{-5}	1.621861×10^{-5}	9.112×10^{-1}	2.6968

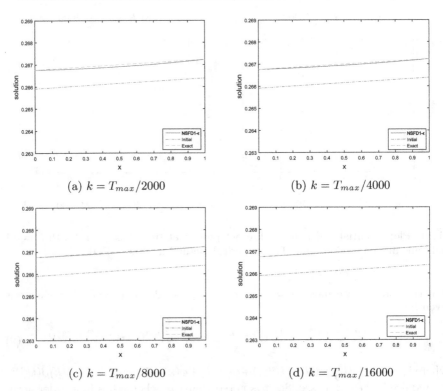

(a) $k = T_{max}/2000$

(b) $k = T_{max}/4000$

(c) $k = T_{max}/8000$

(d) $k = T_{max}/16000$

Fig. 2. Plot of initial and numerical, exact profiles at time $T_{max} = 1.0$ with $h = 0.1$ and some different values of k. The numerical method used is NSFD1-ϵ.

2.3 NSFD2

We propose the following scheme to discretise Eq. (1):

$$\frac{U_j^{n+1} - U_j^n}{\phi(k)} = \left[\frac{U_{j+1}^n - 2U_j^n + U_{j-1}^n}{[\psi(h)]^2} \right] - \alpha U_j^{n+1}(U_j^n)^3 \frac{U_j^n - U_{j-1}^n}{\psi(h)} \beta(1+\gamma)(U_j^n)^5 - \beta\gamma U_j^{n+1} - \beta \left[2U_j^{n+1}(U_j^n)^8 - (U_j^n)^9 \right]. \quad (11)$$

The denominator functions are the same as in NSFD1. We have

$$U_j^{n+1} = \frac{(1-2R)U_j^n + R(U_{j+1}^n + U_{j-1}^n) + \beta\phi(k)[(1+\gamma)(U_j^n)^5 + (U_j^n)^9]}{1 + \alpha r(U_j^n)^3(U_j^n - U_{j-1}^n) + \phi(k)\beta\gamma + 2\phi(k)\beta(U_j^n)^8}. \quad (12)$$

For positivity, the requirement is $1 - 2R \geq 0$ and $1 - \alpha r\gamma \geq 0$. We have the same condition as for NSFD1 scheme.

Boundedness

$$(U_j^{n+1} - \gamma^{\frac{1}{4}})\left(1 + \alpha r(U_j^n)^3(U_j^n - U_{j-1}^n) + \phi(k)\beta\gamma + 2\phi(k)\beta(U_j^n)^8\right) = (1-2R)U_j^n + R(U_{j+1}^n + U_{j-1}^n)$$

$$+ \phi(k)\beta(1+\gamma)((U_j^n)^5 + (U_j^n)^9) - \gamma^{\frac{1}{4}} - \alpha r\gamma^{\frac{1}{4}}(U_j^n)^3(U_j^n - U_{j-1}^n) - \phi(k)\beta\gamma^{\frac{5}{4}} - 2\phi(k)\beta\gamma^{\frac{1}{4}}(U_j^n)^8$$

$$\leq -\alpha r(U_j^n)^3\gamma^{\frac{1}{4}}(U_j^n - U_{j-1}^n) \leq 0. \quad (13)$$

Table 3. L_1 and L_∞ errors, CPU times and rate of convergence (in time) for $\alpha = 1.0, \beta = 1.0, \gamma = 0.01$ at some different time-step size k with spatial mesh size $h = 0.1$ using NSFD2 at $T_{max} = 1.0$.

Time step (k)	L_1 Error	L_∞ Error	R_t	CPU (s)
$T_{max}/2000$	1.967740×10^{-4}	2.981737×10^{-4}	-	0.2936
$T_{max}/4000$	1.967833×10^{-4}	2.981877×10^{-4}	-6.773×10^{-5}	0.5784
$T_{max}/8000$	1.967879×10^{-4}	2.981947×10^{-4}	-3.387×10^{-5}	1.2037
$T_{max}/16000$	1.967902×10^{-4}	2.981981×10^{-4}	-1.644×10^{-5}	2.7648

2.4   NSFD2-ϵ

We construct NSFD2-ϵ by adding the expression $\epsilon_1 U_{j+1}^n + \epsilon_2 U_j^n + \epsilon_3 U_{j-1}^n$ to the right hand side of Eq. (12). This yields

$$U_j^{n+1} = \frac{(1-2R)U_j^n + R(U_{j+1}^n + U_{j-1}^n) + \beta\phi(k)[(1+\gamma)(U_j^n)^5 + (U_j^n)^9]}{1 + \alpha r(U_j^n)^3(U_j^n - U_{j-1}^n) + \phi(k)\beta\gamma + 2\phi(k)\beta(U_j^n)^8} + \epsilon_1 U_{j+1}^n + \epsilon_2 U_j^n + \epsilon_3 U_{j-1}^n \quad (14)$$

Same consistency condition as NSFD1-ϵ was obtained. Further simplification of Eq. (14) produces

$$U_j^{n+1} = \frac{(1-2R)U_j^n + R(U_{j+1}^n + U_{j-1}^n) + \phi(k)\beta((1+\gamma)(U_j^n)^5 + (U_j^n)^9) + \epsilon\chi(U_{j+1}^n - 2U_j^n + U_{j-1}^n)}{\chi}, \quad (15)$$

where $\chi = 1 + \alpha r(U_j^n)^3(U_j^n - U_{j-1}^n) + \phi(k)\beta\gamma + 2\phi(k)\beta(U_j^n)^8$.

We require $0 < \epsilon \ll 1$. It is noted that the scheme is first order accuracy in time and first order accuracy in space. For positively invariant solution $1 - 2R - 2\epsilon\chi \geq 0$ and $1 - \alpha r\gamma \geq 0$. We have that $k \leq 4.414 \times 10^{-3}$ or $k \leq T_{max}/227$.

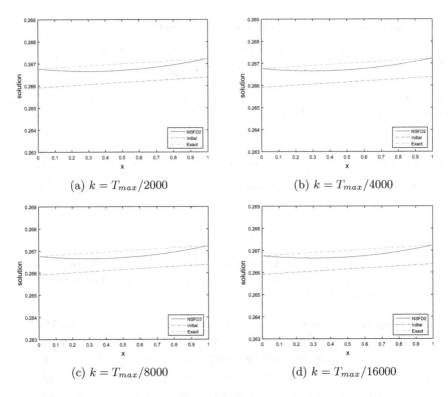

(a) $k = T_{max}/2000$

(b) $k = T_{max}/4000$

(c) $k = T_{max}/8000$

(d) $k = T_{max}/16000$

Fig. 3. Plot of initial and numerical, exact profiles at time $T_{max} = 1.0$ with $h = 0.1$ and some different values of k. The numerical method used is NSFD2.

Boundedness

We assume $0 \leq U_j^n \leq \gamma^{\frac{1}{4}}$ for all relevant values of n and j.

$$(U_j^{n+1} - \gamma^{\frac{1}{4}})_x = (1 - 2R)U_j^n + R(U_{j+1}^n + U_{j-1}^n) + \phi(k)\beta((1+\gamma)(U_j^n)^5 + (U_j^n)^9) + \epsilon\Omega(U_{j+1}^n - 2U_j^n + U_{j-1}^n)$$

$$-\chi \leq \phi(k)\beta(1+\gamma)((U_j^n)^5 + (U_j^n)^9) - \alpha r\gamma^{\frac{1}{4}}(U_j^n)^3(U_j^n - U_{j-1}^n) - \phi(k)\beta\gamma^{\frac{5}{4}} - \phi(k)\beta\gamma^{\frac{1}{4}}(U_j^n)^8$$

$$\leq -\alpha r\gamma^{\frac{1}{4}}(U_j^n)^3(U_j^n - U_{j-1}^n) \leq 0.$$

$$(16)$$

3 Explicit Exponential Finite Difference Method (EEFDM)

In 1985, Bhatacharry [11] conceived the idea of exponential finite difference scheme. This scheme was designed in [11] to solve the heat equation. This scheme conserves the properties (positivity) of many emerging mathematical models in mathematical biology. By rearranging Eq. (1), we have

$$\frac{\partial u}{\partial t} = \beta u \left(1 - u^\delta\right)\left(u^\delta - \gamma\right) - \alpha u^\delta \frac{\partial u}{\partial x} + \frac{\partial^2 u}{\partial x^2} .$$

Table 4. L_1 and L_∞ errors, CPU times and rate of convergence (in time) for $\alpha = 1.0, \beta = 1.0, \gamma = 0.01$ at some different time-step size k with spatial mesh size $h = 0.1$ and $\epsilon = 0.1$ using NSFD2-ϵ at $T_{max} = 1.0$.

Time step (k)	L_1 Error	L_∞ Error	R_t	CPU (s)
$T_{max}/2000$	6.131785×10^{-5}	9.290906×10^{-5}	–	0.2645
$T_{max}/4000$	3.638590×10^{-5}	5.513128×10^{-5}	7.529×10^{-1}	0.5259
$T_{max}/8000$	2.013180×10^{-5}	3.050308×10^{-5}	8.539×10^{-1}	1.1093
$T_{max}/16000$	1.070422×10^{-5}	1.621861×10^{-5}	9.113×10^{-1}	2.6161

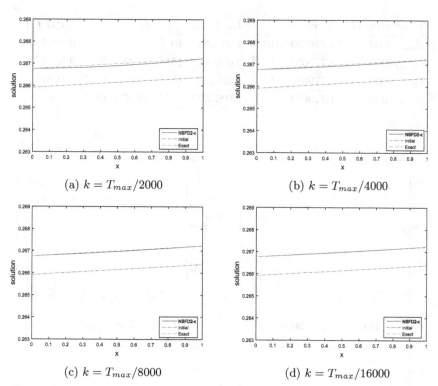

(a) $k = T_{max}/2000$

(b) $k = T_{max}/4000$

(c) $k = T_{max}/8000$

(d) $k = T_{max}/16000$

Fig. 4. Plot of initial and numerical, exact profiles at time $T_{max} = 1.0$ with $h = 0.1$ and some different values of k. The numerical method used is NSFD2-ϵ.

Dividing by u and using standard finite difference approximations for derivatives, we obtain

$$\Lambda U_j^n = \frac{1}{U_j^n} \left[\beta U_j^n \left(1 - (U_j^n)^\delta\right) \left((U_j^n)^\delta - \gamma\right) - \alpha (U_j^n)^\delta \Lambda_x^{(1)} U_j^n + \Lambda_x^{(2)} U_j^n \right], \quad (17)$$

By setting $\delta = 4$, we have

$$U_j^{n+1} = U_j^n \exp\left\{\frac{k}{U_j^n}\left[\beta U_j^n(1-(U_j^n)^4)((U_j^n)^4-\gamma) - \alpha(U_j^n)^4\left(\frac{U_{j+1}^n-U_{j-1}^n}{2h}\right) + \left(\frac{U_{j+1}^n-2U_j^n+U_{j-1}^n}{h^2}\right)\right]\right\}$$

(18)

Table 5. L_1 and L_∞ errors, CPU times and rate of convergence (in time) for $\alpha = 1.0, \beta = 1.0, \gamma = 0.01$ at some different time-step size k with spatial mesh size $h = 0.1$ using EEFDM at $T_{max} = 1.0$.

Time step (k)	L_1 Error	L_∞ Error	R_t	CPU (s)
$T_{max}/2000$	1.779623×10^{-4}	2.696663×10^{-4}	-	0.2964
$T_{max}/4000$	1.779623×10^{-4}	2.696662×10^{-4}	9.521×10^{-8}	0.5549
$T_{max}/8000$	1.779623×10^{-4}	2.696662×10^{-4}	5.049×10^{-8}	1.1924
$T_{max}/16000$	1.779623×10^{-4}	2.696662×10^{-4}	2.596×10^{-8}	2.7481

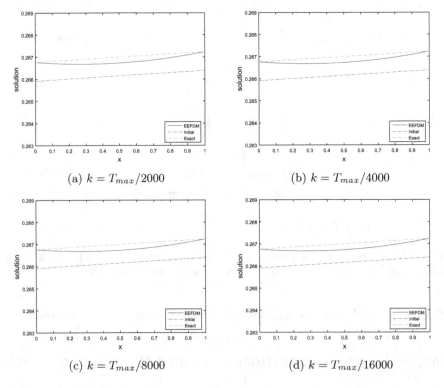

(a) $k = T_{max}/2000$

(b) $k = T_{max}/4000$

(c) $k = T_{max}/8000$

(d) $k = T_{max}/16000$

Fig. 5. Plot of initial, numerical and exact profiles vs x for EEFDM at time $T_{max} = 1.0$ with $h = 0.1$ for different values of k.

3.1   EEFDM-ϵ

We modified the scheme in Eq. (18) following the idea of Ruxun *et al.* [8]. We have

$$U_j^{n+1} = U_j^n \exp\left\{\frac{k}{U_j^n}\left[\beta U_j^n(1 - (U_j^n)^4)((U_j^n)^4 - \gamma) - \alpha(U_j^n)^4\left(\frac{U_{j+1}^n - U_{j-1}^n}{2h}\right) + \left(\frac{U_{j+1}^n - 2U_j^n + U_{j-1}^n}{h^2}\right)\right]\right\}$$

$$+ \epsilon_1 U_{j+1}^n + \epsilon_2 U_j^n + \epsilon_3 U_{j-1}^n. \quad (19)$$

Same consistency condition as NSFD1-ϵ were obtained.

Table 6. L_1 and L_∞ errors, CPU times and rate of convergence (in time) for $\alpha = 1.0, \beta = 1.0, \gamma = 0.01$ at some different time-step size k with spatial mesh size $h = 0.1$ and $\epsilon = 0.1$ using EEFDM-ϵ at $T_{max} = 1.0$.

Time step (k)	L_1 Error	L_∞ Error	R_t	CPU (s)
$T_{max}/2000$	5.936640×10^{-5}	8.995212×10^{-5}	$-$	0.2808
$T_{max}/4000$	3.569349×10^{-5}	5.408214×10^{-5}	7.340×10^{-1}	0.5573
$T_{max}/8000$	1.992047×10^{-5}	3.018287×10^{-5}	8.414×10^{-1}	1.2706
$T_{max}/16000$	1.064557×10^{-5}	1.612975×10^{-5}	9.040×10^{-1}	2.7636

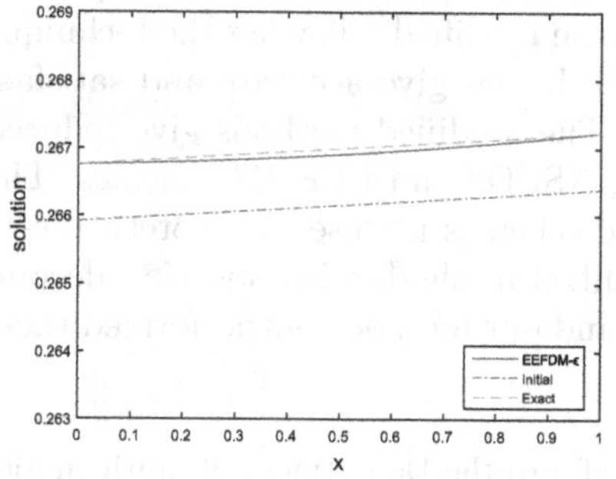

(a) $k = T_{max}/2000$

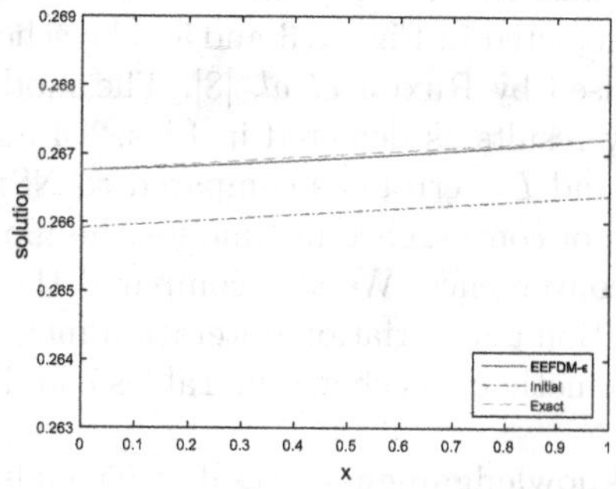

(b) $k = T_{max}/4000$

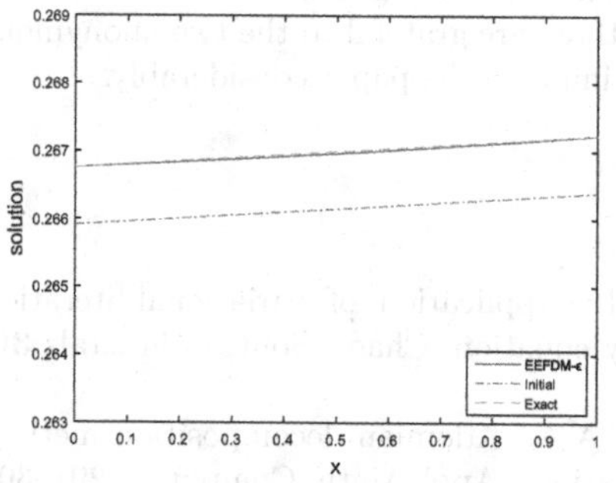

(c) $k = T_{max}/8000$

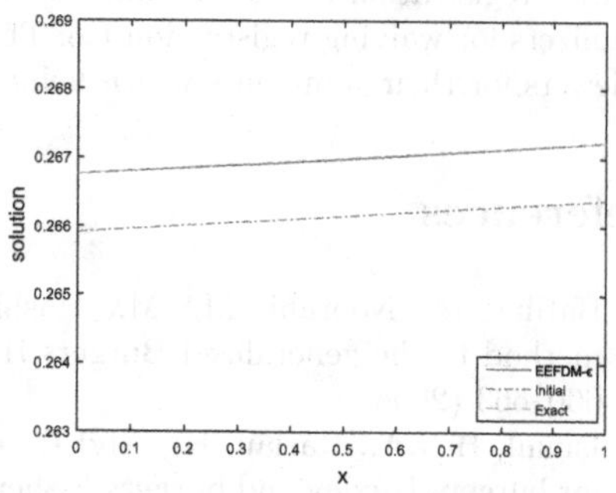

(d) $k = T_{max}/16000$

Fig. 6. Plot of initial and numerical, exact profiles at time $T_{max} = 1.0$ with $h = 0.1$ and some different values of k. The numerical method used is EEFDM-ϵ.

Table 7. Absolute errors from six constructed schemes with the results of [1] and [2] for $\alpha = 1.0, \beta = 1.0$ and $\gamma = 0.01$ using $k = 0.0005$

t	x	NSFD1	NSFD1-ε	NSFD2	NSFD2-ε	EEFDM	EEFDM-ε	VIM [1]	ADM [2]
0.1	0.1	6.8702×10^{-5}	3.2044×10^{-5}	6.8702×10^{-5}	3.2044×10^{-5}	6.5341×10^{-5}	3.1174×10^{-5}	2.1768×10^{-4}	2.1778×10^{-4}
0.1	0.5	1.7270×10^{-4}	8.8190×10^{-5}	1.7270×10^{-4}	8.8190×10^{-5}	1.6656×10^{-4}	8.5868×10^{-5}	2.1721×10^{-4}	2.1731×10^{-4}
0.5	0.1	1.0670×10^{-4}	3.3623×10^{-5}	1.0670×10^{-4}	3.3623×10^{-5}	9.6886×10^{-5}	3.2553×10^{-5}	1.0876×10^{-3}	1.0901×10^{-3}
0.5	0.5	2.9597×10^{-4}	9.3362×10^{-5}	2.9597×10^{-4}	9.3362×10^{-5}	2.6890×10^{-4}	9.0391×10^{-5}	1.0852×10^{-3}	1.0877×10^{-3}

Table 8. Relative errors from six constructed schemes with the results of [1] and [2] for $\alpha = 1.0, \beta = 1.0$ and $\gamma = 0.01$ using $k = 0.0005$

t	x	NSFD1	NSFD1-ε	NSFD2	NSFD2-ε	EEFDM	EEFDM-ε	VIM [1]	ADM [2]
0.1	0.1	2.5823×10^{-2}	1.2044×10^{-2}	2.5823×10^{-2}	1.2044×10^{-2}	2.4560×10^{-2}	1.1717×10^{-2}	8.1822×10^{-2}	8.1860×10^{-2}
0.1	0.5	6.4870×10^{-2}	3.3124×10^{-2}	6.4870×10^{-2}	3.3124×10^{-2}	6.2561×10^{-2}	3.2252×10^{-2}	8.1588×10^{-2}	8.1625×10^{-2}
0.5	0.1	4.0055×10^{-2}	1.2622×10^{-2}	4.0055×10^{-2}	1.2622×10^{-2}	3.6370×10^{-2}	1.2220×10^{-2}	4.0828×10^{-1}	4.0921×10^{-1}
0.5	0.5	1.1102×10^{-1}	3.5022×10^{-2}	1.1102×10^{-1}	3.5022×10^{-2}	1.0087×10^{-1}	3.3908×10^{-2}	4.0711×10^{-1}	4.0804×10^{-1}

4   Conclusion

Three schemes were initially constructed in order to solve the generalized Burgers-Huxley equation and we obtained conditions for which the schemes are positive definite and bounded. There is some phase error in the profiles as depicted in Figs. 1, 3 and 5. The schemes are modified following the technique devised by Ruxun et al. [8]. The modified schemes give accurate and satisfactory results as depicted in Figs. 2, 4 and 6. The modified methods give reduced L_1 and L_∞ errors as compared to NSFD1, NSFD2 and EEFDM schemes. The rate of convergence in time for the modified schemes is close to theoretical rate of convergence. We also compared the modified methods with adomian decomposition and variational iteration methods and our methods outperformed these two methods as shown in Tables 7 and 8.

Acknowledgements. Yusuf O. Tijani is grateful to the Department of Mathematics and Applied Mathematics of Nelson Mandela University for providing some funding to cover registration fees at the university. A. R. Appadu is grateful to the conference organizers for waiving registration fee. The authors are grateful to the two anonymous reviewers for their comments which helped to improve the paper considerably.

References

1. Batiha, B., Noorani, M.S.M., Hashim, I.: Application of variational iteration method to the generalized Burgers-Huxley equation. Chaos, Solitons Fractals **36**, 660–663 (2008)
2. Ismail, H.N.A., Raslan, K., Abd-Rabboh, A.A.: Adomian decomposition method for burgers-Huxley and burgers-Fisher equations. Appl. Math. Comput. **1**, 291–301 (2004)
3. İnan, B.: Finite difference methods for the generalized Huxley and Burgers-Huxley equations. Kuwait J. Sci. **44**, 20–27 (2017)

4. Zibaei, M., Zeinadini, S., Namjoo, M.: Numerical solutions of Burgers-Huxley equation by exact finite difference and NSFD schemes. J. Differ. Equ. Appl. **22**, 1098–1113 (2016)

5. Appadu, A.R., İnan, B., Tijani, Y.O.: Comparative study of some numerical methods for the Burgers-Huxley equation. Symmetry **11**, 1333 (2019). https://doi.org/10.3390/sym11111333

6. Wang, Z.S., Zhuo, X.Y., Lu, Y.K.: Solitary wave solutions of the generalised Burgers-Huxley equation. J. Phys. A: Math. Gen. **23**, 271–274 (1990)

7. Mickens, R.E.: Application of Nonstandard Finite Difference Scheme, pp. 1–261. World Scientific, Singapore (2000)

8. Ruxun, L., Mengping, Z., Ji, W., Xiao-Yuan, L.: The designing approach of difference schemes by controlling the remainder-effect. Int. J. Numer. Meth. Fluids **32**, 523–533 (1999)

9. Appadu, A.R., Dauhoo, M.Z., Rughooputh, S.D.D.V.: Efficient shock-capturing numerical schemes using the approach of minimised integrated square difference error for hyperbolic conservation laws. Proc. Int. Conf. Comput. Its Appl. **32**, 774–789 (2007)

10. Agbavon, K.M., Appadu, A.R., Khumalo, N.: On the numerical solution of Fisher's equation with coefficient of reaction term much smaller than coefficient of reaction term. Adv. Differ. Eqn. **146**, 33 (2019)

11. Bhattacharya, M.C.: An explicit conditionally stable finite difference equation for heat conduction problems. Int. J. Numer. Method Eng. **21**, 239–265 (1985)

A Local Meshless Method
for a Multi-term Time Fractional
Non-linear Diffusion Equation

Akanksha Bhardwaj, Karuna Pati Tripathi, and Alpesh Kumar$^{(\boxtimes)}$ [ID]

Department of Basic Sciences and Humanities, Rajiv Gandhi Institute of Petroleum
Technology, Jais, Amethi 229304, India
{pmth17001,kptripathi,alpeshk}@rgipt.ac.in

Abstract. In the current work, a radial basis function based local mesh-less method is taken into consideration to solve the multi-term time fractional nonlinear diffusion equation. We mentioned the proof of unconditional stability and also theoretically discussed the convergence of the proposed numerical scheme. Some numerical problems are given to show the exactness and efficiency of the developed scheme. The present result indicates that the proposed numerical scheme is very accurate and efficient for modeling and simulating the considered problems.

Keywords: Diffusion equation · RBF · Meshless method · Finite difference method

1 Introduction

In this paper, we consider the time fractional non-linear multi-term diffusion equation

$$\begin{cases} \sum_{i=0}^{l} d_i \, {}_0^c\mathscr{D}_t^{\alpha_i} u(\mathbf{x},t) = \Delta u(\mathbf{x},t) - F(u(\mathbf{x},t)) + f(\mathbf{x},t), & (\mathbf{x},t) \in \Omega \times (0,T], \\ u(\mathbf{x},0) = \xi(\mathbf{x}), & \mathbf{x} \in \Omega \\ u(\mathbf{x},t) = 0, & (\mathbf{x},t) \in \partial\Omega \times (0,T], \end{cases} \tag{1}$$

where $0 < \alpha_l \leq \ldots \leq \alpha_0 < 1$, $d_i \geq 0$, $i = 0,1,\ldots,l$, $l \in \mathbb{N}$, $T > 0$, with convention $d_0 = 1$ and $\xi(\mathbf{x})$ and $f(\mathbf{x},t)$ are sufficiently smooth functions on closed and bounded domain $\Omega \subset \mathbb{R}^2$, with Lipschitz boundary $\partial\Omega$. The non-linear function $F(u(\mathbf{x},t))$ satisfied the assumption $|F(u(\mathbf{x},t))| \leq C|u(\mathbf{x},t)|$, and first derivative of $F(u(\mathbf{x},t))$ with respect to u is bounded. Furthermore for any positive integer z, ${}_0^c\mathscr{D}_t^{\alpha} u(\mathbf{x},t)$ the Caputo's differential operator is defined as follows

$$\begin{cases} {}_0^c\mathscr{D}_t^{\alpha} u(\mathbf{x},t) = \begin{cases} \frac{1}{\Gamma(z-\alpha)} \int_0^t \frac{\partial^z u(\mathbf{x},s)}{\partial s^z} \frac{ds}{(t-s)^{\alpha-(z-1)}}, & z-1 < \alpha < z, \\ \frac{\partial^z u(\mathbf{x},t)}{\partial t^z}, & \alpha = z. \end{cases} \end{cases} \tag{2}$$

Supported by Rajiv Gandhi Institute of Petroleum Technology, Jais, Amethi.

A. Awasthi et al. (Eds.): CSMCS 2020, CCIS 1345, pp. 68–78, 2021.
https://doi.org/10.1007/978-981-16-4772-7_5

2   The Time-Semi Discretization Scheme

For $0 < \alpha < 1$, Caputo's fractional derivative ${}_0^c\mathscr{D}_t^\alpha u(\mathbf{x}, t)$ can be defined as

$$
{}_0^c\mathscr{D}_t^\alpha u(\mathbf{x}, t) = \begin{cases} \dfrac{1}{\Gamma(1-\alpha)} \int\limits_0^t \dfrac{\partial u(\mathbf{x},s)}{\partial s} \dfrac{ds}{(t-s)^\alpha}, & 0 < \alpha < 1, \\[12pt] \dfrac{\partial u(\mathbf{x},t)}{\partial t}, & \alpha = 1. \end{cases} \tag{3}
$$

Let us consider $\delta t = \frac{T}{N}$ denotes the step size in time, where N is any positive integer, and $t_n = n\delta t$, $n = 0, 1, \ldots, N$ be temporal mesh points.

Let us introduce some notations $u^{n-\frac{1}{2}} = \frac{1}{2}(u^n + u^{n-1})$, and $\delta_t u^{n-\frac{1}{2}} = \frac{1}{\delta t}(u^n - u^{n-1})$, where expansion of u^n is $u(\mathbf{x}, t_n)$.

Lemma 1. *For $0 < \alpha < 1$, and $\zeta \in C^2[0, T]$, we have*

$$
\int\limits_0^{t_n} \zeta'(s)(t_n - s)^{-\alpha} ds = \sum_{k=1}^n \frac{\zeta(t_k) - \zeta(t_{k-1})}{\delta t} \int\limits_{t_{k-1}}^{t_k} (t_n - s)^{-\alpha} ds + R^n, \quad 1 \leq n \leq N \tag{4}
$$

and

$$
|R^n| \leq \left(\frac{1}{2(1-\alpha)} + \frac{1}{2} \right) \delta t^{2-\alpha} \max_{0 \leq t \leq t_n} |\zeta''(t)|. \tag{5}
$$

Proof. See [1].

Lemma 2. *Let $0 < \alpha_i < 1$, $a_{i,0} = \frac{1}{\delta t \Gamma(1-\alpha_i)}$ and $b_{i,k} = \frac{\delta t^{1-\alpha_i}}{(1-\alpha_i)} \left[(k+1)^{1-\alpha_i} - (k)^{1-\alpha_i} \right]$, $i = 0, 1, \ldots, l$, $k = 0, 1, 2, \ldots$, then*

$$
\left| \frac{1}{\Gamma(1-\alpha_i)} \int\limits_0^{t_n} \frac{g'(s)}{(t_n - s)^{\alpha_i}} ds - a_{i,0} \left[b_{i,0} g(t_n) - \sum_{k=1}^{n-1} (b_{i,n-k-1} - b_{i,n-k}) g(t_k) - b_{i,n-1} g(0) \right] \right|
$$

$$
\leq \frac{1}{\Gamma(1-\alpha_i)} \left(\frac{1}{2(1-\alpha_i)} + \frac{1}{2} \right) \delta t^{2-\alpha_i} \max_{0 \leq t \leq t_n} |g''(t)| \tag{6}
$$

Proof. It proceeds from Lemma 1.

Lemma 3. *Let $0 < \alpha_i < 1$, and $b_{i,k} = \frac{\delta t^{1-\alpha_i}}{(1-\alpha_i)} \left[(k+1)^{1-\alpha_i} - (k)^{1-\alpha_i} \right]$, $i = 0, 1, \ldots, l$, $k = 0, 1, 2, \ldots$, then*

$$
b_{i,0} > b_{i,1} > b_{i,2} > \ldots > b_{i,k} \to 0, \quad \text{as } k \to \infty.
$$

Proof. See [1].

We define that

$$
w_i(\mathbf{x}, t) = \frac{1}{\Gamma(1-\alpha_i)} \int\limits_0^t \frac{\partial u(\mathbf{x}, s)}{\partial s} \frac{ds}{(t-s)^{\alpha_i}}. \tag{7}
$$

Thus,

$$\sum_{i=0}^{l} d_i w_i^{n-\frac{1}{2}} = \Delta u^{n-\frac{1}{2}} - F(u^{n-1}) + f^{n-\frac{1}{2}} + r_1^{n-\frac{1}{2}}, \quad n \geq 1, \tag{8}$$

where

$$|r_1^{n-\frac{1}{2}}| \leq C_1 \delta t. \tag{9}$$

From (7), we have

$$w_i(\mathbf{x}, t_n) = \frac{1}{\Gamma(1 - \alpha_i)} \int_0^{t_n} \frac{\partial u(\mathbf{x}, t)}{\partial t} \frac{dt}{(t_n - t)^{\alpha_i}}$$

using Lemma 2, we have

$$w_i^n = a_{i,0} \left[b_{i,0} u^n - \sum_{k=1}^{n-1} (b_{i,n-k-1} - b_{i,n-k}) u^k - b_{i,n-1} u^0 \right] + \mathcal{O}(\delta t^{2-\alpha_i}). \tag{10}$$

Now define the operator [1]

$$\mathcal{P}_i(u^n, q) = \left[b_{i,0} u^n - \sum_{k=1}^{n-1} (b_{i,n-k-1} - b_{i,n-k}) u^k - b_{i,n-1} q \right]$$

and using the condition $u^0 = u(\mathbf{x}, 0) = \xi(\mathbf{x}) = \xi$, we have

$$w_i^{n-\frac{1}{2}} = a_{i,0} \mathcal{P}_i(u^{n-\frac{1}{2}}, \xi) + (r_2)_i^{n-\frac{1}{2}} \tag{11}$$

where

$$|(r_2)_i^{n-\frac{1}{2}}| \leq C_2 \delta t^{2-\alpha_i}. \tag{12}$$

now substituting above expression in (8), we have

$$\sum_{i=0}^{l} d_i \left(a_{i,0} \mathcal{P}_i(u^{n-\frac{1}{2}}, \xi) + (r_2)_i^{n-\frac{1}{2}} \right) = \Delta u^{n-\frac{1}{2}} - F(u^{n-1}) + f^{n-\frac{1}{2}} + r_1^{n-\frac{1}{2}}$$

$$\sum_{i=0}^{l} d_i a_{i,0} \mathcal{P}_i(u^{n-\frac{1}{2}}, \xi) = \Delta u^{n-\frac{1}{2}} - F(u^{n-1}) + f^{n-\frac{1}{2}} + R^{n-\frac{1}{2}} \tag{13}$$

where,

$$R^{n-\frac{1}{2}} = -\sum_{i=0}^{l} d_i (r_2)_i^{n-\frac{1}{2}} + r_1^{n-\frac{1}{2}}$$

$$|R^{n-\frac{1}{2}}| \leq \sum_{i=0}^{l} d_i C_2 \delta t^{2-\alpha_i} + C_1 \delta t$$

$$\leq C \delta t$$

where $C = \left[C_1 + \sum\limits_{i=0}^{l} d_i C_2 \right]$.

Now excluding the $R^{n-\frac{1}{2}}$ that is truncation error term, and approximating the analytical value u^n by its approximating value U^n, then the resulted discrete scheme is as following

$$\sum_{i=0}^{l} d_i a_{i,0} \mathcal{P}_i(U^{n-\frac{1}{2}}, \xi) = \Delta U^{n-\frac{1}{2}} - F(U^{n-1}) + f^{n-\frac{1}{2}}, \quad 1 \le n \le N, \qquad (14)$$

or we have the above equation in more specific form as

$$\mathcal{L} U^n = b \qquad (15)$$

where the value of linear differential operator $\mathcal{L}$ and b are:

$$\mathcal{L} U^n = \sum_{i=0}^{l} d_i \frac{a_{i,0} b_{i,0}}{2} U^n - \frac{1}{2} \Delta U^n$$

$$b = \frac{1}{2} \Delta U^{n-1} - \sum_{i=0}^{l} d_i \frac{a_{i,0} b_{i,0}}{2} U^{n-1} + \sum_{i=0}^{l} d_i a_{i,0} \left[\sum_{k=1}^{n-1} (b_{i,n-k-1} - b_{i,n-k}) U^{k-\frac{1}{2}} + b_{i,n-1} U^0 \right] -$$

$$F(U^{n-1}) + f^{n-\frac{1}{2}}$$

2.1 Convergence and Stability Analysis

Next we consider L_2 norm to discuss the stability analysis and convergence of the time discrete numerical scheme.

Lemma 4. *For any $\eta = \{\eta_1, \eta_2, \ldots\}$, $0 < \alpha_i < 1$ and θ, we have*

$$\sum_{j=1}^{n} \mathcal{P}_i(\eta_j, \theta) \eta_j \ge \frac{t_n^{-\alpha_i}}{2} \delta t \sum_{j=1}^{n} \eta_j^2 - \frac{t_n^{1-\alpha_i}}{2(1-\alpha_i)} \theta^2$$

Proof. See [1].

Theorem 1. *Let U^n be the solution of the Eq. (14), belonging to $H_0^1(\Omega)$. Then the time discrete scheme (14) is unconditionally stable and have the following inequality:*

$$\|U^n\|^2 \le C^2 \left(\sum_{i=0}^{l} \frac{2 d_i T}{\delta t \Gamma(2-\alpha_i)} \|\xi\|^2 + \Gamma(1-\alpha_0) T \sum_{j=1}^{n} \|f^{j-\frac{1}{2}}\|^2 \right).$$

Proof. Consider the equation

$$\sum_{i=0}^{l} \frac{d_i}{\delta t \Gamma(1-\alpha_i)} \left[b_{i,0} U^{n-\frac{1}{2}} - \sum_{k=1}^{n-1} (b_{i,n-k-1} - b_{i,n-k}) U^{k-\frac{1}{2}} - b_{n-1} \xi \right] = \Delta U^{n-\frac{1}{2}} - F(U^{n-1}) + f^{n-\frac{1}{2}},$$

$$(16)$$

Multiplying above equation by $U^{n-\frac{1}{2}}$, and integrating over Ω, give

$$\sum_{i=0}^{l} \frac{d_i}{\delta t \Gamma(1-\alpha_i)} \left\{ b_{i,0}\left(U^{n-\frac{1}{2}}, U^{n-\frac{1}{2}}\right) - \sum_{k=1}^{n-1} \left(b_{i,n-k-1} - b_{i,n-k}\right)\left(U^{k-\frac{1}{2}}, U^{n-\frac{1}{2}}\right) - b_{n-1}\left(\xi, U^{n-\frac{1}{2}}\right) \right\}$$

$$= \left(\Delta U^{n-\frac{1}{2}}, U^{n-\frac{1}{2}}\right) - \left(F(U^{n-1}), U^{n-\frac{1}{2}}\right) + \left(f^{n-\frac{1}{2}}, U^{n-\frac{1}{2}}\right), \qquad (17)$$

Now using the fact

$$\left(\Delta U^{n-\frac{1}{2}}, U^{n-\frac{1}{2}}\right) = -\left(\nabla U^{n-\frac{1}{2}}, \nabla U^{n-\frac{1}{2}}\right)$$

$$= -\int_{\Omega} \left(\frac{\nabla U^n + \nabla U^{n-1}}{2}\right)\left(\frac{\nabla U^n + \nabla U^{n-1}}{2}\right) d\Omega$$

$$= -\frac{1}{4}\int_{\Omega}(\nabla U^n + \nabla U^{n-1})^2 d\Omega$$

$$= -\frac{1}{4}\left(\|\nabla U^n + \nabla U^{n-1}\|\right)^2,$$

we have

$$\sum_{i=0}^{l} \frac{d_i}{\delta t \Gamma(1-\alpha_i)} \left\{ b_{i,0}\|U^{n-\frac{1}{2}}\|^2 - \sum_{k=1}^{n-1}(b_{i,n-k-1} - b_{i,n-k})\|U^{k-\frac{1}{2}}\|\|U^{n-\frac{1}{2}}\| - b_{n-1}\|\xi\|\|U^{n-\frac{1}{2}}\| \right\}$$

$$= -\frac{1}{4}\left(\|\nabla U^n + \nabla U^{n-1}\|\right)^2 - \left(F(U^{n-1}), U^{n-\frac{1}{2}}\right) + \left(f^{n-\frac{1}{2}}, U^{n-\frac{1}{2}}\right), \qquad (18)$$

now taking the summation from $n=1$ to m on both the sides, we have

$$\sum_{n=1}^{m}\sum_{i=0}^{l} \frac{d_i}{\delta t \Gamma(1-\alpha_i)} \left\{ b_{i,0}\|U^{n-\frac{1}{2}}\| - \sum_{k=1}^{n-1}(b_{i,n-k-1} - b_{i,n-k})\|U^{k-\frac{1}{2}}\| - b_{n-1}\|\xi\| \right\}\|U^{n-\frac{1}{2}}\|$$

$$\leq -\frac{1}{4}\sum_{n=1}^{m}\left(\|\nabla U^n + \nabla U^{n-1}\|\right)^2 + C\sum_{n=1}^{m}\|U^{n-1}\|\|U^{n-\frac{1}{2}}\| + \sum_{n=1}^{m}\|f^{n-\frac{1}{2}}\|\|U^{n-\frac{1}{2}}\|.$$

$$(19)$$

now using the inequality $|xy| \leq \frac{1}{2\theta}x^2 + \frac{\theta}{2}y^2$, together with $\theta = \frac{d_0 t_m^{-\alpha_0}}{2\Gamma(1-\alpha_0)}$, we have

$$C\sum_{n=1}^{m}\|U^{n-1}\|\|U^{n-\frac{1}{2}}\| \leq C^2\frac{\Gamma(1-\alpha_0)}{d_0 t_m^{-\alpha_0}}\sum_{n=1}^{m}\|U^{n-1}\|^2 + \frac{d_0 t_m^{-\alpha_0}}{4\Gamma(1-\alpha_0)}\sum_{n=1}^{m}\|U^{n-\frac{1}{2}}\|^2.$$

and

$$\sum_{n=1}^{m}\|f^{n-\frac{1}{2}}\|\|U^{n-\frac{1}{2}}\| \leq \frac{\Gamma(1-\alpha_0)}{d_0 t_m^{-\alpha_0}}\sum_{n=1}^{m}\|f^{n-\frac{1}{2}}\|^2 + \frac{d_0 t_m^{-\alpha_0}}{4\Gamma(1-\alpha_0)}\sum_{n=1}^{m}\|U^{n-\frac{1}{2}}\|^2.$$

Now using above relation together with Lemma 4, we have

$$\frac{d_0 t_m^{-\alpha_0}}{2\delta t \Gamma(1-\alpha_0)}\delta t\sum_{n=1}^{m}\|U^{n-\frac{1}{2}}\|^2 - \sum_{i=0}^{l}\frac{d_i t_m^{1-\alpha_i}}{2\delta t \Gamma(2-\alpha_i)}\|\xi\|^2 \leq -\frac{1}{4}\sum_{n=1}^{m}\left(\|\nabla U^n + \nabla U^{n-1}\|\right)^2 +$$

$$c^2\frac{\Gamma(1-\alpha_0)}{d_0 t_m^{-\alpha_0}}\sum_{n=1}^{m}\|U^{n-1}\|^2 + \frac{d_0 t_m^{-\alpha_0}}{4\Gamma(1-\alpha_0)}\sum_{n=1}^{m}\|U^{n-\frac{1}{2}}\|^2 + \frac{\Gamma(1-\alpha_0)}{d_0 t_m^{-\alpha_0}}\sum_{n=1}^{m}\|f^{n-\frac{1}{2}}\|^2 + \frac{d_0 t_m^{-\alpha_0}}{4\Gamma(1-\alpha_0)}\sum_{n=1}^{m}\|U^{n-\frac{1}{2}}\|^2.$$

Now simplifying above relation, we have

$$\sum_{n=1}^{m}\left(\|\nabla U^n + \nabla U^{n-1}\|\right)^2 \le \sum_{i=0}^{l}\frac{2d_i t_m^{1-\alpha_i}}{\delta t \Gamma(2-\alpha_i)}\|\xi\|^2 + 4C^2\Gamma(1-\alpha_0)t_m^{\alpha_0}\sum_{n=1}^{m}\|U^{n-1}\|^2$$

$$+4\Gamma(1-\alpha_0)t_m^{\alpha_0}\sum_{n=1}^{m}\|f^{n-\frac{1}{2}}\|^2$$

$$\sum_{n=1}^{m}\|\nabla U^n\|^2 + \sum_{n=1}^{m}\|\nabla U^{n-1}\|^2 \le \sum_{i=0}^{l}\frac{2d_i t_m^{1-\alpha_i}}{\delta t \Gamma(2-\alpha_i)}\|\xi\|^2 + 4C^2\Gamma(1-\alpha_0)t_m^{\alpha_0}\sum_{n=1}^{m}\|U^{n-1}\|^2$$

$$+4\Gamma(1-\alpha_0)t_m^{\alpha_0}\sum_{n=1}^{m}\|f^{n-\frac{1}{2}}\|^2 \tag{20}$$

switching the index from m to n, we get

$$\|\nabla U^n\|^2 \le \sum_{i=0}^{l}\frac{2d_i t_n^{1-\alpha_i}}{\delta t \Gamma(2-\alpha_i)}\|\xi\|^2 + 4C^2\Gamma(1-\alpha_0)t_n^{\alpha_0}\sum_{j=0}^{n-1}\|U^j\|^2 + 4\Gamma(1-\alpha_0)t_n^{\alpha_0}\sum_{j=1}^{n}\|f^{j-\frac{1}{2}}\|^2 \tag{21}$$

$$\|\nabla U^n\|^2 \le \sum_{i=0}^{l}\frac{2d_i T}{\delta t \Gamma(2-\alpha_i)}\|\xi\|^2 + 4C^2\Gamma(1-\alpha_0)T\sum_{j=0}^{n-1}\|U^j\|^2 + 4\Gamma(1-\alpha_0)T\sum_{j=1}^{n}\|f^{j-\frac{1}{2}}\|^2 \tag{22}$$

Now using Discrete Gronwall Lemma [2], we have

$$\|\nabla U^n\|^2 \le C^2\left(\sum_{i=0}^{l}\frac{2d_i T}{\delta t \Gamma(2-\alpha_i)}\|\xi\|^2 + \Gamma(1-\alpha_0)T\sum_{j=1}^{n}\|f^{j-\frac{1}{2}}\|^2\right) \tag{23}$$

Now using Poincare inequality $\|U^n\|^2 \le \tilde{C}^2\|\nabla U^n\|^2$, we have

$$\|U^n\|^2 \le C^2\left(\sum_{i=0}^{l}\frac{2d_i T}{\delta t \Gamma(2-\alpha_i)}\|\xi\|^2 + \Gamma(1-\alpha_0)T\sum_{j=1}^{n}\|f^{j-\frac{1}{2}}\|^2\right) \tag{24}$$

Theorem 2. *Let u^n, the analytical solution of (13) and U^n, the numerical solution of (14) both belonging to H_0^1, then time semi discrete scheme (14) is convergent with convergence order $\mathcal{O}(\delta t)$.*

Proof. We consider $\mathcal{E}^n = u^n - U^n$ for $n \ge 1$, together with $\mathcal{E}^0 = 0$. Now subtracting (14) from (13), we have

$$\sum_{i=0}^{l}\frac{d_i}{\delta t \Gamma(1-\alpha_i)}\left\{b_{i,0}\mathcal{E}^{n-\frac{1}{2}} - \sum_{k=1}^{n-1}(b_{i,n-k-1} - b_{i,n-k})\mathcal{E}^{k-\frac{1}{2}}\right\} = \Delta\mathcal{E}^{n-\frac{1}{2}} - \left[F(u^{n-1}) - F(U^{n-1})\right] + R^{n-\frac{1}{2}}, \tag{25}$$

Multiplying the above equation by $\mathcal{E}^{n-\frac{1}{2}}$, and taking the integration over Ω, give

$$\sum_{i=0}^{l}\frac{d_i}{\delta t \Gamma(1-\alpha_i)}\left\{b_{i,0}\|\mathcal{E}^{n-\frac{1}{2}}\| - \sum_{k=1}^{n-1}(b_{i,n-k-1} - b_{i,n-k})\|\mathcal{E}^{k-\frac{1}{2}}\|\right\}\|\mathcal{E}^{n-\frac{1}{2}}\|$$

$$= -\frac{1}{4}\left(\|\nabla\mathcal{E}^n + \nabla\mathcal{E}^{n-1}\|\right)^2 - \left(\left[F(u^{n-1}) - F(U^{n-1})\right], \mathcal{E}^{n-\frac{1}{2}}\right) + (R^{n-\frac{1}{2}}, \mathcal{E}^{n-\frac{1}{2}})$$

Now summing the above relation from $n = 1$ to m, we have

$$\sum_{n=1}^{m} \sum_{i=0}^{l} \frac{d_i}{\delta t \Gamma(1-\alpha_i)} \left\{ b_{i,0} \|\mathcal{E}^{n-\frac{1}{2}}\| - \sum_{k=1}^{n-1} (b_{i,n-k-1} - b_{i,n-k}) \|\mathcal{E}^{k-\frac{1}{2}}\| \right\} \|\mathcal{E}^{n-\frac{1}{2}}\|$$

$$\leq -\frac{1}{4} \sum_{n=1}^{m} \left(\|\nabla \mathcal{E}^n + \nabla \mathcal{E}^{n-1}\| \right)^2 - \sum_{n=1}^{m} \left(\left[F(u^{n-1}) - F(U^{n-1}) \right], \mathcal{E}^{n-\frac{1}{2}} \right) + \sum_{n=1}^{m} \|R^{n-\frac{1}{2}}\| \|\mathcal{E}^{n-\frac{1}{2}}\|$$

now using application of Lemma 4, we get

$$\sum_{i=0}^{l} \frac{d_i}{\delta t \Gamma(1-\alpha_i)} \left[\frac{t_m^{-\alpha_i}}{2} \delta t \sum_{n=1}^{m} \|\mathcal{E}^{n-\frac{1}{2}}\|^2 \right] + \frac{1}{4} \sum_{n=1}^{m} \left(\|\nabla \mathcal{E}^n + \nabla \mathcal{E}^{n-1}\| \right)^2$$

$$\leq -\sum_{n=1}^{m} \left(\left[F(u^{n-1}) - F(U^{n-1}) \right], \mathcal{E}^{n-\frac{1}{2}} \right) + \sum_{n=1}^{m} \|R^{n-\frac{1}{2}}\| \|\mathcal{E}^{n-\frac{1}{2}}\|. \qquad (26)$$

Using inequality $|xy| \leq \frac{1}{2\theta}x^2 + \frac{\theta}{2}y^2$, together with $\theta = \frac{d_0 t_m^{-\alpha_0}}{2\Gamma(1-\alpha_0)}$, we have

$$-\sum_{n=1}^{m} \left(\left[F(u^{n-1}) - F(U^{n-1}) \right], \mathcal{E}^{n-\frac{1}{2}} \right) \leq L^2 \frac{\Gamma(1-\alpha_0)}{d_0 t_m^{-\alpha_0}} \sum_{n=1}^{m} \|\mathcal{E}^{n-1}\|^2 + \frac{d_0 t_m^{-\alpha_0}}{4\Gamma(1-\alpha_0)} \sum_{n=1}^{m} \|\mathcal{E}^{n-\frac{1}{2}}\|^2$$

$$\sum_{n=1}^{m} \|R^{n-\frac{1}{2}}\| \|\mathcal{E}^{n-\frac{1}{2}}\| \leq \frac{\Gamma(1-\alpha_0)}{d_0 t_m^{-\alpha_0}} \sum_{n=1}^{m} \|R^{n-\frac{1}{2}}\|^2 + \frac{d_0 t_m^{-\alpha_0}}{4\Gamma(1-\alpha_0)} \sum_{n=1}^{m} \|\mathcal{E}^{n-\frac{1}{2}}\|^2.$$

Using above relation into Eq. (26), we have

$$\frac{d_0 t_m^{-\alpha_0}}{2\Gamma(1-\alpha_0)} \sum_{n=1}^{m} \|\mathcal{E}^{n-\frac{1}{2}}\|^2 + \frac{1}{4} \sum_{n=1}^{m} \left(\|\nabla \mathcal{E}^n + \nabla \mathcal{E}^{n-1}\| \right)^2 \leq L^2 \frac{\Gamma(1-\alpha_0)}{d_0 t_m^{-\alpha_0}} \sum_{n=1}^{m} \|\mathcal{E}^{n-1}\|^2$$

$$+ \frac{d_0 t_m^{-\alpha_0}}{4\Gamma(1-\alpha_0)} \sum_{n=1}^{m} \|\mathcal{E}^{n-\frac{1}{2}}\|^2 + \frac{\Gamma(1-\alpha_0)}{d_0 t_m^{-\alpha_0}} \sum_{n=1}^{m} \|R^{n-\frac{1}{2}}\|^2 + \frac{d_0 t_m^{-\alpha_0}}{4\Gamma(1-\alpha_0)} \sum_{n=1}^{m} \|\mathcal{E}^{n-\frac{1}{2}}\|^2,$$

switching index from m to n, and after simplification we get

$$\|\nabla \mathcal{E}^n\|^2 \leq 4L^2 \Gamma(1-\alpha_0) t_n^{\alpha_0} \sum_{j=0}^{n-1} \|\mathcal{E}^j\|^2 + 4\Gamma(1-\alpha_0) t_n^{\alpha_0} \sum_{j=1}^{n} \|R^{j-\frac{1}{2}}\|^2$$

$$\leq 4L^2 \Gamma(1-\alpha_0) t_n^{\alpha_0} \sum_{j=0}^{n-1} \|\mathcal{E}^j\|^2 + 4n\Gamma(1-\alpha_0) t_n^{\alpha_0} \max_{1 \leq j \leq n} \|R^{j-\frac{1}{2}}\|^2.$$

Now using Poincare inequality, we have

$$\|\mathcal{E}^n\|^2 \leq C_\Omega^2 L^2 \Gamma(1-\alpha_0) t_n^{\alpha_0} \sum_{j=0}^{n-1} \|\mathcal{E}^j\|^2 + C_\Omega^2 \Gamma(1-\alpha_0) T C^2 \delta t^2 \qquad (27)$$

Now application of Discrete Gronwall Lemma [2], with parameters $z_k = 0$, $\delta_0 = C_\Omega^2 C^2 T \Gamma(1-\alpha_0)\delta t^2$, $x_k = L^2 C_\Omega^2 \Gamma(1-\alpha_0) t_n^{\alpha_0}$, and $y_k = \|\mathcal{E}^k\|^2$, we have

$$\|\mathcal{E}^n\|^2 \leq C(T, \alpha_0, C_\Omega)\delta t^2.$$

Therefore we have

$$\|\mathcal{E}^n\| \leq C'(T, \alpha_0, C_\Omega)\delta t,$$

which completes the proof.

3   Spatial Discretization

In local collocation method the computational domain Ω, containing M colloca-
tion points is partitioned into M overlapping sub domains Ω_i, such that $\bigcup\limits_{i=1}^{M} \Omega_i = \Omega$. For each $\mathbf{x}_k^{[i]} \in \Omega_i$, the influence points of $\mathbf{x}_k^{[i]}$ are $\{\mathbf{x}_1^{[i]}, \mathbf{x}_2^{[i]}, \mathbf{x}_3^{[i]}, \ldots, \mathbf{x}_{m_i}^{[i]}\}$ are
m_i closest points of $\mathbf{x}_k^{[i]}$ in sub domain Ω_i.

The numerical approximation of $u(\mathbf{x}, t_n)$ in local interpolation form can be
given as

$$\hat{u}(\mathbf{x}, t_n) = \sum_{j=1}^{m_i} \lambda_j \phi(\|\mathbf{x} - \mathbf{x}_j^{[i]}\|) + \sum_{j=1}^{l} \gamma_j p_j(\mathbf{x}), \tag{28}$$

where $\{\lambda_j\}$ and $\{\gamma_j\}$ are unknown coefficients at n^{th} time level, ϕ is considered
radial basis function, $\|\cdot\|$ is the Euclidean norm and $\{p_j(x)\}_{j=1}^{l}$ denote basis
for the $l = \binom{m-1+d}{m-1}$ dimensional linear space of d-variate polynomials of total
degree $\leq m - 1$. The interpolation condition on sub domain Ω_i

$$\hat{u}(\mathbf{x}_k^{[i]}, t_n) = u(\mathbf{x}_k^{[i]}, t_n), \ \forall \ 1 \leq k \leq m_i, \tag{29}$$

is supported with extra l regularization conditions

$$\sum_{j=1}^{m_i} \lambda_j p_k(\mathbf{x}_j^{[i]}) = 0 \ \forall \ 1 \leq k \leq l. \tag{30}$$

Imposing conditions (29–30) on $\hat{u}(\mathbf{x}, t_n)$, at each stencil we obtain following linear
system

$$\begin{bmatrix} \Phi & P \\ P^t & O \end{bmatrix} \begin{bmatrix} \lambda \\ \gamma \end{bmatrix} = \begin{bmatrix} u\,|_{\Omega_i} \\ O \end{bmatrix} \tag{31}$$

where $\Phi := [\phi\|\mathbf{x}_j^{[i]} - \mathbf{x}_k^{[i]}\|]_{1 \leq j,k \leq m_i}$, $P := [p_k(\mathbf{x}_j^{[i]})]_{1 \leq j \leq m_i, 1 \leq k \leq l}$.
The above system can be written in matrix form as

$$\Lambda_{\Omega_i} = A_{\Omega_i}^{-1} U_{\Omega_i}^n, \tag{32}$$

where $\Lambda_{\Omega_i} = [\lambda_1, \ldots, \lambda_{m_i}, \gamma_1, \ldots, \lambda_l]^{\mathsf{T}}$, $U_{\Omega_i}^n = [u(\mathbf{x}_1^{[i]}, t_n), \ldots, u(\mathbf{x}_{m_i}^{[i]}, t_n), 0, \ldots, 0]^{\mathsf{T}}$,
and A_{Ω_i} is coefficient matrix of the system (31).

Suppose ϕ is a conditionally positive definite function of order m on $\mathbb{R}^d$ and
the points $\Omega_i = \{x_1, x_2, \ldots, x_{n_i}\}$ form $(m-1)$ unisolvent set of centers. Then
the system (31) is uniquely solvable.

For a linear differential operator $\mathscr{D}$, at each stencil $\mathbf{x}_k^{[i]} \in \Omega_i$, we have approximation for $\mathscr{D}u(\mathbf{x}, t_n)$ as;

$$\mathscr{D}\hat{u}(\mathbf{x}_k^{[i]}, t_n) = \sum_{j=1}^{m_i} \lambda_j \mathscr{D}\phi(\|\mathbf{x}_k^{[i]} - \mathbf{x}_j^{[i]}\|) + \sum_{j=1}^{l} \gamma_j \mathscr{D}p_j(\mathbf{x}_k^{[i]}),$$

$$= [\mathscr{D}\phi(\|\mathbf{x}_k^{[i]} - \mathbf{x}_1^{[i]}\|), \ldots, \mathscr{D}\phi(\|\mathbf{x}_k^{[i]} - \mathbf{x}_{m_i}^{[i]}\|), \mathscr{D}p_1(\mathbf{x}_k^{[i]}), \ldots \mathscr{D}p_l(\mathbf{x}_k^{[i]})]\Lambda_{\Omega_i}$$

$$= \mathscr{D}\Psi_{\Omega_i} A_{\Omega_i}^{-1} U_{\Omega_i}^n, \tag{33}$$

where $\Psi_{\Omega_i} = [\phi(\|\mathbf{x}_k^{[i]} - \mathbf{x}_1^{[i]}\|), \ldots, \phi(\|\mathbf{x}_k^{[i]} - \mathbf{x}_{m_i}^{[i]}\|), p_1(\mathbf{x}_k^{[i]}), \ldots p_l(\mathbf{x}_k^{[i]})]$. For each k the local operator $\mathscr{D}\Psi_{\Omega_i} A_{\Omega_i}^{-1}$ is a $1 \times m_i$ row vector.

Now for each collocation points $\mathbf{x}_i \in \Omega$, applying the local collocation method described through Eq. (33) to the linear operator $\mathscr{L}$ defined in Eq. (15), we have

$$\mathscr{L}\Psi_{\Omega_i} A_{\Omega_i}^{-1} U_{\Omega_i}^n = b_i, \quad \mathbf{x}_i \in \Omega. \tag{34}$$

For each arbitrary i, the $\mathscr{L}\Psi_{\Omega_i} A_{\Omega_i}^{-1}$ is a $1 \times m_i$ row vector, that going to store in $M \times M$ matrix, by filling extra spaces by zeros. Thus we have following linear system

$$\mathbf{L}\mathbf{U}^n = \mathbf{b}. \tag{35}$$

The resulting system is sparse having only m_i non zero entries in each rows, and hence can be calculate efficiently.

4 Numerical Experiments

This section included some numerical simulation to verify the theoretical results and demonstrate the reliability and correctness of the proposed method. The second order thin plate spline $r^4 \ln r$ is taken in both the numerical problems and a constant number of collocation points in each subdomain. We used two different definitions of errors to measure the accuracy. They are defined as

$$\text{Maximmu absolute error} \qquad L_\infty = \max_{1 \le i \le M} |u(\mathbf{x}_i, T) - U(\mathbf{x}_i, T)|,$$

$$\text{Root mean square error} \qquad L_{rms} = \sqrt{\frac{1}{M} \sum_{i=1}^{M} |u(\mathbf{x}_i, T) - U(\mathbf{x}_i, T)|^2},$$

where M is total collocation points, $u(\mathbf{x}_i, T)$ and $U(\mathbf{x}_i, T)$ are analytical and numerical solution respectively.

Example 1. Consider the following two-term test problem

$$_0^c\mathscr{D}_t^{\alpha_0} u(x,t) +_0^c \mathscr{D}_t^{\alpha_1} u(x,t) = \Delta u(x,t) - F(u(x,t)) + f(x,t).$$

source term, initial and boundary conditions are computed from the exact solution

$$u(x,t) = (1+t^2)(-x^2 + x).$$

and $f(x,t) = \left(\frac{2t^{2-\alpha_0}}{\Gamma(3-\alpha_0)} + \frac{2t^{2-\alpha_1}}{\Gamma(3-\alpha_1)} \right) (x - x^2) + 2\left(1 + t^2\right) + F(u(x,t)).$

In this problem, we take $F(u(x,t)) = 0$, which means that the given problem is considered as linear problem. We solve this problem in the computational domain $[0,1]$ for different values of δt together with different combinations of α_0 and α_1. Table 1 reports the results with $M = N$, $m = 3$ at $T = 1$ and the computational order of convergence is $(2 - \alpha)$.

The proposed method is compared with the method employed by Qiao and Xu [3]. We found that our proposed method is more correct and very efficient than [3], also the corresponding results for different values of α_0 and $\alpha_1 = 0.2$ are shown in Table 2.

Table 1. The absolute errors and RMS errors for different values of α_0 and α_1 for Example 1.

h	$\alpha_0 = 0.5, \alpha_1 = 0.1$			$\alpha_0 = 0.7, \alpha_1 = 0.3$		
	L_∞	L_{rms}	Rate	L_∞	L_{rms}	Rate
1/10	3.3320e−04	2.2792e−04	–	8.5665e−04	5.8593e−04	–
1/20	1.1935e−04	8.2898e−05	1.5	3.4262e−04	2.3795e−04	1.3
1/40	4.2383e−05	2.9746e−05	1.5	1.3667e−04	9.5908e−05	1.3
1/80	1.4998e−05	1.0587e−05	1.5	5.4596e−05	3.8535e−05	1.3
1/160	5.2987e−06	3.7516e−06	1.5	2.1863e−05	1.5478e−05	1.3

Table 2. Comparison of L_2 errors with $\alpha_1 = 0.2$, for Example 1.

h	$\alpha_0 = 0.50$		$\alpha_0 = 0.75$		$\alpha_0 = 0.95$	
	Present method	Method [3]	Present method	Method [3]	Present method	Method [3]
1/10	2.5369e−04	3.7247e−04	6.5893e−04	9.6734e−04	1.3850e−03	2.0331e−03
1/20	9.1285e−05	1.3154e−04	2.7960e−04	4.0229e−04	6.7602e−04	9.7424e−04
1/30	4.9877e−05	7.1352e−05	1.6872e−04	2.4138e−04	4.4312e−04	6.3399e−04
1/40	3.2416e−05	4.6196e−05	1.1778e−04	1.6786e−04	3.2807e−04	4.6762e−04
1/50	2.3185e−05	3.3963e−05	8.9082e−05	1.2666e−04	2.5974e−04	3.6936e−04

Example 2. Consider the following two-dimensional two-term test problem

$$_0^c\mathscr{D}_t^{\alpha_0} u(x,y,t) +_0^c \mathscr{D}_t^{\alpha_1} u(x,y,t) = \Delta u(x,y,t) - u(x,y,t)^3 + f(x,y,t).$$

source term, initial and boundary conditions are computed from the exact solution

$$u(x,y,t) = t^4 \sin(\pi x + \pi y).$$

and $f(x,y,t) = \sin(\pi x + \pi y) \left(\frac{24t^{4-\alpha_0}}{\Gamma(5-\alpha_0)} + \frac{24t^{4-\alpha_1}}{\Gamma(5-\alpha_1)} + 2\pi^2 t^4 \right) + u(x,y,t)^3.$

We considered this test problem in the computational domain $[0,1]^2$ with 2025 spatial points and $m = 5$ for different δt, α_0 and α_1. To prove the reliability and

Table 3. The absolute errors and RMS errors with different values of α_0, α_1 and δt at $T = 1.0$ s for Example 2.

δt	$\alpha_0 = 0.7, \alpha_1 = 0.4$			$\alpha_0 = 0.9, \alpha_1 = 0.7$		
	L_∞	L_{rms}	Rate	L_∞	L_{rms}	Rate
1/5	2.7105e−02	1.2861e−02	–	4.1120e−02	1.9814e−02	–
1/10	1.4665e−02	6.9194e−03	0.8	2.1751e−02	1.0445e−02	0.9
1/20	7.2565e−03	3.4142e−03	1.0	1.0707e−02	5.1348e−03	1.0
1/40	3.5543e−03	1.6699e−03	1.0	5.2029e−03	2.4933e−03	1.0
1/80	1.7903e−03	8.4149e−04	0.9	2.5673e−03	1.2301e−03	1.0

correctness of the proposed method, the results are reported in Table 3 and this table shows that the proposed method has convergence order one.

5   Conclusion

This work aimed to develop a meshless method based on the radial basis function to numerically solve the multi-term time fractional nonlinear diffusion equation which involves multiple Caputo fractional derivatives in time. The Finite difference method has been used for the discretization in time. Further, we proved that the proposed scheme has unconditional stability property and convergence. A fully discrete scheme was obtained using the meshless local collocation method for spatial variables. To verify the theoretical outcome some test problems are considered. From the presented results one can see that the proposed scheme is robust and reliable to deal with the described problems.

References

1. Sun, Z., Wu, X.: A fully discrete difference scheme for a diffusion-wave system. Appl. Numer. Math. **56**(2), 193–209 (2006)
2. Quarteroni, A., Valli, A.: Numerical Approximation of Partial Differential Equations, vol. 23. Springer, Heidelberg (2008)
3. Qiao, L., Xu, D.: Orthogonal spline collocation scheme for the multi-term time-fractional diffusion equation. Int. J. Comput. Math. **95**(8), 1478–1493 (2018)

On Space-Fractional Diffusion Equations with Conformable Derivative

Kamla Kant Mishra and Shruti Dubey$^{(\boxtimes)}$ ⓘ

Indian Institute of Technology, Madras, Chennai 600036, India
sdubey@iitm.ac.in

Abstract. We study the generalized space fractional diffusion equation on bounded domain Ω. The fractional derivative is considered in conformable sense. In particular, we extend and prove the maximum-minimum principle for the considered parabolic problem. Then, we show the use of this principle in the establishment of uniqueness of the solution and its continuous dependence on the initial and boundary data. We also present the construction of series solution for IBVP of the considered equation with homogeneous and nonhomogeneous boundary values.

Keywords: Conformable fractional derivative · Space-fractional diffusion equation · Initial-boundary-value problems · Maximum (Minimum) principle

1 Introduction

From the last few decades, the theories of fractional calculus have been developed extensively due to its application in the area of science and engineering, see [5, 6,8,9,12,13,22]. In [27], the author proved that there are some complex medium which behaves anomalously due to the diverse characteristic of the elementary units. As a result, the evolution process within such system does not follow the standard laws. Fractional calculus plays a significant role to describe the time evolution of a physical phenomenon within such system. In fact, in the pioneer article [22], the author has pointed out that diffusion process in a complex media does not follow Gaussian statistics, rather it follows some other non-linear statistics, in which the mean square displacement of an unit element depends on logarithm of time or fractional power of time. By considering the statistics $\langle x^2(t) \rangle \sim t^\alpha$, $0 < \alpha < 1$, where $\langle x^2(t) \rangle$ represents the mean square of the displacement of an unit arbitrary element, the author proved that the diffusion phenomenon in such medium is modelled by fractional differential equations. Consequently, the fractional calculus draws a great importance to study the real world physical problems.

Fractional calculus is a generalization of classical calculus which has been generalized in many ways. Among all, the most used versions in the history of

Supported by the Ministry of Human Resource Development (MHRD, now Ministry of Education), GOI.

fractional calculus, are due to Riemann-Liouville and Caputo. A considerable work on Riemann-Liouville and Caputo derivatives is available in literature, see [4,7,24,29,30] and references therein. However, there are some disadvantages as these derivatives do not obey the chain rule property, see [23]. In [11], the author proposed a generalization, known as "Conformable fractional derivative", which is quiet natural and closer to classical derivative and follows the chain rule property. After that, it becomes an attraction among the researcher and some notable works have been studied involving this derivative. For more details, we refer to [1,2,10,14–17,26,28]. Recently, some numerical studies have also been reported for equations with conformable fractional derivative. In [25], authors proposed numerical approximation approach based on differential transform method. In particular, they discussed approximation of conformable time fractional Burger equation using the proposed method. Discretization method using piecewise constant approximation is discussed in [18] while authors in [3], introduced three numerical techniques named conformable variational iteration method, conformable fractional reduced differential transform method and conformable homotopy analysis method for linear and nonlinear conformable partial differential equations.

In this paper, we study the initial-boundary value problem for the space-fractional diffusion equations,

$$u_t(x,t) = kT_x^\alpha u(x,t), \ (x,t) \in \Omega := (0,l) \times (0,T), \tag{1}$$

where T_x^α represents the Conformable spatial fractional differential operator of order $\alpha \in (1,2]$ and k is the diffusion constant. We prove the uniqueness result by utilizing the maximum-minimum principle for the problem (1).

In [19], the author established the uniqueness of classical solution based on maximum-minimum principle of time fractional initial-boundary value problem for diffusion equation with the diffusive term $L(u) := \nabla \cdot (p(x)\nabla u) - q(x)u$, where $p \in C^1(\bar{G})$ and $q \in C(\bar{G})$, over an open bounded domain $G \times (0,T)$, $G \subset \mathbb{R}^n$. Again, the existence and uniqueness of generalized and classical solutions have been investigated for the same problem in [20]. The same author in [21], considered the more generalised version (inhomogeneous equation with inhomogeneous boundary conditions) of the aforesaid problem and showed the equivalence between generalized and classical solution. This was done by using the very well known classical method, known as separation of variable technique. In one dimensional case with constant coefficient, the problem turns out to be a very well investigated diffusion problem, known as Heat equations. To the best of author's knowledge, no work has been reported on the analysis of the conformable spatial-fractional Heat equation. This work provide the opportunity to investigate the said problem.

The organization of this paper is as follows. In the next section, we present the definitions and some of the properties of conformable fractional derivative. In Sect. 3, maximum-minimum principle for the considered problem is proved and applied to show that it possesses at most one solution and solution depends continuously on initial and boundary data. In the last section, analytic solutions of the space fractional diffusion equation with homogeneous and nohomogeneous boundary conditions are provided.

2   Preliminary Results

In this section, we list some definitions, lemmas and weighted spaces which are useful in the sequel. For more details and more properties, we refer [1,11]. Let $\Omega = (0,l) \times (0,T)$, where $l, T \in \mathbb{R}_+$, $C(\overline{\Omega})$ is the set of all continuous functions defined on $\overline{\Omega}$, $C_t^1(0,T)$ is the set of all continuously differentiable functions with respect to t which are defined on $(0,T)$ and $C_x^2(0,l)$ is the set of all two times continuously differentiable functions with respect to x which are defined on $(0,l)$.

Definition 1 [1,11]. *Given a function $f : [a, \infty) \to \mathbb{R}$. Then, the (left) conformable fractional derivative starting from a of order $\alpha \in (0,1]$ is defined by*

$$T_a^\alpha f(x) = \lim_{\epsilon \to 0} \frac{f(x + \epsilon(x - a)^{1-\alpha}) - f(x)}{\epsilon}, \tag{2}$$

if the limit exists.

   In case of the conformable fractional derivative, the point $a \in \mathbb{R}$ appearing in (2) will be called the lower terminal of the left-side conformable derivative. Usually, if the conformable derivative of order α of a function f exists, then we simply say that f is α-differentiable. When $a = 0$, we write T^α instead of T_0^α. If $T_a^\alpha f(x)$ exists on $(a, a + \epsilon)$ then $T_a^\alpha f(a) = \lim_{x \to a^+} (T_a^\alpha f(x))$.

   It may be noted that some authors (see, for example [11]) use the notation T_α^a instead of T_a^α, but, following the tradition from the notations of the classical fractional derivatives, we will write the order above and the lower terminal below.

   The (right) conformable fractional derivative of f of order $\alpha \in (0,1]$ terminating at b is defined as [1,11]

$$T_b^\alpha f(x) = \lim_{\epsilon \to 0} \frac{f(x + \epsilon(b - x)^{1-\alpha}) - f(x)}{\epsilon}. \tag{3}$$

If $T_b^\alpha f(x)$ exists on (a, b), then $T_b^\alpha f(b) = \lim_{x \to b^-} (T_b^\alpha f)(x)$.

   It is worth to note that the conformable fractional derivative of a constant function is zero.

Definition 2 [1,11]. *Given a function $f : [a, \infty) \to \mathbb{R}$ and f is n^{th} differentiable at x. Then, the conformable fractional derivative starting from a of order $\alpha \in (n, n+1]$ is defined by*

$$T_a^\alpha f(x) = \lim_{\epsilon \to 0} \frac{f^{(\lceil \alpha \rceil - 1)}(x + \epsilon(x - a)^{(\lceil \alpha \rceil - \alpha)}) - f^{(\lceil \alpha \rceil - 1)}(x)}{\epsilon}, \tag{4}$$

if the limit exists. $\lceil \alpha \rceil$ is the smallest integer greater than or equal to α.

Consequently, one can show that

$$T^\alpha f(x) = x^{(\lceil\alpha\rceil-\alpha)} f^{\lceil\alpha\rceil}(x).$$

Examples

Conformable fractional derivative of order $\alpha \in (0,1]$

(1) $T^\alpha(f(x)) = T^\alpha(e^{cx}) = cx^{1-\alpha}e^{cx}$.
(2) $T^\alpha(f(x)) = T^\alpha(sin(cx)) = cx^{1-\alpha}cos(cx)$.
(3) $T^\alpha(f(x)) = T^\alpha(\frac{1}{\alpha}x^\alpha) = 1$.
(4) $T^\alpha(f(x)) = T^\alpha(sin(\frac{1}{\alpha}x^\alpha)) = cos(\frac{1}{\alpha}x^\alpha)$.

Next, we give the definition of left and right fractional integral of order α.

Definition 3 [1,11]. *Given a function $f : [a,\infty) \to \mathbb{R}$. Then, the (left) conformable fractional integral starting from a of order $\alpha \in (0,1]$ is defined as*

$$I_a^\alpha f(x) = \int_a^x (t-a)^{\alpha-1} f(t)dt. \tag{5}$$

[1,11] The (right) conformable fractional integral of f of order $\alpha \in (0,1]$ terminating at b is defined as

$$I_b^\alpha f(x) = \int_x^b (b-t)^{\alpha-1} f(t)dt. \tag{6}$$

In our exposition below we will use only left-side conformable derivative but will omit the expression "left-side" for simple writing. Following lemmas present few relation between conformable fractional derivative and integral and also some properties of conformable fractional derivative.

Lemma 1 [1,11]. *If $f : [a,\infty) \to \mathbb{R}$ is a continuous function, then for all $x > a$, $T_a^\alpha I_a^\alpha f(x) = f(x)$.*

Lemma 2 [1,11]. *If $T_a^\alpha f(x)$ is continuous on $[a,b]$, then $I_a^\alpha T_a^\alpha f(x) = f(x) - f(a)$.*

Lemma 3 [1,11]. *Let $f : [a,\infty) \to \mathbb{R}$ be a function which is α differentiable at $x_0 > 0$, then f is continuous at x_0.*

Lemma 4 [1,11]. *Let $\alpha \in (0,1]$ and f,g be α-differentiable functions at a point $x > 0$. Then, following statements hold true:*

(1) $T^\alpha(pf + qg) = pT^\alpha(f) + qT^\alpha(g)$, for all $p,q \in \mathbb{R}$.
(2) $T^\alpha(x^p) = px^{p-\alpha}$, for all $p \in \mathbb{R}$.
(3) $T^\alpha(\mu) = 0$, for all constant functions $f(x) = \mu$.
(4) $T^\alpha(fg) = fT^\alpha g + gT^\alpha f$.
(5) $T^\alpha(\frac{f}{g}) = \frac{gT^\alpha(f)-fT^\alpha(g)}{g^2}$, where $g \neq 0$.
(6) If f is differentiable, then $T^\alpha f(x) = x^{1-\alpha}\frac{df(x)}{dx}$.

Khalil [11] who introduced the notion of conformable fractional derivative, gave the following examples which assure the readers that a function can be α-differentiable at a point but not differentiable at the same point.

Example 1 [11]. Consider a function $f(x) = 2x^{\frac{1}{2}}$. For this function, we have $T^{\frac{1}{2}} f(x) = 1$ for every $x \in \mathbb{R}_+$. Hence $T^{\frac{1}{2}} f(0) = \lim\limits_{x \to 0^+} T^{\frac{1}{2}} f(x) = 1$. Then f is α-differentiable at a point zero, but $f'(0)$ does not exist.

3   Maximum-Minimum Principle

In general, the Eq. (1) has infinite number of solutions. However, deterministic character of real world processes that are modelled as Eq. (1) is ensured by certain conditions describing an initial state of the corresponding process and the observations of its visible parts.

We consider the initial-boundary value problem of the Eq. (1) along with the following initial and boundary conditions:

$$u(x,0) = \phi(x), \quad 0 < x < l, \tag{7}$$

$$u(0,t) = g(t), \quad u(l,t) = h(t), \quad 0 < t < T, \tag{8}$$

where $g \in C[0,T]$, $h \in C[0,T]$, $\phi \in C[0,l]$, $\phi(0) = g(0)$ and $\phi(l) = h(0)$.

Definition 4. *The function $u = u(x,t)$ defined on $\overline{\Omega} = [0,l] \times [0,T]$ is called a classical solution of the problem (1) if it belongs to the space $C(\overline{\Omega}) \cap C_t^1(0,T) \cap C_x^2(0,l)$ and satisfies (1).*

The maximum-minimum principle is a well known principle for the classical PDEs of the elliptic and parabolic type. Here we present the extension of it and establish the maximum-minimum principle for the generalized space-conformable fractional diffusion equation over an open bounded domain $(0,l) \times (0,T)$, $l, T \in \mathbb{R}+$.

Theorem 1. *Let $\overline{\Omega} = \{(x,t) : 0 \leq x \leq l, 0 \leq t \leq T\}$ be a closed rectangle and $\partial\Omega = \{(x,t) \in \Omega : t = 0 \ \ or \ \ x = 0 \ \ or \ \ x = l\}$. Let $u(x, t)$ be a continuous function on $\overline{\Omega}$ which satisfies the following equation:*

$$u_t(x,t) = kT_x^\alpha u(x,t), \quad 1 < \alpha < 2, \quad 0 < x < l, \quad t > 0. \tag{9}$$

Then,

$$\max_{(x,t)\in\overline{\Omega}} \{u(x,t)\} = \max_{(x,t)\in\partial\Omega} \{u(x,t)\}. \tag{10}$$

Proof. Let

$$M = \max_{(x,t)\in\partial\Omega} \{u(x,t)\}. \tag{11}$$

Then, it is enough to show $\max\limits_{(x,t)\in\overline{\Omega}} \{u(x,t)\} \leq M$.

84 K. K. Mishra and S. Dubey

Consider a function

$$v(x,t) = u(x,t) + \epsilon x^2, \tag{12}$$

where ϵ is a positive constant.
Let $D^\alpha = \frac{\partial}{\partial t} - kT_x^\alpha$.
For $(x,t) \in \overline{\Omega} - \partial\Omega$, we have

$$D^\alpha(v(x,t)) = D^\alpha(u(x,t)) - kT_x^\alpha(\epsilon x^2) \tag{13}$$

$$= -k\epsilon T_x^\alpha \tag{14}$$

$$= -k\epsilon x^{1-\alpha} 2x \tag{15}$$

$$= -2k\epsilon x^{2-\alpha} < 0. \tag{16}$$

Hence,

$$D^\alpha(v(x,t)) < 0. \tag{17}$$

Note that

$$v(x,t) = u(x,t) + \epsilon x^2 \tag{18}$$

$$\Rightarrow \quad v(x,t) \le M + \epsilon l^2, \quad \text{for } (x,t) \in \partial\Omega. \tag{19}$$

If $v(x,t)$ attains its maximum at some interior point, say (x_1,t_1), then it implies $D^\alpha v(x_1,t_1) \ge 0$ which contradicts the Eq. (17). Therefore $v(x,t)$ attains its maximum at a point of $\partial\Omega \cup \Gamma$, where $\Gamma = \{(x,t) : t = T\}$. Suppose that $v(x,t)$ attains its maximum at a point $(\overline{x},T) \in \Gamma$, $0 < \overline{x} < l$. Then

$$T_x^\alpha v(x,t) \le 0. \tag{20}$$

Since $v(\overline{x},T) \ge v(\overline{x},T-\delta)$, $0 < \delta < T$, therefore we have,

$$\lim_{\delta \to 0} \frac{v(\overline{x},T-\delta) - v(\overline{x},T)}{-\delta} = v_t(\overline{x},T) \ge 0,$$

which leads to $D^\alpha(v(\overline{x},T)) \ge 0$, a contradiction to (17).
Hence, $\max_{(x,t)\in\overline{\Omega}} v(x,t) = \max_{(x,t)\in\partial\Omega} v(x,t) \le M + \epsilon l^2$.
Therefore, $u(x,t) \le M + \epsilon l^2 - \epsilon x^2$,
$\quad \Rightarrow \quad u(x,t) \le M + \epsilon(l^2 - x^2)$, on $\overline{\Omega}$ and for every $\epsilon > 0$.
Taking $\epsilon \to 0$, we get $u(x,t) \le M$ on $\overline{\Omega}$, which means

$$\max_{\overline{\Omega}} u(x,t) = \max_{\partial\Omega} u(x,t).$$

This completes the proof of the theorem.

Theorem 2. *Let* $\overline{\Omega} = \{(x,t) : 0 \le x \le l, 0 \le t \le T\}$ *be a closed rectangle and* $\partial\Omega = \{(x,t) \in \Omega : t = 0 \text{ or } x = 0 \text{ or } x = l\}$. *Let* $u(x,t)$ *be a continuous function on* $\overline{\Omega}$ *which satisfies the following equation:*

$$u_t(x,t) = kT_x^\alpha u(x,t), \quad 1 < \alpha < 2, \quad 0 < x < l, \quad t > 0. \tag{21}$$

Then,

$$\min_{(x,t)\in\overline{\Omega}} \{u(x,t)\} = \min_{(x,t)\in\partial\Omega} \{u(x,t)\}. \tag{22}$$

Proof. This Theorem can be proved in a similar way to the proof of the Theorem (1) by replacing $u(x,t)$ by $-u(x,t)$.

Remark 1. By the above theorem, the maximum(minimum) of $u(x,t)$ cannot be assumed anywhere inside the rectangle, instead it is attained either on the bottom or lateral sides (unless u is constant).

3.1   Uniqueness and Stability of the Solution

We apply maximum-minimum principle to show that the uniqueness of the solution to considered initial boundary value problem (1), (7)–(8), We also prove that the solution, if it exists, continuously depends on the data given in the problem.

Theorem 3. *The initial value problem (1), (7)–(8) possesses at most one classical solution.*

Proof. Suppose that $u_1(x,t)$ and $u_2(x,t)$ are two solutions of (1) satisfying (7)–(8).
Consider a function $w(x,t) = u_1(x,t) - u_2(x,t)$. Then, $w(x,t)$ solves the following problem:

$$w_t(x,t) - kT_x^\alpha w(x,t) = 0, \quad 0 < x < l, \ \ 0 < t < T, \ \ 1 < \alpha < 2, \qquad (23)$$
$$w(x,0) = 0, \quad 0 \le x \le l, \qquad (24)$$
$$w(0,t) = w(l,t) = 0, \quad 0 \le t \le T. \qquad (25)$$

By maximum-minimum principle Theorems (1) and (2), we get

$$\max_{\overline{\Omega}} w(x,t) = 0 \text{ and } \min_{\overline{\Omega}} w(x,t) = 0, \text{ where } \overline{\Omega} = [0,l] \times [0,T].$$

Therefore, it follows that $w(x,t) = 0$, consequently $u_1(x,t) = u_2(x,t), \ \forall (x,t) \in \overline{\Omega}$. This completes the proof of the theorem.

Consider the IBVP (1), (7)–(8) with $g = h = 0$, that is

$$u_t - kT_x^\alpha u(x,t) = 0, \quad 0 < x < l, \ \ 0 < t < T, \ \ 1 < \alpha < 2, \qquad (26)$$
$$u(x,0) = \phi(x), \quad 0 < x < l, \qquad (27)$$
$$u(0,t) = 0, \quad 0 < t < T, \qquad (28)$$
$$u(l,t) = 0, \quad 0 < t < T. \qquad (29)$$

Theorem 4. *Let $u_j(x,t)$ be two solutions of Eq. (26) with initial data $\phi_j(x)$, $j = 1, 2$. Then,*

$$\max_{0 \le x \le l} |u_1(x,t) - u_2(x,t)| = \max_{0 \le x \le l} |\phi_1(x) - \phi_2(x)|, \quad for \ all \ t \in [0,T]. \quad (30)$$

Proof. Consider the function $w(x,t) = u_1(x,t) - u_2(x,t)$, which satisfies,

$$w_t - kT_x^\alpha w(x,t) = 0, \quad 0 < x < l, \quad 0 < t < T, \quad 1 < \alpha < 2, \tag{31}$$

$$w(x,0) = \phi_1(x) - \phi_2(x), \quad 0 < x < l, \tag{32}$$

$$w(0,t) = 0, \quad 0 < t < T, \tag{33}$$

$$w(l,t) = 0, \quad 0 < t < T. \tag{34}$$

So, by the Theorem (1) and (2),

$$u_1(x,t) - u_2(x,t) \le \max\{\max_{0 \le x \le l}(\phi_1(x) - \phi_2(x)), 0\}$$

$$\Rightarrow u_1(x,t) - u_2(x,t) \le \max_{0 \le x \le l}|\phi_1(x) - \phi_2(x)|.$$

and

$$u_1(x,t) - u_2(x,t) \ge \min\{\min_{0 \le x \le l}(\phi_1(x) - \phi_2(x)), 0\}$$

$$\Rightarrow u_1(x,t) - u_2(x,t) \ge -\max\{\max_{0 \le x \le l}(\phi_1(x) - \phi_2(x)), 0\}$$

$$\Rightarrow u_1(x,t) - u_2(x,t) \ge -\max_{0 \le x \le l}|\phi_1(x) - \phi_2(x)|.$$

Therefore,

$$\max_{0 \le x \le l}|u_1(x,t) - u_2(x,t)| \le \max_{0 \le x \le l}|\phi_1(x) - \phi_2(x)|. \tag{35}$$

This completes the proof of the theorem.

4   Construction of Solutions

In this section, we demonstrate the construction of analytic solution of the IBVP of diffusion equation (1), with the conditions (7)–(8) in the series form using the Fourier method. We present two cases: first for homogeneous and second for nonhomogeneous boundary conditions.

Equation with Homogeneous Boundary Conditions

Consider the following well posed problem of space-fractional diffusion equation

$$u_t - kT_x^\alpha u = 0, \quad 0 < x < l, \quad t > 0, \quad 1 < \alpha < 2, \tag{36}$$

$$u(x,0) = \phi(x), \quad 0 < x < l, \tag{37}$$

$$u(0,t) = u(l,t) = 0, \quad t > 0. \tag{38}$$

Let us consider the variable separable form

$$u(x,t) = X(x)S(t).$$

Substituting it into (36) and separating variables, we get

$$S_t X - kST_x^\alpha(X) = 0, \tag{39}$$

$$\frac{1}{k}S_t\frac{1}{T} = \frac{1}{X}T_x^\alpha(X) = \lambda \quad (say), \tag{40}$$

where λ is a constant. This along with the boundary conditions leads to

$$S_t - k\lambda S = 0, \tag{41}$$

$$T_x^\alpha X - \lambda X = 0, \ X(0) = 0 = X(l). \tag{42}$$

For $\lambda = 0$ and $\lambda > 0$, we get trivial solution $u = 0$.

In case of $\lambda < 0$, say $\lambda = -p^2$, the general solution of fractional differential equation (42) is given by

$$X(x) = c_1 \cos(\frac{p}{\beta}x^\beta) + c_2 \sin(\frac{p}{\beta}x^\beta),$$

where $2\beta = \alpha$ and c_1, c_2 are arbitrary constants. Using the boundary conditions, we obtain

$$X_n(x) = c_n \sin(\frac{1}{l^\beta}x^\beta n\pi), \ n = 1, 2, 3, \ldots \tag{43}$$

For $\lambda = -p^2 = -\frac{\beta^2}{l^{2\beta}}n^2\pi^2$, (41) solves as

$$S_n(t) = d_n e^{-(\frac{\beta}{l^\beta})^2 kn^2\pi^2 t}. \tag{44}$$

Thus, the solutions of space fractional diffusion equation satisfying boundary conditions are given by the functions

$$u_n(x,t) = A_n e^{-(\frac{\alpha/2}{l^{\alpha/2}})^2 kn^2\pi^2 t} \sin\left(\frac{1}{l^{\alpha/2}}x^{\alpha/2}n\pi\right), \ n = 1, 2, 3, \ldots \tag{45}$$

To satisfy the initial condition, we superpose the solutions

$$u(x,t) = \sum_{n=1}^{\infty} A_n e^{-(\frac{\alpha/2}{l^{\alpha/2}})^2 kn^2\pi^2 t} \sin(\frac{1}{l^{\alpha/2}}x^{\alpha/2}n\pi). \tag{46}$$

Then, at $t = 0$, we get

$$\phi(x) = \sum_{n=1}^{\infty} A_n \sin(\frac{1}{l^{\alpha/2}}x^{\alpha/2}n\pi).$$

If initial function $\phi(x)$ can be represented as its Fourier series, then the general solution of (36) is given by (46), where

$$A_n = \frac{2}{l}\int_0^l \phi(x) \sin(\frac{1}{l^{\alpha/2}}x^{\alpha/2}n\pi)dx.$$

Non-homogeneous Boundary Conditions

Consider the following well posed problem for a space-fractional diffusion equation

$$u_t - kT_x^\alpha u = 0, \quad 0 < x < l, \quad t > 0, \quad 1 < \alpha < 2, \qquad (47)$$
$$u(x,0) = f(x), \quad 0 < x < l, \qquad (48)$$
$$u(0,t) = A(t), \quad t \geq 0, \qquad (49)$$
$$u(l,t) = B(t), \quad t \geq 0. \qquad (50)$$

where $A(t)$ and $B(t)$ are smooth functions.

Likewise, the classical parabolic equations (47) is to be transformed into the problem with homogeneous boundary conditions. We first find a function $U(x) \in C^2[0,l]$ such that

$$U''(x) = 0, \; U(0) = A(t), \; U(l) = B(t),$$

This gives,

$$U(x) = \left(\frac{B(t) - A(t)}{l} \right)(x) + A(t).$$

Then, define

$$V(x,t) = u(x,t) - U(x), \; x \in [0,l], \; t \geq 0.$$

Clearly, $V(x,t)$ solves the following problem

$$V_t - kT_x^\alpha V = 0, \quad 0 < x < l, \quad t > 0, \quad 1 < \alpha < 2, \qquad (51)$$
$$V(x,0) = f(x) - U(x), \quad 0 < x < l, \qquad (52)$$
$$V(0,t) = 0, \quad t \geq 0, \qquad (53)$$
$$V(l,t) = 0, \quad t \geq 0. \qquad (54)$$

It is an IBVP with homogeneous boundary conditions and therefore can be solved similar to the previous case. Thus, (47) is solved as $u(x,t) = V(x,t) - \left(\frac{B(t) - A(t)}{l} \right)(x) - A(t)$.

References

1. Abdeljawad, T.: On conformable fractional calculus. J. Comput. Appl. Math. **279**, 57–66 (2015)
2. Abdeljawad, T., AL Horani, M., Khalil, R.: Conformable fractional semigroups of operators. J. Semigroup Theory Appl. **2015** (2015)
3. Acan, O., Firat, O., Keskin, Y.: Conformable variational iteration method, conformable fractional reduced differential transform method and conformable homotopy analysis method for non-linear fractional partial differential equations. Waves Random Complex Media **30**(2), 250–268 (2020)
4. Ahmed, H.M.: Fractional neutral evolution equations with nonlocal conditions. Adv. Differ. Equ. **2013**, 117 (2013)

5. Atangana, A., Noutchie, S.C.O.: Model of break-bone fever via beta-derivatives. BioMed. Res. Int. **2014**, 10, Article ID 523159 (2014)
6. Das, S.: Functional Fractional Calculus, 2nd edn. Springer, Berlin (2011)
7. Dubey, S., Sharma, M.: Solutions to fractional functional differential equations with nonlocal conditions. Fract. Calc. Appl. Anal. **17**(3), 654–673 (2014). https://doi.org/10.2478/s13540-014-0191-3
8. Hilfer, R., Butzer, P.L., Westphal, U.: An Introduction to Fractional Calculus. Applications of Fractional Calculus in Physics, pp. 1–85. World Scientific, River Edge (2010)
9. Hilfer, R.: Applications of Fractional Calculus in Physics. World scientific Singapore (2000)
10. Kaplan, M.: Applications of two reliable methods for solving a nonlinear conformable time-fractional equation. Opt. Quant. Electron. **49**, 312 (2017)
11. Khalil, R., Al Horani, M., Yousef, A., Sababheh, M.: A new definition of fractional derivative. J. Comput. Appl. Math. **264**, 65–70 (2014)
12. Miller, K.S., Ross, B.: An Introduction to the Fractional Calculus and Fractional Differential Equations. Wiley, New York (1993)
13. Podlubny, I.: Fractional Differential Equations: An Introduction to Fractional Derivatives, Fractional Differential Equations, to Methods of Their Solution and Some of Their Applications. Elsevier (1998)
14. Bayour, B., Torres, D.F.M.: Existence of solution to a local fractional nonlinear differential equation. J. Comput. Appl. Math. **312**, 127–133 (2017)
15. Dong, X., Bai, Z., Zhang, W.: Positive solutions for nonlinear eigenvalue problems with conformable fractional differential derivatives. J. Shandong Univ. Sci. Tech. Nat. Sci. (Chin. Ed.) **35**, 85–90 (2016)
16. Song, Q., Dong, X., Bai, Z., Chen, B.: Existence for fractional Dirichlet boundary value problem under barrier strip conditions. J. Nonlinear Sci. Appl. **10**, 3592–3598 (2017)
17. Kadkhoda, N., Jafari, H.: An analytical approach to obtain exact solutions of some space-time conformable fractional differential equations. Adv Differ Equ. **428** (2019)
18. Kartal, S., Gurcan, F.: Discretization of conformable fractional differential equations by a piecewise constant approximation. Int. J. Comput. Math. **96**(9), 1849–1860 (2019)
19. Luchko, Y.: Maximum principle for the generalized time-fractional diffusion equation. J. Math. Anal. Appl. **351**(1), 218–223 (2009)
20. Luchko, Y.: Some uniqueness and existence results for the initial-boundary-value problems for the generalized time-fractional diffusion equation. Comput. Math. Appl. **59**(5), 1766–1772 (2010)
21. Luchko, Y.: Initial-boundary-value problems for the one-dimensional time-fractional diffusion equation. Fract. Calc. Appl. Anal. **15**(1), 141–160 (2012)
22. Metzler, R., Klafter, J.: The random walk's guide to anomalous diffusion: a fractional dynamics approach. Phys. Rep. **339**(1), 1–77 (2000)
23. Diethelm, K.: The Analysis of Fractional Differential Equations: An Application-Oriented Exposition Using Differential Operators of Caputo Type. Springer, Berlin (2010)
24. Sharma, M., Dubey, S.: Asymptotic behavior of solutions to nonlinear nonlocal fractional functional differential equations. J. Nonl. Evol. Equ. Appl. **2015**(2), 21–30 (2015)

25. Singh, B.K., Agrawal, S.: A new approximation of conformable time fractional partial differential equations with proportional delay. Appl. Numer. Math. **157**, 419–433 (2020)
26. Thabet, H., Kendre, S.: Analytical solutions for conformable space-time fractional partial differential equations via fractional differential transform. Chaos Solitons Fractals **109**, 238–245 (2018)
27. Takayama, H.: Cooperative Dynamics in Complex Physical Systems: Proceedings of the Second Yukawa International Symposium, Kyoto, Japan, 24–27 August 1988, vol. 43. Springer, Heidelberg (2012)
28. Zhong, W., Wang, L.: Positive solutions of conformable fractional differential equations with integral boundary conditions. Bound Value Probl. **2018**, 137 (2018)
29. Bazhlekova, E.G., Dimovski, I.H.: Exact solution for the fractional cable equation with nonlocal boundary conditions. Central Eur. J. Phys. **11**, 1304–1313 (2013)
30. Zhou, Y., Jiao, F.: Existence of mild solutions for fractional neutral evolution equations. Comput. Math. Appl. **59**, 1063–1077 (2010)

Compact Finite Difference Method for Pricing European and American Options Under Jump-Diffusion Models

Kuldip Singh Patel[1]($^{(\boxtimes)}$) and Mani Mehra[2]

[1] Dr. SPM International Institute of Information Technology Naya Raipur,
Naya Raipur 493661, Chhattisgarh, India
`kuldip@iiitnr.edu.in`
[2] Indian Institute of Technology Delhi, New Delhi 110016, India
`mmehra@maths.iitd.ac.in`

Abstract. In this article, a compact finite difference method is proposed for pricing European and American options under jump-diffusion models. Partial integro-differential equation and linear complementarity problem governing European and American options respectively are discretized using Crank-Nicolson Leap-Frog scheme. In proposed compact finite difference method, the second derivative is approximated by the value of unknowns and their first derivative approximations which allow us to obtain a tri-diagonal system of linear equations for the fully discrete problem. Further, consistency and stability for the fully discrete problem are also proved. Since jump-diffusion models do not have smooth initial conditions, the smoothing operators are employed to ensure fourth-order convergence rate. Numerical illustrations for pricing European and American options under Merton jump-diffusion model are presented to validate the theoretical results.

Keywords: Compact finite difference method · European and American options · Jump-diffusion models · Operator splitting technique

1 Introduction

F. Black and M. Scholes [4] derived a partial differential equation (PDE) governing the option prices in the stock market with the assumption that the dynamics of the underlying asset are driven by geometric Brownian motion with constant volatility. Though Black-Scholes model is a seminal work in option pricing, numerous studies found that these assumptions are inconsistent with the market price movements. Therefore, various approaches have been considered to overcome the shortcomings of Black-Scholes model. In one of these approaches, Merton [24] incorporated the jumps into the dynamics of underlying asset to determine the volatility skews and it is known as Merton jump-diffusion model.

© Springer Nature Singapore Pte Ltd. 2021
A. Awasthi et al. (Eds.): CSMCS 2020, CCIS 1345, pp. 91–108, 2021.
https://doi.org/10.1007/978-981-16-4772-7_7

In another approach, S. L. Heston [10] considered the volatility to be a stochastic process and this model is known as stochastic volatility model. Apart from these, Dupire [8] considered the volatility to be a deterministic function of time and stock price. Further, Bates [3] combined the jump-diffusion model with stochastic volatility approach to capture the typical features of market option prices. Anderson and Andreasen [1] combined the deterministic volatility function approach with jump-diffusion model and proposed a second-order accurate numerical method for valuation of options. Our interest in present manuscript lies in option pricing under Merton's jump-diffusion model for European and American options.

The prices of European options under jump-diffusion models can be evaluated by solving a partial integro-differential equation (PIDE), whereas a linear complementarity problem (LCP) is solved for the evaluation of American options. Let us introduce some existing literature on numerical methods for the solution of the PIDE and LCP. Cont and Voltchkova [5] used implicit-explicit (IMEX) scheme for pricing European and barrier options and proved the stability and convergence of the proposed scheme. d'Halluin et al. [6] proposed a second-order accurate implicit method for pricing European and American options which uses fast Fourier transform (FFT) for the evaluation of convolution integral. They also proved the stability and the convergence of the fixed-point iteration method. An excellent comparison of various approaches for jump-diffusion models is given in [7]. A three-time levels second-order accurate implicit method using finite difference approximations is proposed for European and American put options under jump-diffusion models in [16] and [17] respectively. Salmi et al. [30] proposed a second-order accurate implicit-explicit (IMEX) time semi-discretization scheme for pricing European and American options under Bates model. They explicitly treated the jump term using the second-order Adams-Bashforth method and rest of the terms are discretized implicitly using the Crank-Nicolson method.

The majority of numerical approaches discussed in [6,16,17,30] to price European and American options under jump-diffusion models are based on second-order discretization methods. In order to increase the accuracy of second order finite difference approximations, inclusion of more grid points in computation stencil is required which results in a computationally expensive scheme. Therefore, finite difference approximations have been developed using compact stencils, commonly known as compact finite difference approximations, at the expense of some complication in their evaluation (see [20] for details). As a result, compact finite difference approximations provide high-order accuracy and better resolution characteristics as compared to finite difference approximations for equal number of grid points. A detailed study about the accuracy and resolution characteristics of various order compact finite difference approximations is presented in [22] and algorithms to compute derivative approximation in one and two-dimensions are also given.

Compact finite difference approximations have been used for option pricing problems [19,32] and in various other applications [15,23,29]. In [19,32], the original equation is considered as an auxiliary equation and each of the derivative of leading term of truncation error is compactly approximated. However, in present manuscript, we consider a different (Hermitian scheme) approach to derive compact scheme for solving PIDE and LCP. An immediate advantage of using Hermitian scheme approach instead of auxiliary equation approach is that earlier approach can be easily extended for solving higher-dimensional PDEs, PIDEs and LCPs arising in other areas of applied Mathematics.

Nevertheless, high-order approximations are not customary tools for option pricing because initial conditions for option pricing are always non-smooth. As a result, it affects the rate of convergence of numerical methods. Various approaches, e.g. co-ordinate transformation [32], and local mesh refinement [19], have been considered for option pricing problems to achieve high-order convergence rate even for non-smooth initial conditions. These approaches suffer with certain drawbacks e.g. it is not always easy to define a coordinate transformation for PIDE and the stability results for using local mesh refinement are not straight forward. Therefore as another approach, we apply smoothing operator to the initial conditions to obtain high-order convergence rate even for non-smooth initial conditions [14].

Our aim in this manuscript is to develop high-order compact scheme for solving PIDEs and LCPs in order to get option price and hedging parameters. The continuous model problem is discussed in next section. In third Section, compact finite difference method for pricing European and American options is proposed. Consistence and stability analysis for European options is also proved in this Section. In fourth section, numerical illustrations are presented to validate the theoretical findings. Finally, conclusions and some future work are given in fifth section.

2   The Mathematical Model

The PIDE for European option price can be written as [16]

$$\frac{\partial u}{\partial \tau}(x,\tau) = \mathbb{L}u, \ (x,\tau) \in (-\infty,\infty) \times (0,T],$$
$$u(x,0) = f(x) \ \forall \ x \in (-\infty,\infty), \tag{1}$$

where

$$\mathbb{L}u = \frac{\sigma^2}{2}\frac{\partial^2 u}{\partial x^2}(x,\tau) + \left(r - \frac{\sigma^2}{2} - \lambda\zeta\right)\frac{\partial u}{\partial x}(x,\tau) - (r+\lambda)u(x,\tau) + \lambda\int_{\mathbb{R}} u(y,\tau)g(y-x)dy, \tag{2}$$

$\tau = T - t$, $x = \ln\left(\frac{S}{S_0}\right)$, $u(x,\tau) = V(S_0 e^x, T-\tau)$, λ is the intensity of the jump sizes, $\zeta = \int_{\mathbb{R}}(e^x - 1)g(x)dx$, and $V(S,t)$ is the option price. Further, the LCP for American options is written as [17]

$$\frac{\partial u}{\partial \tau}(x,\tau) - \mathbb{L}u(x,\tau) \geq 0, \quad u(x,\tau) \geq f(x), \tag{3}$$
$$\left(\frac{\partial u}{\partial \tau}(x,\tau) - \mathbb{L}u(x,\tau)\right)(u(x,\tau) - f(x)) = 0.$$

The initial condition for European and American put options is

$$f(x) = max(K - S_0 e^x, 0) \ \forall \ x \in \mathbb{R}, \tag{4}$$

and the asymptotic behaviour is described as

European put options: $\lim_{x \to -\infty} [u(x, \tau) - (Ke^{-r\tau} - S_0 e^x)] = 0, \ \lim_{x \to \infty} u(x, \tau) = 0$

American put options: $\lim_{x \to -\infty} [u(x, \tau) - (K - S_0 e^x)] = 0, \ \lim_{x \to \infty} u(x, \tau) = 0$

3 Compact Finite Difference Method

In this section, compact finite difference approximations for first and second derivatives will be discussed. Fully discrete problem for PIDE (1) and LCP (3) will also be obtained in this section. Further, consistency and stability of the fully discrete problem will also be proved.

3.1 Compact Finite Difference Approximations for First and Second Derivatives

Let us consider the fourth-order compact finite difference approximations for first and second derivatives [20] of function u as follows

$$\frac{1}{4} u_{x_{i-1}} + u_{x_i} + \frac{1}{4} u_{x_{i+1}} = \frac{1}{\delta x} \left[-\frac{3}{4} u_{i-1} + \frac{3}{4} u_{i+1} \right], \tag{5}$$

$$\frac{1}{10} u_{xx_{i-1}} + u_{xx_i} + \frac{1}{10} u_{xx_{i+1}} = \frac{1}{\delta x^2} \left[\frac{6}{5} u_{i-1} - \frac{12}{5} u_i + \frac{6}{5} u_{i+1} \right], \tag{6}$$

where u_{x_i} and u_{xx_i} represents first and second derivatives approximations of unknown u at grid point x_i. If $\Delta_x u_i$ and $\Delta_{xx} u_i$ represent second-order finite difference approximation for first and second derivative respectively, then we may write

$$\Delta_x u_i = \frac{u_{i+1} - u_{i-1}}{2\delta x}, \quad \Delta_{xx} u_i = \frac{u_{i+1} - 2u_i + u_{i-1}}{\delta x^2}. \tag{7}$$

If the first derivative of unknowns are also considered as variables then Eq. (5) can be written as

$$\frac{1}{4} u_{xx_{i-1}} + u_{xx_i} + \frac{1}{4} u_{xx_{i+1}} = \frac{1}{\delta x} \left[-\frac{3}{4} u_{x_{i-1}} + \frac{3}{4} u_{x_{i+1}} \right]. \tag{8}$$

Eliminating $u_{xx_{i-1}}$ and $u_{xx_{i+1}}$ from Eqs. (6) and (8) and using Eq. (7) we have

$$u_{xx_i} = 2\Delta_{xx} u_i - \Delta_x u_{x_i}. \tag{9}$$

In this way, compact finite difference approximation for second derivative is expressed in terms of the value of the functions and their first derivative approximations. The value of u_{x_i} in Eq. (9) is obtained from Eq. (5). In case of non-periodic boundary conditions, fourth order accurate one-sided compact finite difference approximation for first derivative at boundary point can be obtained from [33]. Moreover, a detailed discussion about the resolution characteristics of above discussed compact finite difference approximations is presented in [27]. It is shown in [27] that resolution characteristics of compact approximation given in (9) are better than the compact approximation given in (6). This is another advantage of splitting the second derivative approximation as above apart from that it gives us a tri-diagonal system of linear equations for the fully discrete problem which will be shown later in this manuscript.

3.2   Localization to Bounded Domain

The domain of the spatial variable is restricted to a bounded interval $\Omega = (-L, L)$ for some fixed real number L to solve the PIDE (1) numerically. For given positive integers M AND N, let $\delta x = 2L/N$ and $\delta \tau = T/M$ and in this way we define $x_n = -L + n\delta x$ $(n = 0, 1, ..., N)$ and $\tau_m = m\delta \tau$ $(m = 0, 1, ..., M)$. Cont and Voltchkova [5] proved that truncation error after localization decreases exponentially point-wise. Further, Matache et al. [21] also proved an exponential bound in L_2-norm on truncation error. Now, the PIDE (1) can be written as

$$\frac{\partial u(x, \tau)}{\partial \tau} = \mathbb{D}u(x, \tau) + \mathbb{I}u(x, \tau), \quad (x, \tau) \in \Omega \times [0, T), \tag{10}$$

where $\mathbb{D}$ corresponds to the differential operator and $\mathbb{I}$ represents the integral operator. The operators $\mathbb{D}$ and $\mathbb{I}$ are as follows

$$\mathbb{D}u(x, \tau) = \frac{\sigma^2}{2} \frac{\partial^2 u}{\partial x^2}(x, \tau) + \left(r - \frac{\sigma^2}{2} - \lambda \zeta\right) \frac{\partial u}{\partial x}(x, \tau) - (r + \lambda)u(x, \tau),$$
$$\mathbb{I}u(x, \tau) = \lambda \int_{\mathbb{R}} u(y, \tau)g(y - x)dy. \tag{11}$$

3.3   Temporal Semi-discretization

A direct application of Crank-Nicolson Leap-Frog scheme for time semi-discretization of Eq. (10) gives us

$$\frac{u^{m+1} - u^{m-1}}{2\delta \tau} = \mathbb{D}\left(\frac{u^{m+1} + u^{m-1}}{2}\right) + \lambda \mathbb{I}(u^m), \quad m \geq 1, \tag{12}$$
$$u^{m+1}(x_{min}) = K(e^{-r\tau_{m+1}} - e^{x_{min}}), \quad u^{m+1}(x_{max}) = 0.$$

Let us suppose u^m and $\tilde{u}^m$ represents the exact and approximate solution of the Eq. (12) respectively with error $e^m := u^m - \tilde{u}^m$. The following theorem has been proved in [13] for the stability of the semi-discrete problem (12).

Theorem 1. *There exist a constant γ such that $\forall\ \delta \tau < \frac{1}{\gamma}$, we have*

$$||e^i||^2 \leq C||e^0||^2, \quad \forall\ 2 \leq i \leq M,$$

where C is a constant depends on r, σ, λ, and T.

3.4   The Fully Discrete Problem

The numerical approximations for the differential operator $\mathbb{D}$ and the integral operator $\mathbb{I}$ are discussed in this section. If $\mathbb{D}_\delta$ represents the discrete approximations for the operator $\mathbb{D}$, then

$$\mathbb{D}_\delta u_n^m = \frac{\sigma^2}{2} u_{xx_n}^m + \left(r - \frac{\sigma^2}{2} - \lambda\zeta\right) u_{x_n}^m - (r + \lambda)u_n^m, \tag{13}$$

where $u_n^m = u(x_n \tau_m)$ and $u_{x_n}^m$, $u_{xx_n}^m$ are the first and second derivative approximations of $u(x_n, \tau_m)$ respectively. Now using Eq. (9) in above Eq. (13), we get

$$\mathbb{D}_\delta u_n^m = \frac{\sigma^2}{2} \left(2\Delta_x^2 u_n^m - \Delta_x u_{x_n}^m\right) + \left(r - \frac{\sigma^2}{2} - \lambda\zeta\right) u_{x_n}^m - (r + \lambda)u_n^m. \tag{14}$$

In this way, second derivative approximation of unknowns are eliminated from the PIDE using the unknowns itself and their first derivative approximation.

Now, the discrete approximation for the integral operator $\mathbb{I}u$ using fourth-order accurate composite Simpson's rule is discussed. Integral operator $\mathbb{I}u(x, \tau)$ given in Eq. (11) is divided into two parts namely on $\Omega = (-L, L)$ and $\mathbb{R}\backslash\Omega$. The value $(\Upsilon(x, \tau, L))$ of integral operator $\mathbb{I}u$ on $\mathbb{R}\backslash\Omega$ can be obtained from

$$\Upsilon(x, \tau, L) = \begin{cases} Ke^{-r\tau}\Phi\left(-\frac{x+\mu_J+L}{\sigma_J}\right) - S_0 e^{x+\frac{\sigma_J^2}{2}+\mu_J}\Phi\left(-\frac{x+\sigma_J^2+\mu_J+L}{\sigma_J}\right), & \text{(European)}, \\ K\Phi\left(-\frac{x+\mu_J+L}{\sigma_J}\right) - S_0 e^{x+\frac{\sigma_J^2}{2}+\mu_J}\Phi\left(-\frac{x+\sigma_J^2+\mu_J+L}{\sigma_J}\right), & \text{(American)}, \end{cases}$$

where $\Phi(y)$ is the cumulative distribution function of standard normal random variable. The discrete approximation $(\mathbb{I}_\delta u)$ for the integral operator $(\mathbb{I}u)$ can be obtained from [25]. If $\mathbb{L}_\delta$ denote the discrete approximation of operator $\mathbb{L}$ (defined in Eq. (2)), then

$$\mathbb{L}_\delta u_n^m = \mathbb{D}_\delta\left(\frac{u_n^{m+1} + u_n^{m-1}}{2}\right) + \mathbb{I}_\delta u_n^m. \tag{15}$$

We find U_n^m (the approximate value of u_n^m) which is the solution of following problem

$$\frac{U_n^{m+1} - U_n^{m-1}}{2\delta\tau} = \mathbb{D}_\delta\left(\frac{U_n^{m+1} + U_n^{m-1}}{2}\right) + \mathbb{I}_\delta U_n^m, \quad 1 \leq m \leq M - 1, 1 \leq n \leq N - 1, \tag{16}$$

Using the values of $\mathbb{D}_\delta U_n^m$ from Eq. (14) in Eq. (16), we obtain

$$\frac{U_n^{m+1} - U_n^{m-1}}{2\delta\tau} = \frac{1}{2}\left[\frac{\sigma^2}{2}\left(2\Delta_x^2 U_n^{m+1} - \Delta_x U_{x_n}^{m+1}\right) + \left(r - \frac{\sigma^2}{2} - \lambda\zeta\right) U_{x_n}^{m+1} - (r + \lambda)U_n^{m+1}\right]$$

$$+ \frac{1}{2}\left[\frac{\sigma^2}{2}\left(2\Delta_x^2 U_n^{m-1} - \Delta_x U_{x_n}^{m-1}\right) + \left(r - \frac{\sigma^2}{2} - \lambda\zeta\right) U_{x_n}^{m-1} - (r + \lambda)U_n^{m-1}\right]$$

$$+ \mathbb{I}_\delta U_n^m, \quad \text{for } 1 \leq m \leq M - 1.$$

Re-arranging the terms in above equation, the following fully discrete problem is obtained

$$\left[1 - \delta\tau\frac{\sigma^2}{2}2\Delta_x^2 - \delta\tau(r+\lambda)\right]U_n^{m+1} = \delta\tau\left[-\frac{\sigma^2}{2}\Delta_x U_{x_n}^{m+1} + \left(r - \frac{\sigma^2}{2} - \lambda\zeta\right)U_{x_n}^{m+1}\right]$$

$$+ \delta\tau\left[\frac{\sigma^2}{2}(2\Delta_x^2 U_n^{m-1} - \Delta_x U_{x_n}^{m-1}) + \left(r - \frac{\sigma^2}{2} - \lambda\zeta\right)U_{x_n}^{m-1} \quad (17)\right.$$

$$\left. - (r+\lambda)U_n^{m+1}\right] + U_n^{m-1} + 2\delta\tau\mathbb{I}_\delta U_n^m.$$

for $1 \leq m \leq M - 1$. Let us introduce the following notation

$$\mathbf{U}^m = (U_1^m, U_2^m, ..., U_{N-1}^m)^T \text{ and } \mathbf{U}_x^m = (U_{x_1}^m, U_{x_2}^m, ..., U_{x_{N-1}}^m)^T,$$

the resulting system of equations corresponding to the difference scheme (17) can be written as

$$A\mathbf{U}^{m+1} = F(\mathbf{U}^m, \mathbf{U}^{m-1}, \mathbf{U}_x^{m-1}, \mathbf{U}_x^{m+1}). \tag{18}$$

The presence of $\mathbf{U}_x^{m+1}$ on the right hand side of the Eq. (18) bind us to use a predictor corrector method to solve the system of equations. Therefore, correcting to convergence approach presented in [26] is used to solve (18).

Computational Complexity: In above discussed approach, number of iterations to achieve desired accuracy are not known in advance. Let number of iterations required by above approach be n_m at a fixed time level m and $n_s := \max_{1 \leq m \leq M} n_m$. We know that a tri-diagonal system of equations is solved with $O(N)$ operations and we have also discussed that matrix-vector multiplication is obtained with $O(N \log N)$ complexity. Therefore, maximum computational complexity of the proposed compact finite difference method will be of order $O((n_s + \log N)NM)$.

Now, the fully discrete problem for American options using compact finite difference method is discussed. Ikonen et al. [11] proposed the operator splitting technique for American put options under Black-Scholes model and it is extended by Toivanen [34] for jump-diffusion models. For detailed explanation about the operator splitting technique, one can see [17] and references therein. A new auxiliary variable ψ is taken such that $\psi = U_\tau - \mathbb{L}U$ and LCP (3) is written as follows:

$$U_\tau - \mathbb{L}U = \psi,$$
$$\psi \geq 0, \quad U \geq f, \quad \psi(U - f) = 0. \tag{19}$$

The above equation is discretized using operator splitting technique as follows:

$$\frac{\tilde{U}_n^{m+1} - U_n^{m-1}}{2\delta\tau} - \left[\mathbb{D}_\delta\left(\frac{\tilde{U}_n^{m+1} + U_n^{m-1}}{2}\right) + \mathbb{I}_\delta U_n^m\right] = \Psi_n^m, \tag{20}$$

$$\frac{U_n^{m+1} - U_n^{m-1}}{2\delta\tau} - \left[\mathbb{D}_\delta\left(\frac{\tilde{U}_n^{m+1} + U_n^{m-1}}{2}\right) + \mathbb{I}_\delta U_n^m\right] = \Psi_n^{m+1}. \tag{21}$$

Now, a pair $(\Psi_n^{m+1}, U_n^{m+1})$ is to be obtained satisfying the Eqs. (20) and (21) and the constraints

$$U_n^{m+1} \geq f(x_n), \quad \Psi_n^{m+1} \geq 0, \quad \Psi_n^{m+1}\left(U_n^{m+1} - f(x_n)\right) = 0. \tag{22}$$

Algorithm to solve Eqs. (20), (21), and (22):

for $m = 0, n = 1, 2, ..., N - 1$
$\frac{\tilde{U}_n^{m+1} - U_n^m}{\delta\tau} = \mathbb{D}_\delta \tilde{U}_n^{m+1} + \mathbb{I}_\delta U_n^m + \Psi_n^m$
end
Solve for $n = 1, 2, ..., N - 1$
$U_n^{m+1} = \max\left(u_0(x_n), \tilde{U}_n^{m+1} - \delta\tau\Psi_n^m\right)$
$\Psi_n^{m+1} = \frac{U_n^{m+1} - \tilde{U}_n^{m+1}}{\delta\tau} + \Psi_n^m$
end
for $m \geq 1, n = 1, 2, ..., N - 1$
$\frac{\tilde{U}_n^{m+1} - U_n^{m-1}}{2\delta\tau} = \mathbb{D}_\delta\left(\frac{\tilde{U}_n^{m+1} + U_n^{m-1}}{2}\right) + \mathbb{I}_\delta U_n^m + \Psi_n^m$
end
Solve for $n = 1, 2, ..., N - 1$
$U_n^{m+1} = \max\left(u_0(x_n), \tilde{U}_n^{m+1} - 2\delta\tau\Psi_n^m\right)$
$\Psi_n^{m+1} = \frac{U_n^{m+1} - \tilde{U}_n^{m+1}}{2\delta\tau} + \Psi_n^m$
end

3.5   Consistency

In order to prove consistency of fully discrete problem (17), the following theorem is a direct application of Taylor series expansion and compact finite difference approximations discussed above (see [26] for more details).

Theorem 2. *For sufficiently small δx and $\delta\tau$, we have*

$$\frac{\partial u}{\partial\tau}(x_n, \tau_m) - \mathbb{L}u(x_n, \tau_m) - \left(\frac{u(x_n, \tau_{m+1}) - u(x_n, \tau_{m-1})}{2\delta\tau} - \mathbb{L}_\delta u(x_n, \tau_m)\right) = O(\delta\tau^2 + \delta x^4), \tag{23}$$

for $m \geq 1$, where $\mathbb{L}$ and $\mathbb{L}_\delta$ are given in Eqs. (2) and (15) respectively and $(x_n, \tau_m) \in (-L, L) \times (0, T]$.

3.6   Stability

The stability of proposed compact finite difference method is proved using von Neumann stability analysis. Consider a single node

$$U_n^m = p^m e^{In\theta}, \tag{24}$$

where $I = \sqrt{-1}$, p^m is the mth power of amplitude at time levels τ_m. We consider the integration term given in Eq. (11) in an equivalent form as follows.

$$\mathbb{I}u(x, \tau) = \lambda \int_{-L}^{L} u(y + x, \tau)g(y)dy.$$

Fourth-order accurate composite Simpson's rule for above equation is then given by

$$\mathbb{I}_\delta u = \delta x \sum_{k=0}^{N} w_k U_{k+n}^m g_k,$$

$$= \delta x \sum_{k=0}^{N} w_k p^m e^{I\theta(k+n)} g_k,$$

$$\equiv p^m e^{I\theta n} G_k,$$

where

$$G_k = \delta x \sum_{k=0}^{N} w_k e^{I\theta k} g_k \quad \text{and} \quad g_k = g(x_k). \tag{25}$$

The following lemma has been proved in [26] which gives that numerical quadrature G_k is fourth order accurate.

Lemma 1. *The numerical quadrature G_k given in Eq. (25) satisfies the following*

$$|G_k| \leq 1 + c\delta x^4,$$

where c is a constant.

For sake of simplicity, we denote $\frac{\sigma^2}{2} = a$ and $\left(r - \frac{\sigma^2}{2} - \lambda\zeta\right) = b$ in the rest of the section. Therefore, the fully discrete problem (17) can be written as follows

$$(1 - 2a\delta\tau\Delta_x^2 + \delta\tau(r+\lambda))U_n^{m+1} = (1 + 2a\delta\tau 2\Delta_x^2 - \delta\tau(r+\lambda))U_n^{m-1} + 2\delta\tau$$

$$\left[\frac{b}{2} - \frac{a}{2}\Delta_x\right] U_{x_n}^{m+1} + 2\delta\tau \left[\frac{b}{2} - \frac{a}{2}\Delta_x\right] U_{x_n}^{m-1} + 2\delta\tau\lambda G_k U_n^m. \tag{26}$$

From [28], the following relations are obtained

$$\Delta_x U_n^m = I\frac{sin(\theta)}{\delta x}U_n^m, \quad \Delta_x^2 U_n^m = \frac{2cos(\theta)-2}{\delta x^2}U_n^m, \quad U_{x_n}^m = I\frac{3sin(\theta)}{\delta x(2+cos(\theta))}U_n^m. \tag{27}$$

Using (27) in the difference scheme (26), we obtain

$$\left[1 - 4a\delta\tau\left(\frac{cos(\theta)-1}{\delta x^2}\right) + \delta\tau(r+\lambda)\right]U_n^{m+1} = \left[1 + 4a\delta\tau\left(\frac{cos(\theta)-1}{\delta x^2}\right) - \delta\tau(r+\lambda)\right]U_n^{m-1}$$

$$+ \delta\tau\left[\left(a\frac{sin(\theta)}{\delta x} + Ib\right)\frac{3sin(\theta)}{\delta x(2+cos(\theta))}\right]U_n^{m+1}$$

$$+ \delta\tau\left[\left(a\frac{sin(\theta)}{\delta x} + Ib\right)\frac{3sin(\theta)}{\delta x(2+cos(\theta))}\right]U_n^{m-1}$$

$$+ 2\delta\tau\lambda G_k U_n^m, \tag{28}$$

which implies

$$\left[1 - \delta\tau a\frac{cos^2(\theta) + 4cos(\theta) - 5}{\delta x^2(2 + cos(\theta))} + \delta\tau(r + \lambda) - I\delta\tau b\frac{3sin(\theta)}{\delta x(2 + cos(\theta))}\right]U_n^{m+1}$$
$$= \left[1 + \delta\tau a\frac{cos^2(\theta) + 4cos(\theta) - 5}{\delta x^2(2 + cos(\theta))} - \delta\tau(r + \lambda) + I\delta\tau b\frac{3sin(\theta)}{\delta x(2 + cos(\theta))}\right]U_n^{m-1}$$
$$+ 2\delta\tau\lambda G_k U_n^m. \tag{29}$$

Now using Eq. (24) in above and divide the above equation by $p^{m-1}e^{In\theta}$, we get the amplification polynomial

$$\Theta(\delta x, \delta\tau, \theta) = \gamma_0 p^2 - 2\gamma_1 p - \gamma_2, \tag{30}$$

where

$$\gamma_0 = \left[1 - \delta\tau\left(a\frac{cos^2(\theta) + 4cos(\theta) - 5}{\delta x^2(2 + cos(\theta))} - (r + \lambda) + Ib\frac{3sin(\theta)}{\delta x(2 + cos(\theta))}\right)\right],$$
$$\gamma_1 = \lambda\delta\tau G_k, \tag{31}$$
$$\gamma_2 = \left[1 + \delta\tau\left(a\frac{cos^2(\theta) + 4cos(\theta) - 5}{\delta x^2(2 + cos(\theta))} - (r + \lambda) + Ib\frac{3sin(\theta)}{\delta x(2 + cos(\theta))}\right)\right].$$

The following lemma is proved in [31] which states the necessary and sufficient condition for a finite difference scheme to be stable.

Lemma 2. *A finite difference scheme is stable if and only if all the roots, p_u, of the amplification polynomial Θ satisfies the following condition:*

- *There is a constant C such that $|p_u| \leq 1 + C\delta\tau$.*
- *There are positive constants a_0 and a_1 such that if $a_0 < |p_u| \leq 1 + C\delta\tau$ then $|p_u|$ is simple root and for any other root p_v, following relation holds*

$$|p_v - p_u| \geq a_1,$$

as $\delta x, \delta\tau \to 0$.

The above Lemma 2 is proved for fully discrete problem (17) in the following theorem.

Theorem 3. *The fully discrete problem (17) is stable in the sense of Von-Neumann for $\delta\tau \leq 1/(2\lambda)$.*

Proof: First, some properties of the coefficients γ_0, γ_1 and γ_2 of amplification polynomial Θ are proved. Using Lemma 1 in Eq. (31) it is observed that $|\gamma_1| < \delta\tau\lambda$. We can write $|\gamma_0| = |(1 - A) - IB|$, where

$$A = a\frac{cos^2(\theta) + 4cos(\theta) - 5}{\delta x^2(2 + cos(\theta))} - (r + \lambda), \quad and \quad B = b\frac{3sin(\theta)}{\delta x(2 + cos(\theta))}.$$

This implies $|\gamma_0|^2 = 1 + A^2 - 2A + B^2$. Since $\frac{cos^2(\theta)+4cos(\theta)-5}{\delta x^2(2+cos(\theta))} < 0$, $a > 0$, $(r + \lambda) > 0 \implies A < 0$, therefore $|\gamma_0| > 1$. Similarly

$$\left|\frac{\gamma_2}{\gamma_0}\right|^2 = \frac{1 + A^2 + 2A + B^2}{1 + A^2 - 2A + B^2}.$$

Again $A < 0 \implies \left|\frac{\gamma_2}{\gamma_0}\right| < 1$. Now, roots of the amplification polynomial Θ can be written as

$$
\begin{aligned}
|p| &= \left| \frac{\gamma_1 \pm \sqrt{\gamma_1^2 - \gamma_0\gamma_2}}{\gamma_0} \right|, \\
&\leq \left|\frac{\gamma_2}{\gamma_0}\right|^{\frac{1}{2}} + 2\left|\frac{\gamma_1}{\gamma_0}\right|, \\
&\leq 1 + 2\delta\tau\lambda.
\end{aligned}
\tag{32}
$$

Hence, first part of the Lemma 2 is proved for constant $C = 2\lambda$. Now for second part of the Lemma 2, let us assume that p_1 and p_2 are two roots of amplification polynomial Θ. Take the constant $a_0 = 1$ which will imply that $p_1 > 1$, then

$$
\begin{aligned}
|p_1 - p_2| &\geq 2|p_1| - |p_1 + p_2|, \\
&\geq 2 - 2\delta\tau\lambda.
\end{aligned}
\tag{33}
$$

If $\delta\tau$ satisfies the given condition, we have $|p_1 - p_2| \geq 1$, and this prove the second part of the Lemma 2 with $a_1 = 1$. This completes the proof.

Remark 1. We would like to compare the stability result presented in this manuscript with the stability results in [26]. Note that in [26], the restriction on $\delta\tau$ depends on jump size λ and risk free interest rate r. However in the present manuscript, the choice of $\delta\tau$ only depend on the jump size λ. Therefore, one can say that stability result present in this manuscript is less restrictive as compared to [26].

4   Numerical Results

In this section, the applicability of the proposed compact finite difference method for pricing European and American options under jump-diffusion models is demonstrated. According to [14], fourth-order convergence cannot be expected for non-smooth initial conditions. Since the initial conditions given in Eqs. (4) has low regularity, the smoothing operator ϕ_4 given in [14] is employed to smoothen the initial conditions and it's Fourier transform is define as

$$\hat{\phi}_4(\omega) = \left(\frac{sin(\omega/2)}{\omega/2}\right)^4 \left[1 + \frac{2}{3}sin^2(\omega/2)\right].$$

As a result, the following smoothed initial condition $(\tilde{u}_0)$ is obtained

$$\tilde{u}_0(x_1) = \frac{1}{\delta x} \int_{-3\delta x}^{3\delta x} \phi_4\left(\frac{x}{\delta x}\right) u_0(x_1 - x)dx,
\tag{34}$$

where u_0 is the actual non-smooth initial condition and x_1 is the grid point where smoothing is required. The smoothed initial conditions obtained from Eq. (34) tends to the original initial conditions as $\delta x \to 0$. The parameters considered for pricing European and American options under Merton jump-diffusion model are listed in Table 1. The parabolic mesh ratio $\left(\frac{\delta \tau}{\delta x^2}\right)$ is fixed as 0.4 in all our computations, although neither the von Neumann stability analysis nor the numerical experiments showed any such restriction. The relative ℓ^2-error $\frac{\|U_{ref}-U\|_{\ell^2}}{\|U_{ref}\|_{\ell^2}}$ is used to determine the numerical convergence rate, where U_{ref} represents the numerical solution on a fine grid ($\delta x = 4.8828125e-04$) and U denotes the numerical solution on coarser grid. Order of convergence is obtained as the slope of the linear least square fit of the individual error points in the loglog plot of error versus number of grid points.

Table 1. The values of parameters for pricing European and American options under jump-diffusion models.

Parameters	λ	T	r	K	σ	μ_J	σ_J	S_0
Values	0.10	0.25	0.05	100	0.15	−0.90	0.45	100

In option pricing, Greeks are important instruments for the measurement of an option position's risks. The rate of change of option price with respect to change in the underlying asset's price is known as Delta whereas the rate of change in the delta with respect to change in the underlying price is called as Gamma. The proposed compact finite difference method is considered for valuation of options and Greeks as well in the following examples.

Table 2. Values of European put options and Greeks under Merton jump-diffusion model with constant volatility using $N = 1536$.

(S, τ)	Option Price		Delta		Gamma	
	In [12]	Our method	In [12]	Our method	In [12]	Our method
(90, 0)	9.285418	9.285416	−0.846715	−0.846716	0.034860	0.034862
(100, 0)	3.149026	3.149018	−0.355663	−0.355661	0.048825	0.048828
(110, 0)	1.401186	1.401182	−0.058101	−0.058103	0.012129	0.012131

Example 1: Merton Jump-Diffusion Model for European Put Options with Constant Volatility

The values of option prices and Greeks for various stock prices are presented in Table 2 and it is observed that proposed compact finite difference method is accurate for valuation of options and Greeks as well. The difference between the reference and numerical solutions as a function of asset price and time with

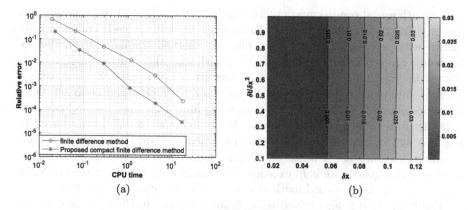

Fig. 1. (a) Efficiency: CPU time and relative error for finite difference method and proposed compact finite differential method (b) Numerical Stability Plot.

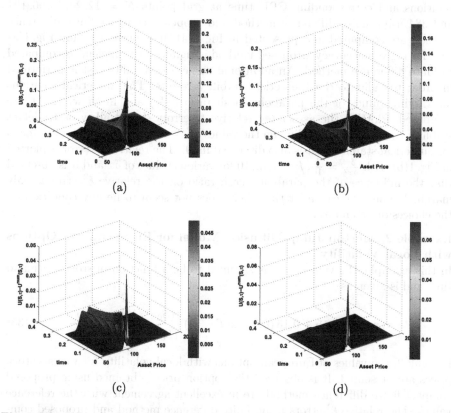

Fig. 2. The difference between reference and numerical solutions as a function of asset price and time using: (a) finite difference method with non-smooth initial condition, (b) proposed compact finite difference method with non-smoothing initial condition, (c) finite difference method with smoothed initial condition and (d) proposed compact finite difference method with smoothed initial condition.

non-smooth initial condition are plotted in Figs. 2(a) and 2(b) respectively. It is observed from the figures that maximum error at strike price is comparatively smaller with proposed compact finite difference method. Similarly, the difference between reference and numerical solutions with smoothed initial conditions are plotted in Figs. 2(c) and 2(d). It is evident from the figures that oscillations in the solution near the strike price are lesser with proposed compact finite difference method. The relative ℓ^2-errors using finite difference method (second-order accurate) and proposed compact finite difference method are plotted in Fig. 3(a) and it can be concluded that proposed method is only second order accurate with non-smooth initial condition. Further, it is observed that numerical order of convergence rate is in excellent agreement with the theoretical order of convergence of the proposed method when initial condition is smoothed.

The PIDE (16) is also solved using finite difference method [16] in order to compare the efficiency of proposed compact finite difference method with finite difference method. The relative ℓ^2 errors between the numerical and reference solutions and corresponding CPU time at grid points $N = 12, 24, 48, 96, 192$ and 384 using finite difference method and proposed compact finite difference method are computed and presented in Fig. 1(a). It is observed from Fig. 1(a) that for a given accuracy, proposed method is significantly efficient as compared to finite difference method. An additional numerical stability test is performed in order to validate the theoretical stability results. The numerical solutions for varying values of the parabolic mesh ratio ($\frac{\delta\tau}{\delta x^2}$) and mesh width δx are computed. Plotting the associated relative ℓ^2 errors should allow us to detect stability restrictions depending on the values of $\delta\tau$ and δx. The similar approach for numerical stability test is also discussed in [9]. The relative ℓ^2 error is plotted in Fig. 1(b) with $\frac{\delta\tau}{\delta x^2} = \frac{k}{10}$, $k = 1, ..., 10$ for various values of δx and it is observed that the influence of the parabolic mesh ratio on the relative ℓ^2 error is only marginal. Thus, we can infer that there does not seem to be any condition on the choices of $\delta\tau$ and δx.

Example 2: Merton Jump-Diffusion Model for European Put Options with Local Volatility

In this example, the volatility σ is assumed to be a function of stock price and time and is given as

$$\sigma(x, \tau) = 0.15 + 0.15 \left(0.5 + 2(T - \tau)\right) \frac{\left((S_0 e^x / 100) - 1.2\right)^2}{\left(S_0 e^x / 100\right)^2 + 1.44}. \tag{35}$$

In Table 3, the values of European options with local volatility for various stock prices are presented. It is observed that option prices obtained using proposed compact finite difference method are in excellent agreement with the reference values. The relative ℓ^2-errors using finite difference method and proposed compact finite difference method are plotted in Fig. 3(b) and it is observed that proposed method is only second order accurate with non-smooth initial condition. The numerical order of convergence rate agrees with the theoretical order of convergence rate of the proposed method when the initial condition is smoothed.

Table 3. Values of European put options with local volatility under Merton jump-diffusion model using $N = 1536$.

	$(S, \tau) = (90, 0)$	$(S, \tau) = (100, 0)$	$(S, \tau) = (110, 0)$
Reference values [18]	9.317323	3.183681	1.407745
Proposed method	9.317322	3.183682	1.407743

Example 3: Merton Jump-Diffusion Model for American Put Options with Constant Volatility

The values of American options for various stock prices are presented in Table 4 and it is observed that proposed compact finite difference method is also accurate for valuation of American options. The relative ℓ^2-errors using finite difference method and proposed compact finite difference method are plotted in Fig. 3(c) and it is observed that proposed method is only second order accurate with non-smooth initial condition. The numerical order of convergence rate is 3.2 with smoothed initial condition which does not represents the theoretical order of convergence rate. The reason could be the lack of regularity of the problem due to the free boundary feature which needs further research to be resolved [2].

Table 4. Values of American put options under Merton jump-diffusion model with $N = 1536$.

	$(S, \tau) = (90, 0)$	$(S, \tau) = (100, 0)$	$(S, \tau) = (110, 0)$
Reference values [18]	10.003866	3.241207	1.419790
Proposed method	10.003862	3.241208	1.419791

Example 4: Merton Jump-Diffusion Model for American Put Options with Local Volatility

In Table 5, the values of American options with non-constant volatility are presented for various stock prices. It can be concluded that proposed method is also accurate for valuation of American options with non-constant volatility. The relative ℓ^2-errors using finite difference method and proposed compact finite difference method are plotted in Fig. 3(d).

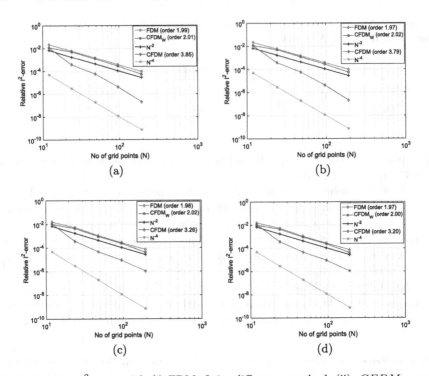

Fig. 3. Relative ℓ^2 error with (i) FDM: finite difference method, (ii). $CFDM_W$: proposed compact finite difference method without smoothing the initial condition, (iii). CFDM: proposed compact finite difference method with smooth initial condition for (a) European put option with constant volatility, (b) European put option with local volatility, (c) American put option with constant volatility, and (d) American put option with local volatility.

Table 5. Values of American put options with local volatility under Merton jump-diffusion model using $N = 1536$.

	$(S, \tau) = (90, 0)$	$(S, \tau) = (100, 0)$	$(S, \tau) = (110, 0)$
Reference values [18]	10.008881	3.275957	1.426403
Proposed method	10.008880	3.275955	1.426403

5   Conclusion and Future Work

In this article, a compact finite difference method has been proposed for pricing European and American options under Merton jump-diffusion model with constant and local volatilities. Consistency and stability of fully discrete problem have also been proved. The effect of non-smooth initial condition on the numerical convergence rate is discussed and it is shown that smoothing of initial condition helps us to achieve high-order numerical convergence rate. Moreover, Greeks (Delta and Gamma) are computed for European options and it is shown that proposed compact finite difference method is accurate for valuation of options and Greeks as well. It would be interesting to extend the proposed compact finite difference method for stochastic volatility jump-diffusion models (a.k.a. Bates model) as a future work.

References

1. Andersen, L., Andreasen, J.: Jump-diffusion process: volatility smile fitting and numerical methods for option pricing. Rev. Deriv. Res. **4**, 231–262 (2000)
2. Bastani, A.F., Ahmadi, Z., Damircheli, D.: A radial basis collocation method for pricing American options under regime-switching jump-diffusions. Appl. Numer. Math. **65**, 79–90 (2013)
3. Bates, D.: Jump and stochastic volatility: exchange rate process implicit in deutsche mark options. Rev. Finan. Stud. **9**, 69–107 (1996)
4. Black, F., Scholes, M.: Pricing of options and corporate liabilities. J. Polit. Econ. **81**, 637–654 (1973)
5. Cont, R., Voltchkova, E.: A finite difference scheme for option pricing in jump-diffusion and exponential Levy models. SIAM J. Numer. Anal. **43**, 1596–1626 (2005)
6. d'Halluin, Y., Forsyth, P.A., Veztal, K.R.: Robust numerical methods for contingent claims under jump-diffusion process. IMA J. Numer. Anal. **25**, 87–112 (2005)
7. Duffy, D.J.: Numerical analysis of jump diffusion models: a partial differential equation approach. Technical Report, Datasim (2005)
8. Dupire, B.: Pricing with a smile. RISK **39**, 18–20 (1994)
9. During, B., Fournie, M.: High-order compact finite difference scheme for option pricing in stochastic volatility models. J. Comput. Appl. Math. **236**, 4462–4473 (2012)
10. Heston, S.L.: A closed form solution for options with stochastic volatility with applications to bond and currency options. Rev. Finan. Stud. **6**, 327–343 (1993)
11. Ikonen, S., Toivanen, J.: Operator splitting method for American option pricing. Appl. Math. Lett. **17**, 809–814 (2004)
12. Kadalbajoo, M.K., Kumar, A., Tripathi, L.P.: A radial basis function based implicit-explicit method for option pricing under jump-diffusion models. Appl. Numer. Math. **110**, 159–173 (2016)
13. Kadalbajoo, M.K., Kumar, A., Tripathi, L.P.: An efficient numerical method for pricing option under jump-diffusion model. Int. J. Adv. Eng. Sci. Appl. Math. **7**, 114–123 (2015)
14. Kreiss, H.O., Thomee, V., Widlund, O.: Smoothing of initial data and rates of convergence for parbolic difference equations. Commun. Pure Appl. Math. **23**, 241–259 (1970)
15. Kumar, V.: High-order compact finite-difference scheme for singularly-perturbed reaction-diffusion problems on a new mesh of Shishkin type. J. Optim. Theory Appl. **143**, 123–147 (2009)
16. Kwon, Y., Lee, Y.: A second-order finite difference method for option pricing under jumps-diffusion models. SIAM J. Numer. Anal. **49**, 2598–2617 (2011)
17. Kwon, Y., Lee, Y.: A second-order tridigonal method for American option under jumps-diffusion models. SIAM J. Sci. Comput. **43**, 1860–1872 (2011)
18. Lee, J., Lee, Y.: Stability of an implicit method to evaluate option prices under local volatility with jumps. Appl. Numer. Math. **87**, 20–30 (2015)

19. Lee, S.T., Sun, H.W.: Fourth order compact scheme with local mesh refinement for option pricing in jump-diffusion model. Numer. Methods Partial Differ. Eq. **28**, 1079–1098 (2011)
20. Lele, S.K.: Compact finite difference schemes with spectral-like resolution. J. Comput. Phys. **103**, 16–42 (1992)
21. Matache, A.M., Schwab, C., Wihler, T.P.: Fast numerical solution of parabolic integro-differential equations with applications in finance. SIAM J. Sci. Comput. **27**, 369–393 (2005)
22. Mehra, M., Patel, K.S.: Algorithm 986: a suite of compact finite difference schemes. ACM Trans. Math. Softw. **44**, 1–31 (2017)
23. Mehra, M., Patel, K.S., Shukla, A.: Wavelet-optimized compact finite difference method for convection-diffusion equations. Int. J. Nonlinear Sci. Numer. (2020, in press). https://doi.org/10.1515/ijnsns-2018-0295
24. Merton, R.C.: Option pricing when underlying stocks return are discontinous. J. Finan. Econ. **3**, 125–144 (1976)
25. Patel, K.S., Mehra, M.: Fourth-order compact finite difference scheme for American option pricing under regime-switching jump-diffusion models. Int. J. Appl. Comput. Math. **3**, 547–567 (2017)
26. Patel, K.S., Mehra, M.: Fourth-order compact scheme for option pricing under the Merton's and Kou's jump-diffusion models. Int. J. Theor. Appl. Finan. **21**, 1–26 (2018)
27. Patel, K.S., Mehra, M.: A numerical study of Asian option with high-order compact finite difference scheme. J. Appl. Math. Comput. **57**, 467–491 (2018)
28. Patel, K.S., Mehra, M.: High-order compact finite difference scheme for pricing Asian option with moving boundary condition. Differ. Equ. Dyn. Syst. **27**, 39–56 (2019)
29. Patel, K.S., Mehra, M.: Fourth-order compact scheme for space fractional advection-diffusion reaction equations with variable coefficients. J. Comput. Appl. Math. **380**, 112963 (2020)
30. Salmi, S., Toivanen, J., Sydow, L.V.: An IMEX-scheme for pricing options under stochastic volatility models with jumps. SIAM J. Sci. Comput. **36**, B817–B834 (2014)
31. Strikewerda, J.C.: Finite Difference Schemes and Partial Differential Equations. SIAM (2004)
32. Tangman, D.Y., Gopaul, A., Bhuruth, M.: Numerical pricing of options using high-order compact finite difference schemes. J. Comput. Appl. Math. **218**, 270–280 (2008)
33. Tian, Z.F., Liang, X., Yu., P.: A higher order compact finite difference algorithm for solving the incompressible Navier-Stokes equations. Int. J. Numer. Meth. Eng. **88**, 511–532 (2011)
34. Toivanen, J.: Numerical valuation of European and American options under Kuo's jump diffusion model. SIAM J. Sci. Comput. **30**, 1949–1970 (2008)

New Exact Solutions for Double Sine-Gordon Equation

Subin P. Joseph$^{(\boxtimes)}$ ⓘ

Government Engineering College, Wayanad 670 644, Kerala, India
subinpj@gecwyd.ac.in

Abstract. Finding exact solutions of nonlinear partial differential equations is one of the very difficult topics in mathematical physics. In this paper, we find new exact solutions to the general Sine-Gordon equation. These equations are used in different fields such as electromagnetic waves propagating in semiconductor quantum super lattices, fluxion dynamics in Josephson junctions and nonlinear optics. Different ansatz methods are applied to obtain the required solutions. Solutions are obtained in terms of periodic functions and hyperbolic functions.

Keywords: Sine-Gordon equation · Exact solutions · Periodic solutions

1 Introduction

One of the important nonlinear partial differential equations attracting intensive study in obtaining exact solutions is the Sine-Gordon equation. Solving such nonlinear partial differential equations have attracted several researchers due to the importance of these equations in mathematical physics. Researchers have worked on solving several variants of Sine-Gordon equation. They have derived certain particular solutions for these equations [1,4,5,8,9,11,12,14–17].

Sine-Gordon equation can be classified as a hyperbolic partial differential equation. There are several methods available to solve hyperbolic partial differential equations numerically, such as composite methods [2,3]. Sine-Gordon equations have found applications in different physical situations including electromagnetic waves propagating in semiconductor quantum super lattices, fluxion dynamics in Josephson junctions, nonlinear optics and charge density waves [6,7,10,13].

The double Sine-Gordon equation is given by

$$\frac{\partial^2 u}{\partial t^2} - \frac{\partial^2 u}{\partial x^2} = \gamma \sin(nu) + \delta \sin(2nu),$$

Supported by TEQIP-II Four Funds, Government Engineering College, Wayanad.

A. Awasthi et al. (Eds.): CSMCS 2020, CCIS 1345, pp. 109–121, 2021.
https://doi.org/10.1007/978-981-16-4772-7_8

where $u = u(x, t)$ and γ, δ, n are real parameters. When $\delta = 0$, we get the single Sine-Gordon equation

$$\frac{\partial^2 u}{\partial t^2} - \frac{\partial^2 u}{\partial x^2} = \gamma \sin(nu).$$

In this paper a general double Sine-Gordon equation given by

$$\alpha \frac{\partial^2 u}{\partial t^2} + \beta \frac{\partial^2 u}{\partial x^2} = \gamma \sin(nu) + \delta \sin(2nu) \tag{1}$$

is considered. A traveling wave ansatz method is applied to obtain the required solutions for the Eq. (1) in the next section. The computations are done using a computer algebra system. Some of the existing solutions are shown to be particular solutions of the derived solutions.

2 Exact Solutions

The traveling wave transformation is used to obtain the exact solutions of the double Sine-Gordon equation (1). Assume that the solution is in the form

$$u = u(bx + ct).$$

Substituting this in the Eq. (1) we get an ordinary differential equation

$$\left(\beta b^2 + \alpha c^2\right) u'' - \gamma \sin(nu) - \delta \sin(2nu) = 0.$$

This equation was solved in [17] by two different methods to obtain certain solutions. Another family of solutions are derived in [15]. In this paper we derive several other new exact solutions. The transformation

$$u = \frac{2}{n} \arctan v \tag{2}$$

is used to derive the required solutions. Then the Eq. (1) becomes

$$v \left(n(\gamma - 2\delta)v^2 + 2\alpha \left(\frac{\partial v}{\partial t}\right)^2 + 2\beta \left(\frac{\partial v}{\partial x}\right)^2 + \gamma n + 2\delta n \right)$$

$$= (v^2 + 1) \left(\alpha \frac{\partial^2 v}{\partial t^2} + \beta \frac{\partial^2 v}{\partial x^2} \right).$$

Now assuming traveling wave solution $v = v(\zeta)$, where $\zeta = bx + ct$, this equation becomes

$$v(\zeta) \left(2v'(\zeta)^2 \left(\beta b^2 + \alpha c^2\right) + n \left(\gamma + 2\delta + (\gamma - 2\delta)v(\zeta)^2\right)\right)$$
$$- \left(v(\zeta)^2 + 1\right) v''(\zeta) \left(\beta b^2 + \alpha c^2\right). \tag{3}$$

Any solution to this ordinary differential equation will lead to a solution to the double Sine-Gordon equation. To obtain the required solutions several ansatz forms are assumed for the function v and the solutions are derived with the help of computer algebra system.

2.1   Sin-Cos Ansatz

Assume the solutions in the form

$$v(\zeta) = A_0 + A_1 \sin \zeta. \tag{4}$$

Substituting this ansatz in the Eq. (3), this will be a solution if the following algebraic equations are satisfied.

$$2A_1^2 A_0 \left(\beta b^2 + \alpha c^2\right) + A_0^3 n(\gamma - 2\delta) + A_0 \gamma n + 2A_0 \delta n = 0,$$
$$3A_0 A_1^2 n(\gamma - 2\delta) = 0$$
$$2A_1^3 \left(\beta b^2 + \alpha c^2\right) + A_0^2 A_1 \left(\beta b^2 + \alpha c^2\right) + A_1 \left(\beta b^2 + \alpha c^2\right) \tag{5}$$
$$+ 3A_0^2 A_1 n(\gamma - 2\delta) + A_1 \gamma n + 2A_1 \delta n = 0$$
$$A_1^3 \left(-\beta b^2 - \alpha c^2\right) + A_1^3 n(\gamma - 2\delta) = 0.$$

Solving this algebraic system simultaneously,

$$A_0 = 0,$$
$$A_1 = \pm \frac{\sqrt{\gamma}}{\sqrt{2\delta - \gamma}}, \tag{6}$$
$$c = \pm \frac{\sqrt{n(\gamma - 2\delta) - b^2 \beta}}{\sqrt{\alpha}}.$$

Then the original solutions for the Eq. (1) are given by

$$u_1 = \pm \frac{2}{n} \arctan \left(\frac{\sqrt{\gamma} \sin \left(bx \pm \frac{t\sqrt{n(\gamma - 2\delta) - b^2 \beta}}{\sqrt{\alpha}} \right)}{\sqrt{2\delta - \gamma}} \right). \tag{7}$$

Similarly using the ansatz

$$v(\zeta) = A_0 + A_1 \cos \zeta, \tag{8}$$

we get the solutions

$$u_2 = \pm \frac{2}{n} \arctan \left(\frac{\sqrt{\gamma} \cos \left(bx \pm \frac{t\sqrt{n(\gamma - 2\delta) - b^2 \beta}}{\sqrt{\alpha}} \right)}{\sqrt{2\delta - \gamma}} \right). \tag{9}$$

Letting $n = 3, b = -4, \alpha = 1, \beta = -3, \delta = 3, \gamma = 1$, the solution (7) becomes

$$\frac{2}{3} \arctan \left(\frac{\sin \left(\sqrt{33}t - 4x \right)}{\sqrt{5}} \right).$$

Graphical representation of this solution is given in Fig. 1. Letting $n = 5, b = -4, \alpha = 1, \beta = -3, \delta = 3, \gamma = 1$, the solution (9) becomes

$$-\frac{2}{5} \arctan\left(\frac{\cos\left(\sqrt{23}t + 4x\right)}{\sqrt{5}}\right).$$

Graphical representation of this solution is given in Fig. 2.

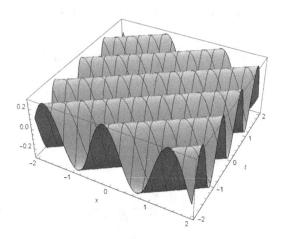

Fig. 1. A solution of Sin-Gordon equation given by (7) with $n = 3, b = -4, \alpha = 1, \beta = -3, \delta = 3, \gamma = 1$

2.2 Sinh-Cosh Ansatz

Assume the solutions in the ansatz form

$$v(\zeta) = A_0 + A_1 \sinh \zeta. \tag{10}$$

Substituting this ansatz in the Eq. (3), this will be a solution if the following algebraic equations are satisfied.

$$
\begin{aligned}
&2A_1^2 A_0 \left(\beta b^2 + \alpha c^2\right) + A_0^3 n(\gamma - 2\delta) + A_0\gamma n + 2A_0\delta n = 0, \\
&2A_0 A_1^2 \left(\beta b^2 + \alpha c^2\right) + 2A_0 A_1^2 \left(-\beta b^2 - \alpha c^2\right) + 3A_0 A_1^2 n(\gamma - 2\delta) = 0, \\
&2A_1^3 \left(\beta b^2 + \alpha c^2\right) + A_0^2 A_1 \left(-\beta b^2 - \alpha c^2\right) + A_1 \left(-\beta b^2 - \alpha c^2\right) \\
&\qquad\qquad\qquad + 3A_0^2 A_1 n(\gamma - 2\delta) + A_1\gamma n + 2A_1\delta n = 0, \\
&2A_1^3 \left(\beta b^2 + \alpha c^2\right) + A_1^3 \left(-\beta b^2 - \alpha c^2\right) + A_1^3 n(\gamma - 2\delta) = 0.
\end{aligned}
\tag{11}
$$

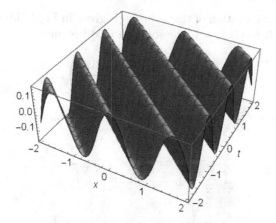

Fig. 2. A solution of Sin-Gordon equation given by (9) with $n = 5, b = -4, \alpha = 1, \beta = -3, \delta = 3, \gamma = 1$

Solving this algebraic system simultaneously,

$$A_0 = 0,$$

$$A_1 = \pm \frac{\sqrt{\gamma}}{\sqrt{\gamma - 2\delta}}, \tag{12}$$

$$c = \pm \frac{\sqrt{-b^2\beta - \gamma n + 2\delta n}}{\sqrt{\alpha}}.$$

Then the original solutions for the Eq. (1) are given by

$$u_3 = \pm \frac{2}{n} \arctan \left(\frac{\sqrt{\gamma} \sinh \left(bx \pm \frac{t\sqrt{-b^2\beta - \gamma n + 2\delta n}}{\sqrt{\alpha}} \right)}{\sqrt{\gamma - 2\delta}} \right). \tag{13}$$

Similarly using the ansatz

$$v(\zeta) = A_0 + A_1 \cosh \zeta, \tag{14}$$

we get the solutions

$$u_4 = \pm \frac{2}{n} \arctan \left(\frac{\sqrt{\gamma} \cosh \left(bx \pm \frac{t\sqrt{-b^2\beta - \gamma n + 2\delta n}}{\sqrt{\alpha}} \right)}{\sqrt{2\delta - \gamma}} \right). \tag{15}$$

Letting $n = 5, b = -4, \alpha = 1, \beta = -3, \delta = 3, \gamma = 1$, the solution (13) becomes

$$\frac{2}{5} \arctan \left(\frac{\sinh \left(\sqrt{13}t - 4x \right)}{\sqrt{7}} \right).$$

Graphical representation of this solution is given in Fig. 3. Letting $n = 5, b = -4, \alpha = 1, \beta = -3, \delta = 3, \gamma = 1$, the solution (15) becomes

$$\arctan\left(\frac{\cosh\left(\sqrt{35}t + 3x\right)}{\sqrt{2}}\right).$$

Graphical representation of this solution is given in Fig. 4.

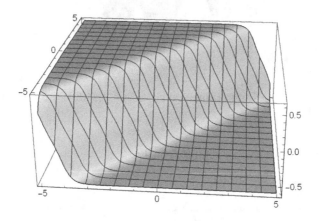

Fig. 3. A solution of Sin-Gordon equation given by (13) with $n = 5, b = -4, \alpha = 1, \beta = -3, \delta = 3, \gamma = 1$

2.3   Sec-Csc Ansatz

Let us take the solutions in the ansatz form

$$v(\zeta) = A_0 + A_1 \sec \zeta. \tag{16}$$

Substituting this ansatz in the Eq. (3), this will be a solution if the following algebraic equations are satisfied.

$$A_0^3 n(\gamma - 2\delta) + A_0 \gamma n + 2A_0 \delta n = 0,$$
$$A_1 A_0^2 \left(\beta b^2 + \alpha c^2\right) + A_1 \left(\beta b^2 + \alpha c^2\right) + 3A_1 A_0^2 n(\gamma - 2\delta) + A_1 \gamma n + 2A_1 \delta n = 0,$$
$$2A_0 A_1^2 \left(\beta b^2 + \alpha c^2\right) + 4A_0 A_1^2 \left(-\beta b^2 - \alpha c^2\right) = 0,$$
$$-2A_0 A_1^2 \left(\beta b^2 + \alpha c^2\right) - 2A_0 A_1^2 \left(-\beta b^2 - \alpha c^2\right) + 3A_0 A_1^2 n(\gamma - 2\delta) = 0, \tag{17}$$
$$2A_1^3 \left(-\beta b^2 - \alpha c^2\right) + 2A \left(\beta b^2 + \alpha c^2\right) = 0,$$
$$A_1^3 \left(-\left(\beta b^2 + \alpha c^2\right)\right) + 2A_0^2 A_1 \left(-\beta b^2 - \alpha c^2\right)$$
$$+ 2A_1 \left(-\beta b^2 - \alpha c^2\right) + A_1^3 n(\gamma - 2\delta) = 0.$$

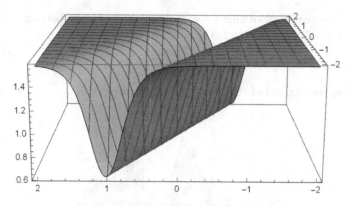

Fig. 4. A solution of Sin-Gordon equation given by (15) with $n = 5, b = -4, \alpha = 1, \beta = -3, \delta = 3, \gamma = 1$

Solving this algebraic system simultaneously,

$$A_0 = 0,$$

$$A_1 = \pm \frac{\sqrt{-\gamma - 2\delta}}{\sqrt{\gamma}},$$

$$c = \pm \frac{\sqrt{-b^2\beta - n(\gamma + 2\delta)}}{\sqrt{\alpha}}.$$

(18)

Then the original solutions for the Eq. (1) are given by

$$u_5 = \pm \frac{2}{n} \arctan \left(\frac{\sqrt{-\gamma - 2\delta} \sec \left(\frac{t\sqrt{-b^2\beta - n(\gamma + 2\delta)}}{\sqrt{\alpha}} + bx \right)}{\sqrt{\gamma}} \right).$$

(19)

Similarly using the ansatz

$$v(\zeta) = A_0 + A_1 \csc \zeta,$$

(20)

we get the solutions

$$u_6 = \pm \frac{2}{n} \arctan \left(\frac{\sqrt{-\gamma - 2\delta} \csc \left(\frac{t\sqrt{-b^2\beta - n(\gamma + 2\delta)}}{\sqrt{\alpha}} + bx \right)}{\sqrt{\gamma}} \right).$$

(21)

Letting $n = 2, b = 4, \alpha = -1, \beta = 3, \delta = 3, \gamma = -3$, the solution (19) becomes

$$\arctan \left(\sec \left(3\sqrt{6}t + 4x \right) \right).$$

Graphical representation of this solution is given in Fig. 5. Letting $n = 2, b = -3, \alpha = 1, \beta = -3, \delta = -3, \gamma = 1$, the solution (21) becomes

$$\arctan\left(\sqrt{5}\csc\left(\sqrt{37}t - 3x\right)\right).$$

Graphical representation of this solution is given in Fig. 6.

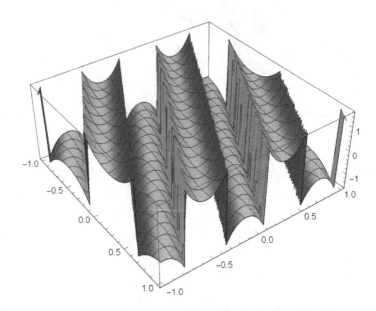

Fig. 5. A solution of Sin-Gordon equation given by (19) with $n = 2, b = 4, \alpha = -1, \beta = 3, \delta = 3, \gamma = -3$

2.4   Csch-Sech Ansatz

Let us take the solutions in the ansatz form

$$v(\zeta) = A_0 + A_1 \operatorname{csch}\zeta. \tag{22}$$

Substituting this ansatz in the Eq. (3), this will be a solution if the following algebraic equations are satisfied.

$$A_0^3 n(\gamma - 2\delta) + A_0\gamma n + 2A_0\delta n = 0,$$
$$A_1 A_0^2 \left(-\beta b^2 - \alpha c^2\right) + A_1 \left(-\beta b^2 - \alpha c^2\right) + 3A_1 A_0^2 n(\gamma - 2\delta) + A_1\gamma n + 2A_1\delta n = 0,$$
$$2A_0 A_1^2 \left(\beta b^2 + \alpha c^2\right) + 4A_0 A_1^2 \left(-\beta b^2 - \alpha c^2\right) = 0,$$
$$2A_0 A_1^2 \left(\beta b^2 + \alpha c^2\right) + 2A_0 A_1^2 \left(-\beta b^2 - \alpha c^2\right) + 3A_0 A_1^2 n(\gamma - 2\delta) = 0, \tag{23}$$
$$2A_1^3 \left(\beta b^2 + \alpha c^2\right) + 2A_1^3 \left(-\beta b^2 - \alpha c^2\right) = 0,$$
$$2A_1^3 \left(\beta b^2 + \alpha c^2\right) + A_1^3 \left(-\beta b^2 - \alpha c^2\right) + 2A_0^2 A_1 \left(-\beta b^2 - \alpha c^2\right)$$
$$+ 2A_1 \left(-\beta b^2 - \alpha c^2\right) + A_1^3 n(\gamma - 2\delta) = 0.$$

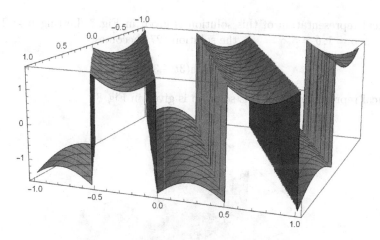

Fig. 6. A solution of Sin-Gordon equation given by (21) with $n = 2, b = -3, \alpha = 1, \beta = -3, \delta = -3, \gamma = 1$

Solving this algebraic system simultaneously,

$$A_0 = 0,$$

$$A_1 = \pm\frac{\sqrt{\gamma + 2\delta}}{\sqrt{\gamma}},$$

$$c = \pm\frac{\sqrt{n(\gamma + 2\delta) - b^2\beta}}{\sqrt{\alpha}}. \tag{24}$$

Then the original solutions for the Eq. (1) are given by

$$u_7 = \pm\frac{2}{n}\arctan\left(\frac{\sqrt{\gamma + 2\delta}\operatorname{csch}\left(bx \pm \frac{t\sqrt{n(\gamma + 2\delta) - b^2\beta}}{\sqrt{\alpha}}\right)}{\sqrt{\gamma}}\right). \tag{25}$$

Similarly using the ansatz

$$v(\zeta) = A_0 + A_1\operatorname{sech}\zeta, \tag{26}$$

we get the solutions

$$u_8 = \pm\frac{2}{n}\arctan\left(\frac{\sqrt{-\gamma - 2\delta}\operatorname{sech}\left(\frac{bx \pm t\sqrt{n(\gamma + 2\delta) - b^2\beta}}{\sqrt{\alpha}}\right)}{\sqrt{\gamma}}\right). \tag{27}$$

Letting $n = 4, b = 2, \alpha = 1, \beta = -3, \delta = 3, \gamma = 1$, the solution (25) becomes

$$\frac{1}{2}\arctan\left(\sqrt{7}\operatorname{csch}\left(2\sqrt{10}t + 2x\right)\right).$$

Graphical representation of this solution is given in Fig. 7. Letting $n = 2, b = -3, \alpha = 1, \beta = 1, \delta = 5, \gamma = -1$, the solution (27) becomes

$$\arctan(3\operatorname{sech}(3t - 3x)).$$

Graphical representation of this solution is given in Fig. 8.

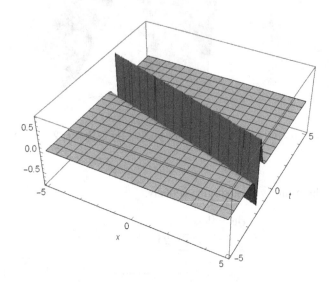

Fig. 7. A solution of Sin-Gordon equation given by (25) with $n = 4, b = 2, \alpha = 1, \beta = -3, \delta = 3, \gamma = 1$

2.5   A Rational Function Ansatz

Let us take the solutions in the rational ansatz form

$$v(\zeta) = A + \frac{A_1}{B + B_1 \cot(\zeta)}. \tag{28}$$

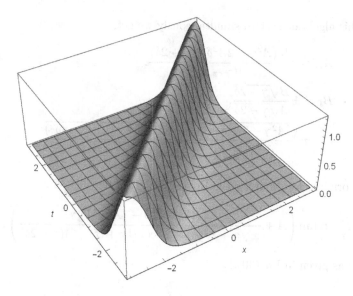

Fig. 8. A solution of Sin-Gordon equation given by (27) with $n = 2, b = -3, \alpha = 1, \beta = 1, \delta = 5, \gamma = -1$

Substituting this ansatz in the Eq. (3), this will be a solution if the following algebraic equations are satisfied.

$$A^3 B^4 n(\gamma - 2\delta) + A^2 A_1 B \left(3B^2 n(\gamma - 2\delta) - 2B_1{}^2 \left(\beta b^2 + \alpha c^2\right)\right)$$
$$+ A_1 B \left(A_1{}^2 n(\gamma - 2\delta) - 2B_1{}^2 \left(\beta b^2 + \alpha c^2\right) + B^2 n(\gamma + 2\delta)\right)$$
$$+ A \left(-2A_1{}^2 B_1{}^2 \left(\beta b^2 + \alpha c^2\right) + 3A_1{}^2 B^2 n(\gamma - 2\delta) + B^4 n(\gamma + 2\delta)\right) = 0,$$

$$B_1 \left(A_1{}^3 \left(2\beta b^2 + 2\alpha c^2 + n(\gamma - 2\delta)\right) - 2 \left(A^2 + 1\right) A_1 B_1{}^2 \left(\beta b^2 + \alpha c^2\right)\right.$$
$$+ 4AB^3 n \left(A^2 \gamma - 2A^2 \delta + \gamma + 2\delta\right) + 2AA_1{}^2 B \left(2\beta b^2 + 2\alpha c^2 + 3n(\gamma - 2\delta)\right)$$
$$A_1 B^2 \left(2 \left(A^2 + 1\right) b^2 \beta + 2\alpha \left(A^2 + 1\right) c^2 + 9A^2 n(\gamma - 2\delta) + 3\gamma n + 6\delta n\right)\right) = 0,$$

$$B_1 n(2AB + A_1) \left(A^2 B(\gamma - 2\delta) + AA_1(\gamma - 2\delta) + B(\gamma + 2\delta)\right) = 0, \tag{29}$$

$$B_1 \left(2 \left(A^2 + 1\right) A_1 B^2 \left(\beta b^2 + \alpha c^2\right) + 4AA_1{}^2 B \left(\beta b^2 + \alpha c^2\right) + 2A_1{}^3 \left(\beta b^2 + \alpha c^2\right)\right.$$
$$4ABB_1{}^2 n \left(A^2 \gamma - 2A^2 \delta + \gamma + 2\delta\right) + A_1 B_1{}^2 \left(-2 \left(A^2 + 1\right) \beta b^2 - 2\alpha \left(A^2 + 1\right) c^2\right.$$
$$+ 3A^2 n(\gamma - 2\delta) + \gamma n + 2\delta n)) = 0,$$

$$B_1 \left(2A^2 A_1 B \left(\beta b^2 + \alpha c^2\right) + 2AA_1{}^2 \left(\beta b^2 + \alpha c^2\right) + 2A_1 B \left(\beta b^2 + \alpha c^2\right)\right.$$
$$+ A^3 B_1{}^2 n(\gamma - 2\delta) + AB_1{}^2 n(\gamma + 2\delta)\right) = 0.$$

Solving this algebraic system simultaneously we get,

$$A_1 = -\frac{B\left(A^2\gamma - 2A^2\delta + \gamma + 2\delta\right)}{A(\gamma - 2\delta)},$$

$$B_1 = \pm\frac{B\sqrt{\gamma + 2\delta}}{A\sqrt{\gamma - 2\delta}},$$

(30)

$$c = \pm\frac{\sqrt{A^2\gamma - 2A^2\delta + \gamma + 2\delta}\sqrt{8\beta b^2\delta + n\left(\gamma^2 - 4\delta^2\right)}}{2\sqrt{2}\sqrt{A}\sqrt{-\frac{\alpha\delta(A^2\gamma - 2A^2\delta + \gamma + 2\delta)}{A}}}.$$

Then the original solutions for the Eq. (1) are given by

$$u_9 = \pm\frac{2}{n}\arctan\left(A + \frac{2\left(A^2 - 1\right)\delta - \left(A^2 + 1\right)\gamma}{\pm\sqrt{\gamma + 2\delta}\sqrt{\gamma - 2\delta}\cot\left(bx + ct\right) + A(\gamma - 2\delta)}\right),$$

(31)

where c is as given in Eq. (30).

3   Discussion

Several new exact solutions for generalized double Sine-Gordon equation are derived in this paper. These solutions are derived using the trigonometric and hyperbolic ansatz forms. The required calculations can be done using any of the computer algebra system such as Maple or Mathematica. We can also obtain several exact solutions to single sine-Gordon equation (1) by putting $\delta = 0$ in the exact solutions derived in this paper. The tan-cot solutions and tanh-coth solutions are excluded in this paper as they were previously derived in the papers [15, 17]. These solutions were derived by them using a variable separated ordinary differential equation method. The ansatz method used in this paper are easier and powerful method in deriving the exact solutions of Double-Sine Gordon equations. Such ansatz method can be applied to obtain exact solutions of several other nonlinear partial differential equations that appear in mathematical physics.

References

1. Alquran, M., Al-Khaled, K.: The tanh and sine-cosine methods for higher order equations of Korteweg- de Vries type. Phys. Scr. **84**, 025010 (2011)
2. Appadu, A.R.: Some applications of the concept of minimized integrated exponential error for low dispersion and low dissipation. Int. J. Numer. Meth. Fluids **68**(2), 244 (2011)
3. Appadu, A.R., Nguetchue, S.N.N.: The technique of MIEELDLD as a measure of the shock-capturing property of numerical methods for hyperbolic conservation laws. Progr. Comput. Fluid Dyn. Int. J. **15**(4), 247 (2015)
4. Bin, H., Qing, M., Yao, L., Weiguo, R.: New exact solutions of the double sine-Gordon equation using symbolic computations. Appl. Math. Comput. **186**, 1334 (2007)

5. Burt, P.B.: Exact, multiple soliton solutions of the double sine Gordon equation. Proc. R. Soc. Lond. A **359**, 479 (1978)
6. Gani, V., Kudryavtsev, A.: Kink-antikink interactions in the double sine-Gordon equation and the problem of resonance frequencies. Phys. Rev. E **60**, 3305 (1999)
7. Goldobin, E., Sterck, A., Koelle, D.: Josephson vortex in a ratchet potential theory. Phys. Rev. E **63**, 03111 (2001)
8. Huang, Y., Li, B.: Exact traveling wave solutions for the modified double sine-Gordon equation. J. Math. Res. **7**(2), 182 (2015)
9. Joardar, A., Kumar, D., Woadud, K.: New exact solutions of the combined and double combined sinhcosh- Gordon equations via modified Kudryashov method. Int. J. Phys. Res. **6**(1), 25 (2018)
10. Lou, S., Ni, G.: Deforming some special solutions of the sine-Gordon equation to that of the double sine-Gordon equation. Phys. Lett. A **140**, 33 (1989)
11. Panigrahi, M., Dash, P.: Mixing exponential and double sine-Gordon equation. Phys. Lett. A **321**, 330 (2004)
12. Peng, Y.: Exact solutions for some nonlinear partial differential equations. Phys. Lett. A **314**, 401 (2003)
13. Salerno, M., Quintero, N.: Soliton ratchets. Phys. Rev. E **65**, 025602 (2002)
14. Sun, Y.: New travelling wave solutions for sine-Gordon equation. J. Appl. Math. 2014, Article ID 841416 (2014)
15. Sun, Y.: New exact traveling wave solutions for double sine-Gordon equation. Appl. Math. Comput. **258**, 100 (2015)
16. Wang, M., Li, X.: Exact solutions to the double sine-Gordon equation. Chaos, Solitons Fractals **27**, 477 (2006)
17. Wazwaz, A.: The tanh method and a variable separated ode method for solving double sine-Gordon equation. Phys. Lett. A **350**, 367 (2006)

Numerical Study of Mixed Convection in Single and Double Lid Driven Cavity Using LBM

Srijit Sen, D. Arumuga Perumal$^{(\boxtimes)}$, and Ajay Kumar Yadav

Department of Mechanical Engineering, National Institute of Technology Karnataka,
Surathkal, Mangalore 575025, India
perumal@nitk.edu.in

Abstract. The lattice Boltzmann method (LBM) has been gaining popularity over the last two decades and the method has been extended from simple fluid flow problems to problems involving heat transfer. In the present work, an attempt is made to model cases involving mixed convection. Two types of problems are considered in this study; the first one dealing with mixed convection in a single-sided lid-driven cavity and the second one dealing with mixed convection in a double-sided lid-driven cavity in parallel and anti-parallel configurations at constant Prandtl number and various values of Richardson number. For the first problem, a square domain is considered with a moving lid at a lower temperature while the stationary wall at the bottom at a higher temperature. The cavity side walls are treated with an adiabatic boundary condition. In LBM, a forcing term dependent on temperature difference is utilized to vary the value of y-velocity in order to satisfy the effects of gravity on mixed convection. A grid independence study is conducted to show that the results are independent of the grid chosen, and good agreement with literature is achieved. The second problem is an extension of the first one; the cavity bottom wall is first given a velocity in the opposite direction, and then in the same direction, and the velocity streamlines, temperature contours and local Nusselt number variation in the top wall for these cases are plotted. The developed method helps in the visualization of various phenomena such as splitting of flow into two halves for the parallel configuration and formation of secondary vortices with high Reynolds number.

Keywords: Mixed convection · LBM · Lid driven cavity

1 Introduction

Mixed convection flows occur in many technological and industrial applications in nature, e.g., solar receivers exposed to wind currents, electronic devices cooled by fans, nuclear reactors cooled during emergency shutdown, heat exchangers placed in a low-velocity environment, flows in the ocean and in the atmosphere,

© Springer Nature Singapore Pte Ltd. 2021
A. Awasthi et al. (Eds.): CSMCS 2020, CCIS 1345, pp. 122–133, 2021.
https://doi.org/10.1007/978-981-16-4772-7_9

and so on [1–5]. It is known that mixed convection is a combination of natural convection and forced convection. Natural convection is a type of flow in which fluid movement is generated by buoyancy force. The colder part of the fluid being heavier than the warmer part gets pulled downwards, whereas the warmer part moves upwards. On coming in contact with the heating source the colder part becomes warm and a cycle is thus established. Forced convection is a type of heat transfer in which transport of heat and movement of fluid is generated by an external source like a fan, pump, moving plate and so on. In mixed convection, both phenomena work simultaneously and the relative strength of each phenomenon is determined by the Richardson number, which is the ratio of Grashoff number and the square of Reynolds number. A heated lid driven cavity proves to be a perfect problem to study mixed convection as the moving plate(s) provides a source for forced convection and the effect of gravity brings into picture buoyancy driven natural convection.

CFD studies of mixed convection in a heated single lid driven cavity have been studied by many authors. Moallemi et al. [2] and Iwatsu et al. [6] modeled mixed convection in a heated single lid driven cavity. Some attempts have been made to model mixed convection using LBM. Darzi et al. [7] validated a specific case presented in Moallemi et al.'s [2] work and extended the results to study mixed convection in inclined lid driven cavity at various angles of inclination. Lamarti et al. [8] validated Iwatsu's results and studied mixed convection in a heated lid driven cavity using Multi Relaxation time (MRT) LBM. However, traditionally a MRT approach is more computationally intensive than a Single Relaxation time (SRT) approach. Hence in this work we use a SRT approach with a forcing term demonstrated by Guo et al. [4] is used to first verify Moallemi et al.'s [2] results for Pr = 1.0 and to then observe the effects of mixed convection in a double lid driven cavity in parallel and anti-parallel configuration.

2   Problem and Solution Methodology

2.1   Problem Specification

Fluid flow and heat transfer are modeled via LBM in the two dimensional cavity with length equivalent to the grid size chosen. In all cases the top lid is assigned a lower temperature $(T_c = 0)$ and the bottom lid is assigned a higher temperature $(T_h = 1)$. The vertical walls are assumed to be insulated. In the first scenario, which is used for validation, the top lid moves to the right with velocity u_{top} and the bottom lid remains stationary (Fig. 1(a)). In the second scenario, the top and bottom lid move in opposite directions with velocity u_{top} (Fig. 1(b)). In the third scenario, the top and the bottom lid move in the same direction with velocity u_{top} (Fig. 1(c)). The flow is assumed to be in the laminar regime with density changing continuously by virtue of the Boussinesq Approximation.

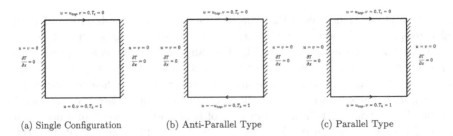

(a) Single Configuration   (b) Anti-Parallel Type   (c) Parallel Type

Fig. 1. Description of boundary conditions for three mixed convection cases

2.2 Non Dimensional Mixed Convection Equations

The governing equations of the problem in non dimensional form are the conservation of mass, momentum and thermal energy:

$$\frac{\partial U}{\partial X} + \frac{\partial V}{\partial Y} = 0 \tag{1}$$

$$U\frac{\partial U}{\partial X} + V\frac{\partial U}{\partial Y} = -\frac{\partial P}{\partial X} + \frac{1}{Re}\left(\frac{\partial^2 U}{\partial X^2} + \frac{\partial^2 U}{\partial Y^2}\right) \tag{2}$$

$$U\frac{\partial V}{\partial X} + V\frac{\partial V}{\partial Y} = -\frac{\partial P}{\partial Y} + \frac{1}{Re}\left(\frac{\partial^2 V}{\partial X^2} + \frac{\partial^2 V}{\partial Y^2}\right) + \frac{Gr}{Re^2}\Theta \tag{3}$$

$$U\frac{\partial \Theta}{\partial X} + V\frac{\partial \Theta}{\partial Y} = \frac{1}{RePr}\left(\frac{\partial^2 \Theta}{\partial X^2} + \frac{\partial^2 \Theta}{\partial Y^2}\right) \tag{4}$$

In the above equations, it can be noted that an additional term, namely the buoyancy term gets added to the y- momentum equation due to the effect of external force due to buoyancy being felt only in the y-direction i.e. gravitation force direction and not in the x-direction. As is evident from the above set of equations, the parameters that govern the simulation are Reynolds number, Grashoff number and Prandtl number. The coupling between heat transfer and fluid flow is via the buoyancy term in the y-momentum equation. The Richardson number (Ri) as defined above is a more intuitive parameter that can replace Grashoff number as it takes into account both Reynolds number and Grashoff number, giving a relative comparison of natural and mixed convection. Four cases are tested: case (i) Ri = 0.01, Re = 1000, case (ii) Ri = 0.4, Re = 500, case (iii) Ri = 1, Re = 1000 & case (iv) Ri = 4, Re = 500. Prandtl number is taken as 1 for all 4 cases.

2.3 Lattice Boltzmann Methodology

The lattice Boltzmann method is an alternate method to CFD for solving fluid flow problems. It is a middleman between the continuum and molecular approaches of flow modeling. The method has found applications in a variety of domains thanks to the speed of the method and the ease of adoption of concepts of parallelism. It also has been extended to multiphase and heat transfer problems showing the method's versatility.

The basic assumption of the LBM is that a system can be described as a probability distribution over particles. Each particle has a probability density f associated with it which represents that the particle is at a position x with velocity c (taken as 1) at time t. The Boltzmann equation is as follows [9]:

$$f_i\left(x + \mathbf{c}_i \delta t, t + \delta t\right) = \frac{1}{\tau}\left(f_i^{eq} - f_i\right) \qquad (5)$$

Where τ is the relaxation time and $\mathbf{c}_i$ is the velocity in the i-th direction, f_i^{eq} is usually computed from statistical particle equilibrium distribution. The relaxation time τ is taken as

$$\tau = 3\nu + 0.5 \qquad (6)$$

The kinematic viscosity ν is computed as

$$\nu = \frac{u_{top}L}{Re} \qquad (7)$$

The space and time discretization are done as usual but the velocities are discretized in a different manner. Particles are given velocities in a set number of directions. In the current study a D2Q9 model is used in which the particles are assigned velocities as shown in Fig. 2. The discrete particle velocities are defined as,

$$\mathbf{c}_i = \begin{cases} (0,0) & \text{if } i = 0 \\ \left(cos\left(\frac{\pi\,(i-1)}{4}\right), sin\left(\frac{\pi\,(i-1)}{4}\right)\right) & \text{if } i = 1,2,3,4 \\ \left(\sqrt{2}\,cos\left(\frac{\pi\,(i-1)}{4}\right), \sqrt{2}\,sin\left(\frac{\pi\,(i-1)}{4}\right)\right) & \text{if } i = 5,6,7,8 \end{cases} \qquad (8)$$

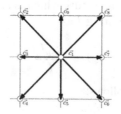

Fig. 2. D2Q9 lattice model

Equation 5 is solved in two steps [9]:

1. Collision

$$f_i^{'}(x,t) = f_i(x,t) + \frac{1}{\tau}(f_i^{eq}(x,t) - f_i(x,t)) \qquad (9)$$

2. Streaming

$$f_i\left(x + \mathbf{c}_i \delta t, t + \delta t\right) = f_i'(x, t) \tag{10}$$

The equilibrium distribution function is chosen as,

$$f_i^{eq} = \rho w_i \left[1 + \frac{3(\mathbf{c}_i \cdot \mathbf{u})}{c^2} + \frac{9(\mathbf{c}_i \cdot \mathbf{u})^2}{2c^4} - \frac{3(\mathbf{u} \cdot \mathbf{u})}{2c^2}\right] \tag{11}$$

Where $\mathbf{u}$ is the macroscopic velocity vector. The lattice weights are given by $w_0 = 4/9$, $w_1 = w_2 = w_3 = w_4 = 1/9$, $w_5 = w_6 = w_7 = w_8 = 1/36$. Now, the density ρ is defined as follows:

$$\rho = \sum_{n=0}^{8} f_i \tag{12}$$

The velocity used to compute macroscopic velocity also needs to be modified by adding a force term to take into account the effect of buoyancy. In mixed convection, the effect of gravity is taken into account by means of a force term. The force term can be modeled in three ways according to Mohamad et al. [3] out of which the approach used by Guo et al. [4] is chosen due to its ability to satisfy the Navier Stokes equation.

$$\mathbf{F} = \rho \mathbf{g} \beta \left(T - T_{ref}\right) \tag{13}$$

$$F_i = w_i \left(1 - \frac{1}{2\tau}\right) \left[\frac{\mathbf{c}_i - \mathbf{u}}{c_s^2} + \frac{\mathbf{c}_i(\mathbf{c}_i \cdot \mathbf{u})}{c_s^4}\right] \mathbf{F} \tag{14}$$

$$\mathbf{u} = \frac{1}{\rho} \sum_{n=0}^{8} \mathbf{c}_i f_i + \frac{\mathbf{F}\tau}{\rho} \tag{15}$$

Where $\mathbf{g}$ is gravitational acceleration in y-direction, β is thermal expansion coefficient and T_{ref} is reference temperature which is assumed as 0.5. $\mathbf{F}_i$ gets added to Eq. (5). To find $\mathbf{F}$, we compute $\mathbf{g}\beta$ as

$$\mathbf{g}\beta = \frac{u_{top}^2 Ri}{L} \tag{16}$$

This modifies the Eq. (9) as follows:

$$f_i'(x, t) = f_i(x, t) + \frac{1}{\tau}(f_i^{eq}(x, t) - f_i(x, t)) + w_i\left(1 - \frac{1}{2\tau}\right)\left[\frac{\mathbf{c}_i - \mathbf{u}}{c_s^2} + \frac{\mathbf{c}_i(\mathbf{c}_i \cdot \mathbf{u})}{c_s^4}\right]\mathbf{F} \tag{17}$$

For modeling temperature a D2Q9 approach is used for temperature probability distribution function g and associated relaxation time τ_s, with the equation to be solved being:

$$g_i\left(x + \mathbf{c}_i \delta t, t + \delta t\right) = \frac{1}{\tau_s}\left(g_i^{eq} - g_i\right) \tag{18}$$

where

$$g_i^{eq} = Tw_i \left[1 + \frac{3(\mathbf{c}_i \cdot \mathbf{u})}{c^2} \right] \tag{19}$$

$$\tau_s = \frac{3\nu}{Pr} + 0.5 \tag{20}$$

For velocity, the bounce back boundary condition is applied at the stationary boundaries for velocity whereas the moving boundaries are assigned equilibrium distribution functions [10]. For temperature, at the top and bottom wall the constant temperature boundary condition [11] is applied whereas at the boundaries a second order accurate scheme is applied which at left and right boundaries respectively as follows:

$$g_i(0) = \frac{4}{3}g_i(1) - \frac{1}{3}g_i(2) \tag{21}$$

$$g_i(n) = \frac{4}{3}g_i(n-1) - \frac{1}{3}g_i(n-2) \tag{22}$$

3 Validation

The credibility of the LBM code can be proven by comparing the results provided by Moallemi et al. [2]. In this section problem 1 as described above is considered. As speed of sound in LBM is taken as $1/sqrt(3)$, taking $u_{top} = 0.1$ brings Mach number below 0.3 makes sure the flow is incompressible. T_{ref} is assumed to be

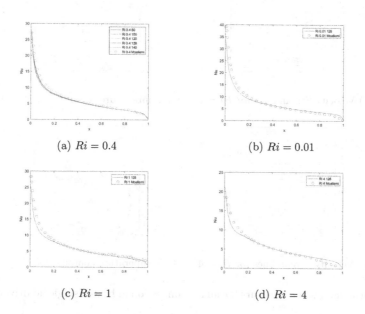

(a) $Ri = 0.4$ (b) $Ri = 0.01$

(c) $Ri = 1$ (d) $Ri = 4$

Fig. 3. Plot of Nusselt number at hot wall v/s non-dimensional length

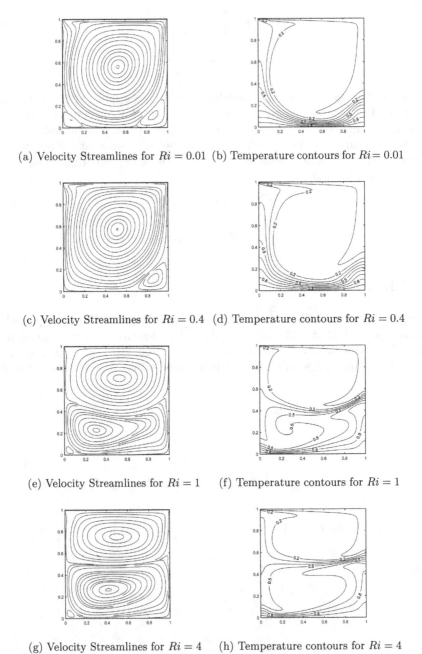

(a) Velocity Streamlines for $Ri = 0.01$ (b) Temperature contours for $Ri = 0.01$

(c) Velocity Streamlines for $Ri = 0.4$ (d) Temperature contours for $Ri = 0.4$

(e) Velocity Streamlines for $Ri = 1$ (f) Temperature contours for $Ri = 1$

(g) Velocity Streamlines for $Ri = 4$ (h) Temperature contours for $Ri = 4$

Fig. 4. Streamlines and contours for mixed convection in heated single lid driven cavity

0.5. The variation of Nusselt number (ratio of temperature difference between boundary node and adjacent node, and nodal distance) at the top wall with normalized length is plotted in Fig. 3.

Figure 3(a) clearly proves that the simulation results are independent of lattice sizes chosen as the grids with 100, 120, 128 and 140 elements are close to each other. Based on the results, 128×128 is chosen as the standard for all simulations. The simulations for Ri = 0.01, 0.40, 1.0 and 4.0 show good agreement with the results provided by Moallemi [2] as shown in Fig. 3 (a) to (d). The streamline patterns and temperature contours are shown in the Fig. 4(a)–(h).

4   Results and Discussion

Figure 5 shows the velocity streamline patterns and temperature contours for mixed convection in heated double lid driven cavity in anti-parallel configuration. In this configuration, a primary vortex is formed at the centre of the cavity for all cases. However in case (i) and case (iii), where the Reynolds number is high, two secondary vortices appear near the trailing edge of the plate. It is also observed that the flow structure is also similar classical single lid-driven cavity flow. Figure 6 shows the velocity streamline patterns and temperature contours for mixed convection in heated double lid driven cavity in parallel configuration. In this case, the flow splits with two primary vortices being formed on the top and bottom of the flow domain. In case (i) and (iii), secondary vortices appear at the right wall, where the flow gets split into two parts by virtue of higher Reynolds number. The top half and bottom half of case (i) seems to be a mirror of each other whereas in the other cases, the flow structure at the top suppresses the flow structure at the bottom. The effect of convection is noticed in the contours. It is also seen that, a very small Ri can cause good heat transfer effect than a high Ri.

In the anti-parallel configuration, the left wall has a higher average temperature than the right wall for all values of Ri. In the parallel configuration, certain interesting observations can be made. In the top half section, the left side has a higher average temperature than the right side, whereas at the bottom half section the right side has a higher average temperature than the left side. A considerable difference in temperature is noticed between the two halves along the imaginary boundary between the halves, which becomes more prominent as the right wall is approached. There is a stronger vertical gradient in temperature along the right wall near the boundary, in comparison with the left wall, where the change is much more gradual. From Fig. 6(a), it is observed that the primary vortices are formed & separated by free shear layer in between them.

Figure 7 (a)–(b) shows the variation of Nusselt number at the top wall with length. Figure 7(a) shows the anti-parallel configuration. For various Ri the flow trend is not changed greatly. It is noticed that the cold wall Nusselt number in the anti-parallel configuration in general is higher than that of the parallel configuration. For case (iii) and case (iv) in the anti-parallel case, a point of

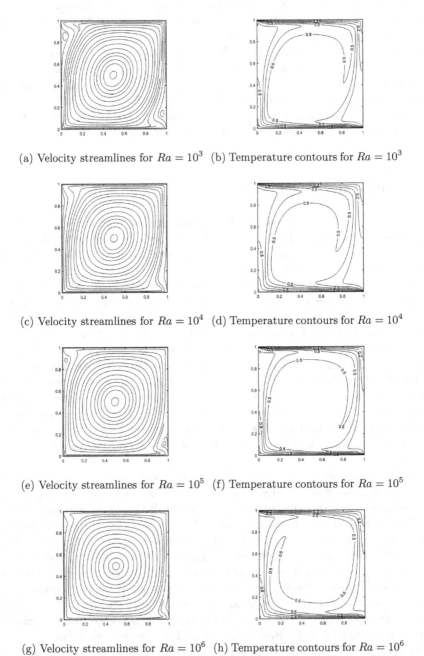

(a) Velocity streamlines for $Ra = 10^3$ (b) Temperature contours for $Ra = 10^3$

(c) Velocity streamlines for $Ra = 10^4$ (d) Temperature contours for $Ra = 10^4$

(e) Velocity streamlines for $Ra = 10^5$ (f) Temperature contours for $Ra = 10^5$

(g) Velocity streamlines for $Ra = 10^6$ (h) Temperature contours for $Ra = 10^6$

Fig. 5. Streamlines and contours for mixed convection in a heated double lid driven cavity in anti-parallel configuration

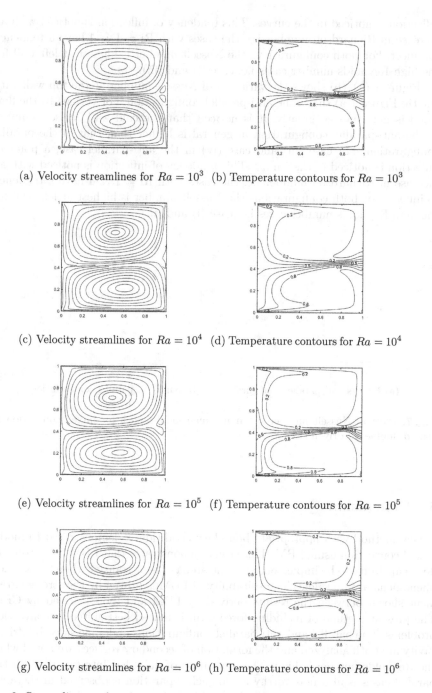

(a) Velocity streamlines for $Ra = 10^3$ (b) Temperature contours for $Ra = 10^3$

(c) Velocity streamlines for $Ra = 10^4$ (d) Temperature contours for $Ra = 10^4$

(e) Velocity streamlines for $Ra = 10^5$ (f) Temperature contours for $Ra = 10^5$

(g) Velocity streamlines for $Ra = 10^6$ (h) Temperature contours for $Ra = 10^6$

Fig. 6. Streamlines and contours for mixed convection in a heated double lid driven cavity in parallel configuration

inflection is noticed in the curves. This tendency of inflection is noticed with an increase in Richardson number i.e. the cases with Ri = 1 & 4 have a tendency to inflect. For both configurations the Nusselt number is highest at left wall for the high Reynolds number cases i.e. case (i) and (iii).

Figure 7 (a)–(b) shows the variation of Nusselt number at the top wall with length. Figure 7(a) shows the anti-parallel configuration. For various Ri the flow trend is not changed greatly. It is noticed that the cold wall Nusselt number in the anti-parallel configuration in general is higher than that of the parallel configuration. For case (iii) and case (iv) in the anti-parallel case, a point of inflection is noticed in the curves. This tendency of inflection is noticed with an increase in Richardson number i.e. the cases with Ri = 1 & 4 have a tendency to inflect. For both configurations the Nusselt number is highest at left wall for the high Reynolds number cases i.e. case (i) and (iii).

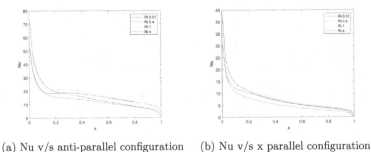

(a) Nu v/s anti-parallel configuration   (b) Nu v/s x parallel configuration

Fig. 7. Plots of Nusselt number v/s non dimensional length for mixed convection in heated double lid driven cavity

5   Conclusion

A new method with appropriate boundary condition is demonstrated to model mixed convection using LBM. This study demonstrates that LBM has a tremendous capability to be implemented in the arena of heat transfer where convection phenomena are prevalent. The suitability of LBM is proven by the strong agreement shown between the results derived by LBM and those derived by CFD. The present method of including force term is indeed best suited to convection problems. The parallel and anti-parallel configurations of double sided lid driven cavity are both analyzed and the formation of secondary vortices are noted when the Reynolds number is high. The separation of fluid flow into two parts in the parallel cases is also noteworthy as no such separation is observed in the anti-parallel cases. Future scope of work could include studying mixed convection in flow over objects which could be useful for aerospace applications.

References

1. Houat, S., Bouayed, Z.E.: The lattice Boltzmann method for mixed convection in a cavity. Energy Procedia **139**, 186–191 (2017)
2. Moallemi, M.K., Jang, K.S.: Prandtl number effects on laminar mixed convection heat transfer in a lid-driven cavity. Int. J. Heat Mass Transf. **35**, 1881–1892 (1992)
3. Mohamad, A.A., Kuzmin, A.: A critical evaluation of forcing term in lattice Boltzmann method, natural convection problem. Int. J. Heat Mass Transf. **53**, 990–996 (2010)
4. Guo, Z., Zheng, C., Shi, B.: Discrete lattice effects on the forcing term in the lattice Boltzmann method. Phys. Rev. E **65**, 0463081–6 (2002)
5. Basak, T., Roy, S., Sharma, P.K., Pop, I.: Analysis of mixed convection flows within a square cavity with uniform and non-uniform heating of bottom wall. Int. J. Therm. Sci. **48**, 891–912 (2009)
6. Iwatsu, R., Hyun, J.M., Kuwahara, K.: Mixed convection in a driven cavity with a stable vertical temperature gradient. Int. J. Heat Mass Transf. **36**, 1601–1608 (1993)
7. Darzi, A.A.R., Farhadi, M., Sedighi, K., Fattahi, E.: Mixed convection simulation of inclined lid driven cavity using lattice Boltzmann method. IJST Trans. Mech. Eng. **35**, 73–83 (2011)
8. Lamarti, H., Mahdaoui, M., Bennacer, R., Chahboun, A.: Numerical simulation of mixed convection heat transfer of fluid in a cavity driven by an oscillating lid using lattice Boltzmann method. Int. J. Heat Mass Transf. **137**, 615–629 (2019)
9. Perumal, D.A., Dass, A.K.: A Review on the development of lattice Boltzmann computation of macro fluid flows and heat transfer. Alex. Eng. J. **54**, 955–971 (2015)
10. Perumal, D.A., Dass, A.K.: Application of lattice Boltzmann method for incompressible viscous flows. Appl. Math. Model. **37**, 4075–4092 (2013)
11. Karki, P., Yadav, A.K., Perumal, D.A.: Study of adiabatic obstacles on natural convection in a square cavity using lattice Boltzmann method. ASME J. Therm. Sci. **11**, 034502 1–16 (2019)

A New (3, 3) Low Dispersion Upwind Compact Scheme

Subhajit Giri and Shuvam Sen$^{(\boxtimes)}$ (ID)

Department of Mathematical Sciences, Tezpur University, Tezpur 784028, India
{subhajit,shuvam}@tezu.ernet.in

Abstract. In this work, we propose a new upwind compact scheme with appropriately designed new boundary closures. The scheme is obtained by minimizing weighted dispersion error and is asymptotically stable. As the formulation leads to an implicit tridiagonal system for approximating spatial derivative it is computationally efficient for long time simulation. The scheme thus derived is tested in conjunction with explicit and implicit time advancing strategies. Verification and validation studies help establish the newly developed method.

Keywords: Dispersion relation · Upwind · Compact

1 Introduction

Wave propagation problems often require solutions that are accurate in the far-field and for longer periods. In such situations, it is imperative to simulate flows resolving a wide range of spatial and temporal scales. For example, the challenging areas of direct numerical simulation (DNS) and large eddy simulation (LES) of turbulence, aeroacoustics, and fluid-structure interactions (FSI) could be cited. The severe computational requirements of such processes might be mitigated by adopting a highly accurate dispersion error-free numerical method. In this context, compact schemes offer an attractive choice because of their spectral like resolution [1]. These schemes offer higher order approximations to differential operators using compact stencils and implicitly relate various function values and their derivatives at discrete nodes. Compact discretizations are known to carry higher spectral resolution compared to the explicit methods. Although implicit they often lead to a diagonally dominant banded system. Indeed compact schemes leading to the tridiagonal system are most favoured because of their obvious computational advantages. Although compact schemes employ a stencil with fewer grid points, their implicit nature can involve a large number of points in the domain thereby making such schemes attractive.

Traditionally compact schemes are of central type [1,2]. As such these schemes carry no dissipation error but do carry significant dispersion error [3]. Central compact schemes applied to problems with periodic boundary conditions are indeed efficient. However, for practical problems, periodic boundary conditions are often absent and one-sided approximations are required for boundary

© Springer Nature Singapore Pte Ltd. 2021
A. Awasthi et al. (Eds.): CSMCS 2020, CCIS 1345, pp. 134–145, 2021.
https://doi.org/10.1007/978-981-16-4772-7_10

points. This forced upwinding near boundaries render instability to the entire discretization process [4]. Thus many a time convection dominated flow requires extra filtering or added numerical dissipation [3]. On the other hand, upwind biased compact schemes are seen to be robust and are used for the Navier-Stokes equation with great success [3–6]. The upwind biased nature of the compact scheme invariably introduces numerical dissipation and is found enough to control aliasing error [3]. Here it is important to remember that good quality numerical solutions schemes should not only resolve all scales present in the flow but also adequately capture the physical propagation speed of the individually resolved scales. Failure might lead to an extreme form of dispersion error often seen as unphysical q-waves. In this context importance of dispersion relation preservation (DRP) in conjunction with high accuracy approximations for acoustic problems are well documented [7].

In the last two decades development of upwind compact schemes to simulate fluid flow problems has seen significant attention. Among them, the works of Zhong [3], Sengupta et al. [4], and Bhumkar et al. [6] deserve special mention. The higher order compact finite-difference schemes developed by Zhong [3] were found to be stable and were less dissipative than a straightforward upwind scheme developed using an upwind-biased grid stencil. But in this work, the author did not attempt to optimize the scheme developed for interior as well as boundary closures. Sengupta et al. [4] analyzed various upwind compact schemes for spatial discretization and highlighted the importance of boundary closure for the overall stability of the scheme. The authors further suggested special boundary treatment to avoid the stability shortcomings of the schemes. Bhumkar et al. [6] stressed the importance of dispersion relation preserving nature of upwind compact schemes for good quality numerical simulation. They optimally reduced dispersion error and worked with varied stencils of lengths three to thirteen. But the authors dealt with wavenumber range $[0, 7\pi/8]$ instead of requisite range $[0, \pi]$. Further, the work made little effort to derive stable and compatible boundary closures.

Issue of stability of various inner and boundary schemes was deliberated by Gustafsson, Kreiss, and Sundström [8]. The technique referred to as G-K-S stability theory provides conditions that schemes must satisfy to ensure stability. But its application to fully discrete higher order schemes with multistage time integration is highly involved [3,9]. On the other hand, application to a semi-discrete hyperbolic system is easier. Unfortunately, a disturbing feature of this stability definition is that the solution need not remain bounded for all time, even though the actual solution remains bounded. The definition only ensures that the error remains uniformly bounded by an exponential amount for all time [9,10]. Thus simulation resulting from such a scheme might lead to unstable modes in the numerical solution to dominate after a sufficiently long time and was amply demonstrated by Carpenter et al. [9]. Carpenter et al. [9] showed that the asymptotic stability of the upwind schemes with numerical boundary closures is necessary for the stability of long time numerical integration. This procedure requires that the eigenvalues of the spatial discretization matrices contain no positive real parts. Numerical computations often reveal that the matrices for compact upwind schemes with boundary conditions carry a full set of eigenvalues thereby elevating any further need of eigenvalue analysis.

In this work, we develop a new upwind compact scheme that employs a stencil of size three and is third order accurate. The scheme is termed $(3, 3)$ as it discretizes spatial derivative at a nodal point using functional value at three grid points and its gradients also at those three points. The scheme thus developed is supplemented by newly developed boundary closures which render the scheme globally third order accurate. As our main motivation is to arrive at a scheme efficient for long time simulation in situations involving convection and diffusion we carry out asymptotically stability analysis of the scheme. Finally, numerical investigation help establish the efficiency of the newly proposed algorithm. All computations are done using in-house C-codes run on a system supported by Intel Core i3 processor with 4 GB RAM.

2   Upwind Compact Spatial Discretization

The model equation often used in deriving the upwind schemes is the linear wave equation

$$\frac{\partial u}{\partial t} + c\frac{\partial u}{\partial x} = 0, \quad a \le x \le b, \quad t > 0, \quad c > 0. \tag{1}$$

This equation is complemented with the Dirichlet boundary condition

$$u(a,t) = g(t), \tag{2}$$

and initial condition

$$u(x,0) = f(x). \tag{3}$$

Traditionally first order spatial derivative in Eq. (1) at an interior grid point say jth node with uniform grid spacing h can compactly be approximated as

$$\sum_{l=-M}^{M} b_l u'_{j+l} = \frac{1}{h}\sum_{l=-N}^{N} a_l u_{j+l}, \tag{4}$$

where u'_j is the numerical approximation of $(\partial u/\partial x)_j$. Compact schemes are known to attain higher spectral resolution on a coarser mesh. The scheme here uses a total of $2M + 1$ and $2N + 1$ grid points on left and right respectively leading to a banded system of equations with bandwidth $2M + 1$. In this study, we are interested to estimate gradients using only the adjacent grid points. Such a choice is inherently advantageous as it leads to a tridiagonal system and is computationally efficient. Thus in our case $M = 1 = N$ leading to $(3, 3)$ system. The system is given by the Eq. (4) and is often expressed in linear algebraic form

$$M_1 u' = \frac{1}{h} M_2 u \tag{5}$$

where $\boldsymbol{u} = (u_0, u_1, ..., u_n)^T$. We strive to evaluate the coefficients a_l and b_l of the upwind schemes in such a manner that the order of the schemes is one less than the maximum achievable order of the central stencil. Thus opting to go with third order accuracy and hence we are left with a free parameter called ϱ. This free parameter is set as the coefficient of the leading truncation term i.e.

$$b_{-1}u'_{j-1} + u'_j + b_1 u'_{j+1} = \frac{1}{h}\left(a_{-1}u_{j-1} + a_0 u_j + a_1 u_{j+1}\right) - \frac{\varrho}{4!}h^3\left(\frac{\partial^4 u}{\partial x^4}\right)_j + ...,$$

$$j = 1, 2, ..., n-1. \quad (6)$$

Equation (6) contains five unknowns, namely b_{-1}, b_1, a_{-1}, a_0, a_1. For uniqueness b_0 is set to unity. One needs five equations to obtain these coefficients. By using the Taylor series expansion and equating the coefficients upto third order on both sides of Eq. (6) we get,

$$a_{-1} + a_0 + a_1 = 0, \quad (7)$$
$$-a_{-1} + a_1 - b_{-1} - b_1 = 1, \quad (8)$$
$$a_{-1} + a_1 + 2b_{-1} - 2b_1 = 0, \quad (9)$$
$$-a_{-1} + a_1 - 3b_{-1} - 3b_1 = 0, \quad (10)$$
$$a_{-1} + a_1 + 4b_{-1} - 4b_1 = \varrho. \quad (11)$$

In terms of ϱ the other coefficients are given by

$$b_{\pm 1} = \mp\frac{\varrho}{4} + \frac{1}{4}, \quad a_{\pm 1} = -\frac{\varrho}{2} \pm \frac{3}{4}, \quad a_0 = \varrho. \quad (12)$$

We intend to choose ϱ in such a manner that the associated upwind scheme carries minimum dispersion error. Subsequent to the work of Haras and Ta'asan [11] we start by taking $u_j = e^{I\omega(jh)}$ in the Eq. (4) and obtain

$$I\omega_{eq}h(b_{-1}e^{-I\omega h} + 1 + b_1 e^{I\omega h}) = (a_{-1}e^{-I\omega h} + a_0 + a_1 e^{I\omega h}) \quad (13)$$

where ω and ω_{eq} are the exact and approximate wavenumber respectively. In general, ω_{eq} is a complex quantity and its difference from ω could be minimized over wavenumber domain $[-\pi, \pi]$. Subsequently, the expression for the real part of ω_{eq} denoted here as $Re[\omega_{eq}h]$ is used to define error function E as

$$E = \int_{-\pi}^{\pi} (\omega h - Re[\omega_{eq}h])^2 |u_0(\omega h)|^2 d(\omega h). \quad (14)$$

Here u_0 is the weight function and we are inclined to work with $u_0(\omega h) = e^{-\omega^2 h^2}$ as such a choice entails a higher emphasis on smaller values of ωh. Thus the error function in terms of ϱ is

$$E = \int_{-\pi}^{\pi} \left[\frac{2\varrho \sin x(\varrho - \varrho\cos x) + 3\sin x(2 + \cos x)}{\varrho^2 \sin^2 x + (2 + \cos x)^2} - x\right]^2 e^{-2x^2} dx. \quad (15)$$

We minimize the dispersion error function E with respect to ϱ and obtain $\varrho = 0.8300949493$ as the point of minima. The corresponding value of the other coefficients is given in Table 1 leading to compact upwind discretization for interior nodes. On the other hand choice, $\varrho = 0$ leads to central fourth order Padé scheme.

Table 1. Third order low dispersion upwind compact scheme.

Parameter	$j = 0$	$1 \leq j < n$	$j = n$
b_{-1}	–	0.4575237373	2.1351328557
b_1	2.1351328557	0.0424762627	–
a_{-3}			0.0225221426
a_{-2}			−0.6351328557
a_{-1}	–	−1.1650474747	−1.9324335721
a_0	−2.5450442852	0.8300949493	2.5450442852
a_1	1.9324335721	0.3349525254	–
a_2	0.6351328557	–	–
a_3	−0.0225221426	–	–

2.1   Boundary Formulation

Considering that there are $n + 1$ grid points $j = 0, 1, ..., n$ laid out in one direction, it is imperative to develop independent and adequate boundary closures for the two extreme nodes. Our scheme being $(3, 3)$ the discretization developed earlier could be implemented at all other nodes. We present below the procedure adopted to obtain closure at $j = 0$. This approximation is proposed to be obtained from a relation of the form

$$u_0' + b_1 u_1' = \frac{1}{h}(a_0 u_0 + a_1 u_1 + a_2 u_2 + a_3 u_3) \tag{16}$$

to preserve the overall tridiagonal nature and third order truncation error of the system. Introducing additional free parameter and writing the modified differential equation as discussed earlier for the interior nodes the constraints satisfying third order accuracy here are

$$a_0 + a_1 + a_2 + a_3 = 0, \tag{17}$$
$$a_1 + 2a_2 + 3a_3 - b_1 = 1, \tag{18}$$
$$a_1 + 4a_2 + 9a_3 - 2b_1 = 0, \tag{19}$$
$$a_1 + 8a_2 + 27a_3 - 3b_1 = 0, \tag{20}$$
$$a_1 + 16a_2 + 81a_3 - 4b_1 = \varrho. \tag{21}$$

In terms of ϱ, the other coefficients are given by

$$b_1 = 3 - \frac{\varrho}{2}, \quad a_0 = -\frac{17}{6} + \frac{\varrho}{6}, \quad a_1 = \frac{3}{2} + \frac{\varrho}{4}, \quad a_2 = \frac{3}{2} - \frac{\varrho}{2}, \quad a_3 = -\frac{1}{6} + \frac{\varrho}{12}. \tag{22}$$

Subsequently the error function E in terms of ϱ for the scheme in Eq. (16) is

$$E = \int_{-\pi}^{\pi} \left[\frac{(8\varrho^2 - 94\varrho + 348)\sin x + (24 - 4\varrho - \varrho^2)\sin 2x + (2\varrho - 4)\sin 3x}{6(6-\varrho)^2 + 24(6-\varrho)\cos x + 24} - x \right]^2 e^{-2x^2} \, dx.$$

(23)

As earlier minimization of the error function with respect to ϱ leads to $\varrho = 1.7297342886$. The corresponding value of the other coefficients is given in Table 1. Note that the closure at $j = n$ is the mirror image of the above procedure and hence its derivation is avoided. Nevertheless, the coefficients could be found in Table 1.

2.2   Stability

As compact finite difference schemes require additional approximations at grid points near the boundaries of the computational domain its analysis should invariably include boundary closures. In this work, we carry out asymptotic stability analysis of the upwind scheme in conjunction with Dirichlet boundary closures by computing the eigenvalues of the matrices obtained by spatial discretization of the wave equation. As we discuss the upwind scheme Neumann boundary condition is not deliberated on [3, 4, 9]. In periodic domain the scheme is automatically stable. The asymptotic stability, which requires that the eigenvalues of the spatial discretization matrices contain no positive real parts, is necessary for the stability of long time integration of the equation. The newly developed low dispersion unwind compact scheme having global third order accuracy can be expressed in compact form as

$$M_1 u' = \frac{1}{h} M_2 u$$

where M_1 and M_2 are $(n+1) \times (n+1)$ matrices with M_1 being tridiagonal. In these two matrices, the first and the last row correspond to the first and the last column of Table 1 whereas the other elements directly correspond to the middle column of the same table. Using the boundary condition the semi-discrete form of the prototype PDE given by Eq. (1) can be expressed as

$$\frac{\partial \tilde{u}}{\partial t} + \frac{c}{h} \tilde{C} \tilde{u} + \frac{c}{h} \tilde{M}_1^{-1} B g(t) = 0$$

(24)

where $\tilde{u} = (u_1, u_2, ..., u_n)^T$. $\tilde{C} = \tilde{M}_1^{-1} \tilde{M}_2$ with $\tilde{M}_1$ and $\tilde{M}_2$ reduced from M_1 and M_2 on account of boundary condition Eq. (2) being applied. For completeness we report below the matrices $\tilde{M}_1$ and $\tilde{M}_2$ as also the vector B.

$$\tilde{M}_1 = \begin{pmatrix} \frac{1}{b_{-1}} - b_1^{(0)} & \frac{b_1}{b_{-1}} & & & \\ b_{-1} & 1 & b_1 & & \\ & \cdot & \cdot & \cdot & \\ & & b_{-1} & 1 & b_1 \\ & & & b_{-1}^{(n)} & 1 \end{pmatrix}_{n \times n},$$

$$\tilde{M}_2 = \begin{pmatrix} \frac{a_0}{b_{-1}} - a_1^{(0)} & \frac{a_1}{b_{-1}} - a_2^{(0)} & -a_3^{(0)} \\ a_{-1} & a_0 & a_1 \\ & & \cdot & \cdot & \cdot \\ & & a_{-1} & a_0 & a_1 \\ & a_{-3}^{(n)} & a_{-2}^{(n)} & a_{-1}^{(n)} & a_0^{(n)} \end{pmatrix}_{n \times n},$$

$$B = \begin{pmatrix} \frac{a_{-1}}{b_{-1}} - a_0^{(0)} \\ 0 \\ \cdot \\ 0 \\ 0 \end{pmatrix}_n.$$

In the above expressions superscript is used to denote the corresponding boundary nodes related to the first and last column in Table 1. The first row of the vector B documents dependence of the discretization on the boundary condition. The asymptotic stability condition for the semi-discrete equations requires that all the eigenvalues of the matrix $-\tilde{C}$ contain no positive real parts. The same computed on an 81×81 grid is depicted in Fig. 1(a). It is heartening to see that all the eigenvalues of the matrix have negative real parts rendering our newly developed scheme asymptotically stable. Figure 1(b) shows the eigenvalue spectrum for the fourth order Padé scheme ($\varrho = 0$) with fourth order boundary closure. This figure presented for the sake of comparison clearly shows that there are eigenvalues with positive real part rendering the scheme asymptotically unstable.

3   Numerical Examples

3.1   Problem 1: Propagation of Sinusoidal Wave

Following Carpenter et al. [9] we consider the propagation of sinusoidal wave $u(x, t) = \sin 2\pi(x - t)$ in the bounded domain $[-1, 1]$. The boundary and initial conditions are

$$u(-1, t) = \sin 2\pi(-1 - t), \quad t > 0 \tag{25}$$
$$\text{and} \quad u(x, 0) = \sin 2\pi x, \quad -1 \le x \le 1 \tag{26}$$

respectively. We have used 41 grid points for computing the solution up to time $t = 60$. For this problem, time is advanced with the fourth order four stage explicit R-K scheme. For spatial discretization apart from newly developed (3, 3) scheme we also use Padé approximation discussed earlier. Simulations are run for CFL numbers (N_c) 0.25 and 0.5. In Figs. 2(a) and (b) we have plotted time evolution of L^2-norm error for new (3, 3) scheme. From these figures, we see that error quickly settles down to a periodic profile with a small amplitude implying asymptotic stability of the scheme. In Figs. 3(a) and (b) we have plotted L^2-norm error for $N_c = 0.25$ and 0.5 computed using fourth order central Padé

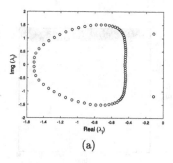
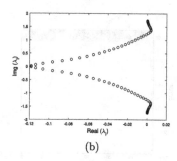

(a)   (b)

Fig. 1. Eigen value spectra (a) New (3, 3) scheme for $\varrho = 0.8300949493$, (b) Fourth order central Padé scheme for $\varrho = 0$.

scheme with fourth order boundary closure. Although theoretically, the schemes carry higher order of accuracy an unbounded error growth is registered for the scheme documenting asymptotically unstable nature of the scheme. This test case establishes the efficiency of the strategy advocated here. A comparison of the CPU time and error reported at $t = 60$ is presented in Table 2. Padé scheme is found to consume CPU time almost four times that of newly developed (3, 3) scheme. This may be attributed to its higher error leading to more iterations for convergence.

Table 2. Error and CPU time at $t = 60$.

Scheme	New (3, 3)		Padé	
	Error	CPU Time (s)	Error	CPU Time (s)
$N_c = 0.25$	5.3e–3	6.8	3.4e–1	23.4
$N_c = 0.50$	4.5e–3	3.8	1.0e–2	12.7

3.2   Problem 2: Convection of Wave Combination

Next we consider convection of combination of two waves of wavenumbers $2\pi k_1$ and $2\pi k_2$ [12]. The initial condition is given by

$$u(x,0) = e^{-\frac{(x-x_m)^2}{b^2}} \times [\cos(2\pi k_1(x - x_m)) + \cos(2\pi k_2(x - x_m))] \qquad (27)$$

where $x_m = 90$, $b = 20$, $k_1 = 0.125$ and $k_2 = 0.0625$. Solution is computed up to $t = 300$ for $N_c = 0.5$ and 1.0. In this problem, time discretization is carried out using the implicit two stage fourth order Gauss-Legendre scheme (IRK24) [13]. This serves as a test case for the newly developed scheme in conjunction with implicit time discretization. Numerical solutions are shown in Fig. 4. L^2-norm error between numerical and exact solutions at $t = 300$ is found to be approximately 4.66×10^{-2} for both the cases. CPU time for this problem with N_c values 0.5 and 1.0 is 1.2 s and 0.6 s respectively.

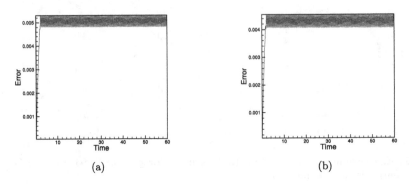

Fig. 2. Time evolution of L^2-norm error for new (3, 3) scheme at (a) $N_c = 0.25$, (b) $N_c = 0.5$.

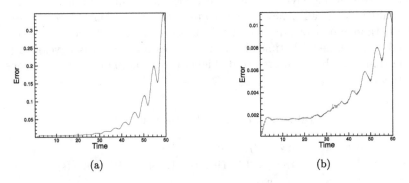

Fig. 3. Time evolution of L^2-norm error for central Padé scheme at (a) $N_c = 0.25$, (b) $N_c = 0.5$.

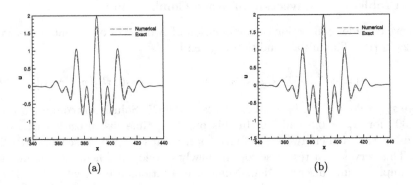

Fig. 4. Numerical solution at $t = 300$ for (a) $N_c = 0.5$, (b) $N_c = 1.0$.

3.3   Problem 3: Convection-Diffusion of Gaussian Pulse

Finally, we study unsteady two-dimensional convection-diffusion equation with zero source term given by

$$a\frac{\partial\psi}{\partial t} - \frac{\partial^2\psi}{\partial x^2} - \frac{\partial^2\psi}{\partial y^2} + c\frac{\partial\psi}{\partial x} + d\frac{\partial\psi}{\partial y} = 0. \tag{28}$$

We consider a Gaussian pulse in a square domain $[0,2] \times [0,2]$ following Sen [14] whose analytical solution is

$$\psi(x,y,t) = \frac{1}{4t+1} \exp\left[-\frac{(ax - ct - 0.5a)^2}{a(4t+1)} - \frac{(ay - dt - 0.5a)^2}{a(4t+1)} \right]. \tag{29}$$

Initially, the Gaussian pulse is centered at $(0.5, 0.5)$ with pulse height 1. Dirichlet boundary conditions are used for this problem along all boundaries. The usual procedure to approximate the diffusion terms ψ_{xx} and ψ_{yy} is to use explicit central differencing. Such a technique lead to loss of high accuracy of the discretized governing equation, which is achieved by the compact schemes on the convective terms. Further, as we emphasize dispersion error reduction it is important to employ a suitable discretization for the diffusion terms. Recently Sen [14] developed a central compact fourth order discretization for second order derivative. This approximation was found to carry good numerical dispersion and dissipation characteristics. Further, it uses functional values and their gradients in a three-point stencil. Hence the strategy developed by Sen [14] is seen to be particularly suitable in this context. Of course, to compute ψ_x and ψ_y, we employ the newly developed $(3,3)$ scheme. Time advancing is carried out with the implicit Crank-Nicolson method. For this simulation convection coefficients are fixed at $c = d = 80$ with $a = 100$. Computations are done for three different grids 21×21, 41×41, and 81×81. Errors in L_1, L_2, and L_∞ norms at time $t = 0.5$ and $t = 1.0$ are shown in Table 3. In this table, we also present CPU time. With grid size decreasing by a factor of two the associated algebraic system increasing by a factor of four. Additionally, as the temporal step is reduced by a factor of four, CPU time as expected increases by a factor of sixteen. In Fig. 5 we compare the analytical solution with the solution computed using the newly developed scheme in the region $0.8 \le x, y \le 1.8$. An excellent comparison can be seen in this figure.

Table 3. L_1, L_2 and L_∞-norm error and CPU time with $\delta t = h^2 = k^2$.

Time		21×21	41×41	81×81
$t = 0.5$	L_1	6.0710e−2	1.4420e−4	3.7067e−6
	L_2	2.1452e−1	6.1790e−4	1.5732e−5
	L_∞	2.3199e+0	8.6617e−3	1.8969e−4
	CPU Time	1.1	20.9	326.3
$t = 1.0$	L_1	1.5141e−2	7.5830e−5	7.8776e−6
	L_2	3.6316e−2	2.4850e−4	2.5428e−5
	L_∞	2.1846e−1	2.3459e−3	1.7324e−4
	CPU Time	2.4	40.0	643.6

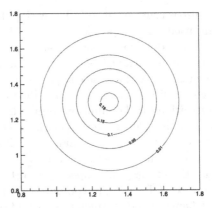

Fig. 5. Comparison of numerical (solid, blue) and analytical (dashed, red) contour at $t = 1.0$. (Color figure online)

4   Conclusion

In this work, we have developed a new $(3, 3)$ dispersion relation preserving third order optimized upwind compact scheme. The scheme is obtained by minimizing phase error over the entire wavenumber range. Subsequently, the boundary closures with optimum dispersion accuracy are also developed. Overall the scheme is found to be asymptotically stable. Three numerical tests are envisaged to illustrate the importance of dispersion relation preserving character and stability of the newly developed spatial discretization. They duly demonstrate the efficiency and accuracy of the scheme proposed.

Acknowledgement. The second author is thankful to Science & Engineering Research Board, India for assistance under Mathematical Research Impact Centric Support (File Number: MTR/2017/000038).

References

1. Lele, S.K.: Compact finite difference schemes with spectral-like resolution. J. Comput. Phys. **103**, 16–42 (1992)
2. Vishal, M.R., Gaitonde, D.V.: On the use of higher-order finite-difference schemes on curvilinear and deforming meshes. J. Comput. Phys. **181**, 155–185 (2002)
3. Zhong, X.: High-order finite-difference schemes for numerical simulation of hypersonic boundary-layer transition. J. Comput. Phys. **144**, 662–709 (1998)
4. Sengupta, T.K., Ganeriwal, G., De, S.: Analysis of central and upwind compact schemes. J. Comput. Phys. **192**, 677–694 (2003)
5. Rai, M.M., Moin, P.: Direct numerical simulation of transition and turbulence in a spatially evolving boundary layer. J. Comput. Phys. **109**, 169–192 (1993)
6. Bhumkar, Y.G., Sheu, T.W.H., Sengupta, T.K.: A dispersion relation preserving optimized upwind compact difference scheme for high accuracy flow simulations. J. Comput. Phys. **278**, 378–399 (2014)
7. Tam, C.K.W., Webb, J.C.: Dispersion-relation-preserving finite difference schemes for computational acoustics. J. Comput. Phys. **107**, 262–281 (1993)
8. Gustafsson, B., Kreiss, H.-O., Sundström, A.: Stability theory of difference approximation for mixed initial boundary value problems II. Math. Comput. **26**, 649–686 (1972)
9. Carpenter, M.H., Gottlieb, D., Abarbanel, S.: Stable and accurate boundary treatments for compact high-order finite-difference schemes. Appl. Numer. Math. **12**, 55–87 (1993)
10. Carpenter, M.H., Gottlieb, D., Abarbanel, S.: The stability of numerical boundary treatments for compact high-order finite-difference schemes. J. Comput. Phys. **108**, 272–295 (1993)
11. Haras, Z., Ta'asan, S.: Finite difference schemes for long time integration. J. Comput. Phys. **114**, 265–279 (1994)
12. Giri, S., Sen, S.: A new class of diagonally implicit Runge-Kutta methods with zero dissipation and minimized dispersion error. J. Comput. Appl. Math. **376**, 112841 (2020)
13. Butcher, J.: Numerical Methods for Ordinary Differential Equations. Wiley, West Sussex (2008)
14. Sen, S.: A new family of (5,5) CC-4OC schemes applicable for unsteady Navier-Stokes equations. J. Comput. Phys. **251**, 251–271 (2013)

Solution of Variable-Order Space Fractional Bioheat Equation by Chebyshev Collocation Method

Rupali Gupta$^{(\boxtimes)}$ and Sushil Kumar

Applied Mathematics and Humanities Department, S. V. National Institute
of Technology Surat, Gujarat 395007, India
{d18ma007,sushilk}@amhd.svnit.ac.in

Abstract. This study presents the mathematical model of space fractional variable-order bioheat equation and proposes a numerical method to find its solution. We use the Chebyshev collocation method, which converts the differential equation into a set of algebraic equations, and then we solve it for unknowns. The effect of temperature distribution within living biological tissue has been studied. The results of the temperature profile in the case of variable-order are demonstrated graphically. An investigation of different types of fractional-orders on maximum temperature is also presented. In the end, the comparison between various types of fractional-order shows the usefulness of variable-order over other cases.

Keywords: Bioheat equation · Caputo type variable-order fractional derivative · Chebyshev polynomial · Collocation method

1 Introduction

Thermal regulation of body temperature has been used for a long period. The effect of temperature alterations is studied in mainly three categories: hyperthermia (raising the temperature) [18], cryobiology (subfreezing the temperature) [15] and hypothermia (lowering the temperature) [36]. In hyperthermia, the temperature lies approximately 42 °C to heat the particular volume of the tumor for around 30–40 min.

To ensure that extreme temperatures are contained in tumors and not in the corresponding healthy tissues, one should have a profound knowledge of the transport of heat in a human body to anticipate the temperature profiles that can be achieved during therapy. The diffusion of heat in living tissues is known as bioheat transfer. The mathematical model for the heat transfer in biological tissues, used by many researchers, is derived in 1948 by Pennes [30] and referred to as the bioheat equation. This model is widely known for its simplicity and effectiveness.

In [7], authors implemented the Galerkin and implicit backward Euler approach to solve the multi-segmental bioheat model, coupled with the arterial fluid

© Springer Nature Singapore Pte Ltd. 2021
A. Awasthi et al. (Eds.): CSMCS 2020, CCIS 1345, pp. 146–158, 2021.
https://doi.org/10.1007/978-981-16-4772-7_11

dynamics of a human body. Giordano *et al.* [16] derived a linear bioheat transfer operator in cylindrical, spherical, and Cartesian coordinate systems and depicted the advantage of these in the heat transfer model. Deka *et al.* [10] used the differential transform method for the steady and unsteady heat equation. The one-dimensional bioheat model with constant and space-dependent heat generation function was solved both analytically and numerically in [3]. They applied separation of the variable in the first case and explicit finite difference technique for numerical solution. The 1-D time-dependent bioheat equation in the Cartesian coordinate solved analytically in [12], an extension of [12] is found in [11], where the authors find the exact solution in spherical and cylindrical coordinates.

Recently, fractional derivatives have attracted the attention of many research scientists, and the ubiquitous appearance can be seen in various applications like physics [27], viscoelasticity [1], biology [26], fluid mechanics [4] and many areas in engineering [17,29]. The fractional calculus has been used to enhance the simulation precision of several anomalies in the sciences. The reason behind the enormous application of the fractional derivative usage is its non-local property [19], which means that the current state's data is affected by all the data of previous states.

It is generally known that finding an analytical solution to such problems is not an easy task. Therefore, computational and approximation methods are widely accepted techniques to achieve the solution. In the field of heat transfer analysis, the temperature distribution is the primary concern. In [28], the authors used a fractional calculus approach to examine the periodic transfer of heat in peripheral tissue areas. Fahmy [14] studied the effect of fractional derivative parameters on temperature profile using the combination of the general boundary element method and the radial basis function collocation method for the dual-phase bioheat model. Ezzat *et al.* [13] used a different approach to model the fractional-order bioheat equation. The authors considered the fractional derivative as an indicator of bioheat efficiency in the bioheat model. In [34], the role of fractional-orders has been investigated by Singh *et al.* in the fractional single-phase lagging heat conduction model using the Riemann-Liouville space fractional derivative and applied the finite-difference and homotopy perturbation methods for approximation of bioheat equation. Damor *et al.* [9] employed the same model with Caputo fractional time derivative and solved by utilizing a quadrature formula for time derivative and central difference formula for approximating the space derivative. They also discussed the stability of the method and the thermal damage of tissues. The analysis is based on time-fractional derivative whereas, the space derivative is considered as integer order. An extension of [9] is given in [31], where authors discussed three cases by introducing the heat generation function as time depended, space depended, and sinusoidal function. They also analyzed the case of the locally variable initial condition and time-depended boundary conditions.

Samko and Ross give the generalization of the constant fractional derivative in [32], where the order of derivatives depends upon the space only or time only or space and time both. The generalization has been implemented in biology, physics, engineering, and many other fields [2,8,35]. Mathematical structures that have been built on this new term are more sensitive to the solution. It also plays a significant part in the accuracy of the solution. Bhrawy and Zaky [6] implemented the shifted Chebyshev collocation method to solve a differential equation with Dirichlet boundary conditions and Caputo's variable-order fractional derivative is utilized. Shekari et al. [33] solved 2-D variable-order time-fractional PDE by using a meshless approach. In [22], Heydari et al. has solved the fractional diffusion-wave equation with all kind of boundary conditions by using Chebyshev wavelets. In [23], authors used Legendre wavelet to solve the 2-D transient dual-phase lag bioheat model.

In all the work mentioned above, the analysis of the transfer of heat in thermo-fluid is based on constant fractional derivative whereas, the investigation on space and time depended fractional derivative is significantly less. The main target of this paper is to present a mathematical problem of bioheat equation with space fractional variable-order. By using the Chebyshev collocation method, we examine the temperature profile for different variable-orders. From the results and graphs, we make conclusions.

2 Mathematical Model

Due to simplicity, ease of use and effectiveness in studying the transfer of heat in tissues, the Pennes' bioheat Model is widely known among many researchers [30]. In this article, we present the space fractional variable-order Pennes' bioheat model by replacing the space derivative to variable-order Caputo's fractional derivative.

$$\rho C \frac{\partial^\beta T}{\partial t^\beta} = K_t \frac{\partial^{\alpha(r,t)} T}{\partial r^{\alpha(r,t)}} + W_b C_b \left(T_a - T\right) + Q_{met} + Q_{ext}, \quad 0 < \beta \le 1, \ 1 < \alpha \le 2,$$

$$\mathrm{T}\left(r,t\right)|_{t=0} = T,$$
$$\frac{\partial T\left(r,t\right)}{\partial r}\bigg|_{r=0} = 0, \tag{1}$$
$$\mathrm{T}\left(r,t\right)|_{r=R} = T_w,$$

where T, r, t, ρ, C, K_t represent the tissue's local temperature, space coordinate, time, density, specific heat and thermal conductivity of the tissues respectively. Subscript b stands for blood, W_b represent blood perfusion rate, T_a stands for the arterial blood temperature which is considered as constant, R is the tissue length. Q_{met} is the metabolic heat created by natural process of the body. Any reduction in metabolic heat will lead to an increase in input energy for hyperthermia and conversly [24]. The relation between metabolic heat and tissue temperature is given by

$$Q_{met} = Q_{mo}\left[1 + d\left(\frac{T - T_0}{T_0}\right)\right], \tag{2}$$

where $d = 0.1 T_0$.

Q_{ext} represent the heat source which is produced by an external electromagnetic field, given by

$$Q_{ext} = \rho SP e^{a(y-0.01)}, \tag{3}$$

where $y = R - r$, notations S, a, P and y represent the antenna constants, transmitted power, distance of the tissue from the outer surface respectively. For simplicity, we define the dimensionless variables as given below [34]

$$x = \frac{r}{R}, \tau = \left(\frac{K_t}{\rho C R^{\alpha(x,\tau)}}\right)^{\frac{1}{\beta}} t, \theta = \left(\frac{T-T_0}{T_0}\right), \theta = \left(\frac{T_a-T_0}{T_0}\right), \theta = \left(\frac{T_w-T_0}{T_0}\right), \tag{4}$$

$$\xi_m = \frac{Q_{mo}R^2}{T_0 K_t}, \xi_f = \sqrt{\frac{W_b C_b}{K_t}} R, \xi_r = \frac{\rho SP}{T_0 K_t} R^2, a_0 = 0.04 \times a, b_0 = R \times a.$$

By using Eq. (4), Eqs. (1)–(3) can be written as

$$\frac{\partial^\beta \theta}{\partial \tau^\beta} = K_t \frac{\partial^{\alpha(x,\tau)} \theta}{\partial x^{\alpha(x,\tau)}} + \Lambda \theta + \Theta,$$

$$\theta(x,\tau)|_{\tau=0} = 0,$$

$$\left.\frac{\partial \theta(x,\tau)}{\partial \tau}\right|_{x=0} = 0, \tag{5}$$

$$\theta(x,\tau)|_{x=1} = \theta_w,$$

where $\Lambda = \xi_m d - \xi_f^2$, $\Theta(x) = \xi_m - \xi_f^2 \theta_a + \xi_r \exp(a_0 - b_0 x)$.

3   Preliminaries

Here we give some basic definitions of variable-order fractional derivatives, shifted Chebyshev polynomial and its derivative.

Definition 1. *The Riemann-Liouville fractional derivative of order $\alpha(x,t)$ is defined as [5]*

$$^{RL}_0 D_x^{\alpha(x,t)} u(x,t) = \begin{cases} \frac{1}{\Gamma(n-\alpha(x,t))} \frac{d^n}{dx^n} \int_0^x \frac{u(s,t)}{(x-s)^{\alpha(x,t)+1-n}} ds, & n-1 < \alpha(x,t) < n \in \mathbb{N}, \\ \frac{d^n}{dx^n} u(x,t), & \alpha(x,t) = n, \end{cases}$$

Definition 2. *The Caputo fractional derivative of order $\alpha(x,t)$ is defined as [21]*

$$^{C}_0 D_x^{\alpha(x,t)} u(x,t) = \begin{cases} \frac{1}{\Gamma(n-\alpha(x,t))} \int_0^x \frac{1}{(x-s)^{\alpha(x,t)+1-n}} \frac{\partial^n u(s,t)}{\partial s^n} ds, & n-1 < \alpha(x,t) < n, \\ \frac{d^n}{dx^n} u(x,t), & \alpha(x,t) = n, \end{cases}$$

with the help of definition given above, we can write the following formula [20]

$$^{C}_0 D_x^{\alpha(x,t)} x^\gamma = \begin{cases} \frac{\Gamma(\gamma+1)}{\Gamma(\gamma+1-\alpha(x,t))} x^{\gamma-\alpha(x,t)}, & \gamma \geq \lceil \alpha(x,t) \rceil, \\ 0, & \gamma < \lceil \alpha(x,t) \rceil. \end{cases} \tag{6}$$

Definition 3. *The analytical closed form of shifted Chebyshev polynomial $T_n^*(x)$ is given by [25]*

$$T_n^*(x) = \sum_{i=0}^{n} \frac{(-1)^i 2^{2n-2i} n (2n-i-1)!}{i!(2n-2i)!} \left(\frac{x}{a}\right)^{n-i}, n = 1,2,3....$$

Specially, $T_n^*(0) = (-1)^n$ and $T_n^*(a) = 1$.

Theorem 1. *The Caputo variable-order fractional derivative of shifted Chebyshev polynomial is given by*

$$
{}_0^C D_x^{\alpha(x,t)} T_n^*(x) = \begin{cases} \displaystyle\sum_{i=0}^{n-\lceil \alpha(x,t)\rceil} \xi_{n,i}^{\alpha(x,t)} x^{n-i-\alpha(x,t)}, & n \geq \lceil \alpha(x,t)\rceil, \\ 0, & n < \lceil \alpha(x,t)\rceil, \end{cases}
$$

where

$$\xi_{n,i}^{\alpha(x,t)} = \frac{(-1)^i 2^{2n-2i} n (2n-i-1)!(n-i)!}{i!(2n-2i)!\Gamma(n-i-\alpha(x,t))} \left(\frac{1}{a}\right)^{n-i}.$$

Proof. By using Definition 3 and Eq. (6), we can write

$$
{}_0^C D_x^{\alpha(x,t)} T_n^*(x) = \sum_{i=0}^{n-\lceil \alpha(x,t)\rceil} \frac{(-1)^i 2^{2n-2i} n (2n-i-1)!(n-i)!}{i!(2n-2i)!\Gamma(n-i-\alpha(x,t))} \left(\frac{1}{a}\right)^{n-i} x^{n-i-\alpha(x,t)}, \quad n \geq \lceil \alpha(x,t)\rceil,
$$

$$
{}_0^C D_x^{\alpha(x,t)} T_n^*(x) = 0, \quad n = 0,1,.....\lceil \alpha(x,t)\rceil - 1,
$$

which gives the desired result.   □

4 Proposed Numerical Technique

We use Chebyshev collocation method for the numerical solution of bioheat model given in Eq. (5). Since Chebyshev polynomials generate an orthogonal basis set so we use it to approximate the temperature.

We can write

$$\theta(x,\tau) \cong \sum_{j=0}^{n}\sum_{i=0}^{m} \phi_i(x) c_{ij} \psi_j(\tau) = \Phi\, C\, \Psi, \tag{7}$$

where ϕ, ψ and C are $1 \times (m+1)$, $(n+1) \times 1$ and $(m+1) \times (n+1)$ matrices respectively and denoted as

$$\Phi(x) = [\phi_0(x)\, \phi_1(x)\, \phi_2(x) .....\phi_m(x)],$$

$$\Psi(\tau) = [\psi_0(\tau)\, \psi_1(\tau)\, \psi_2(\tau) .....\psi_n(\tau)]^T,$$

and

$$C = \begin{bmatrix} c_{00} & c_{01} & \cdots & c_{0n} \\ c_{10} & c_{11} & \cdots & c_{1n} \\ c_{20} & c_{21} & \cdots & c_{2n} \\ \vdots & \vdots & \ddots & \vdots \\ c_{m0} & c_{m1} & \cdots & c_{mn} \end{bmatrix}.$$

From Eq. (7), it can be written as

$$\,_0^C D_\tau^\beta \theta\left(x,\tau\right) = \,_0^C D_\tau^\beta \left(\Phi\left(x\right) C \Psi\left(\tau\right)\right) = \left(\Phi\left(x\right) C \,_0^C D_\tau^\beta \Psi\left(\tau\right)\right). \tag{8}$$

$$\,_0^C D_\tau^{\alpha(x,\tau)} \theta\left(x,\tau\right) = \,_0^C D_\tau^{\alpha(x,\tau)} \left(\Phi\left(x\right) C \Psi\left(\tau\right)\right) = \left(\,_0^C D_\tau^{\alpha(x,\tau)} \Phi\left(x\right) C \Psi\left(\tau\right)\right). \tag{9}$$

Substituting the values from Eqs. (7), (8) and (9) in Eq. (5), we get

$$\left(\Phi\left(x\right) C \,_0^C D_\tau^\beta \Psi\left(\tau\right)\right) = \left(\,_0^C D_\tau^{\alpha(x,\tau)} \Phi\left(x\right) C \Psi\left(\tau\right)\right) + \Lambda\left(\Phi\left(x\right) C \Psi\left(\tau\right)\right) + \Theta\left(x\right). \tag{10}$$

$$\begin{aligned}
\Phi\left(x\right) C \Psi\left(\tau\right)\big|_{\tau=0} &= 0, \\
\Phi'\left(x\right) C \Psi\left(\tau\right)\big|_{x=0} &= 0, \\
\Phi\left(x\right) C \Psi\left(\tau\right)\big|_{x=1} &= \theta_w.
\end{aligned} \tag{11}$$

To get the values of unknowns, we need to collocate Eq. (10). For this, we use uniform nodes $x_i, i = 1, 2..., m-1$ and $\tau_j = 1, 2, ..., n$ as a collocating points.

We get $n(m-1)$ equations from Eq. (10) which is of the form,

$$A_{(m-1),(m+1)} C_{(m+1),(n+1)} B_{(n+1),n}^T = M_{(m-1),(m+1)} C_{(m+1),(n+1)} N_{(n+1),n}^T$$
$$+ \Lambda\left(A_{(m-1),(m+1)} C_{(m+1),(n+1)} N_{(n+1),n}^T\right) + \Theta\left(x\right). \tag{12}$$

The initial and boundary conditions can be written as

$$A_{(m-1),(m+1)} C_{(m+1),(n+1)} I_{1(1,(n+1))}^T = 0,$$

$$B_{1(1,(m+1))} C_{(m+1),(n+1)} R_{(n+1),(n+1)}^T = 0,$$

$$B_{2(1,(m+1))} C_{(m+1),(n+1)} R_{(n+1),(n+1)}^T = \theta_w,$$

where

$$A = \begin{bmatrix} \phi_0\left(x_1\right) & \phi_1\left(x_1\right) & \cdots & \phi_m\left(x_1\right) \\ \phi_0\left(x_2\right) & \phi_1\left(x_2\right) & \cdots & \phi_m\left(x_2\right) \\ \vdots & \vdots & \ddots & \vdots \\ \phi_0\left(x_{m-1}\right) & \phi_1\left(x_{m-1}\right) & \cdots & \phi_m\left(x_{m-1}\right) \end{bmatrix}, B = \begin{bmatrix} \,_0^C D_\tau^\beta \psi_0\left(\tau_1\right) & \,_0^C D_\tau^\beta \psi_1\left(\tau_1\right) & \cdots & \,_0^C D_\tau^\beta \psi_n\left(\tau_1\right) \\ \,_0^C D_\tau^\beta \psi_0\left(\tau_2\right) & \,_0^C D_\tau^\beta \psi_1\left(\tau_2\right) & \cdots & \,_0^C D_\tau^\beta \psi_n\left(\tau_2\right) \\ \vdots & \vdots & \ddots & \vdots \\ \,_0^C D_\tau^\beta \psi_0\left(\tau_n\right) & \,_0^C D_\tau^\beta \psi_1\left(\tau_n\right) & \cdots & \,_0^C D_\tau^\beta \psi_n\left(\tau_n\right) \end{bmatrix},$$

$$C = \begin{bmatrix} c_{00} & c_{01} & \cdots & c_{0n} \\ c_{10} & c_{11} & \cdots & c_{1n} \\ c_{20} & c_{21} & \cdots & c_{2n} \\ \vdots & \vdots & \ddots & \vdots \\ c_{m0} & c_{m1} & \cdots & c_{mn} \end{bmatrix}, M = \begin{bmatrix} D_\tau^{\alpha(x,\tau)} \phi_0\left(x_1\right) & D_\tau^{\alpha(x,\tau)} \phi_1\left(x_1\right) & \cdots & D_\tau^{\alpha(x,\tau)} \phi_m\left(x_1\right) \\ D_\tau^{\alpha(x,\tau)} \phi_0\left(x_2\right) & D_\tau^{\alpha(x,\tau)} \phi_1\left(x_2\right) & \cdots & D_\tau^{\alpha(x,\tau)} \phi_m\left(x_2\right) \\ \vdots & \vdots & \ddots & \vdots \\ D_\tau^{\alpha(x,\tau)} \phi_0\left(x_{m-1}\right) & D_\tau^{\alpha(x,\tau)} \phi_1\left(x_{m-1}\right) & \cdots & D_\tau^{\alpha(x,\tau)} \phi_m\left(x_{m-1}\right) \end{bmatrix},$$

$$N = \begin{bmatrix} \psi_0\left(\tau_1\right) & \psi_1\left(\tau_1\right) & \cdots & \psi_n\left(\tau_1\right) \\ \psi_0\left(\tau_2\right) & \psi_1\left(\tau_2\right) & \cdots & \psi_n\left(\tau_2\right) \\ \vdots & \vdots & \ddots & \vdots \\ \psi_0\left(\tau_n\right) & \psi_1\left(\tau_n\right) & \cdots & \psi_n\left(\tau_n\right) \end{bmatrix}, R = \begin{bmatrix} \psi_0\left(\tau_0\right) & \psi_1\left(\tau_0\right) & \cdots & \psi_n\left(\tau_0\right) \\ \psi_0\left(\tau_1\right) & \psi_1\left(\tau_1\right) & \cdots & \psi_n\left(\tau_1\right) \\ \vdots & \vdots & \ddots & \vdots \\ \psi_0\left(\tau_n\right) & \psi_1\left(\tau_n\right) & \cdots & \psi_n\left(\tau_n\right) \end{bmatrix},$$

$$I_1 = \begin{bmatrix} \psi_0\left(0\right) \\ \psi_1\left(0\right) \\ \vdots \\ \psi_n\left(0\right) \end{bmatrix}, B_1 = \begin{bmatrix} \phi_0'\left(0\right) \\ \phi_1'\left(0\right) \\ \vdots \\ \phi_m'\left(0\right) \end{bmatrix}_1, B_2 = \begin{bmatrix} \phi_0\left(1\right) \\ \phi_1\left(1\right) \\ \vdots \\ \phi_m\left(1\right) \end{bmatrix}.$$

We use Kronecker product to write Eq. (12) with initial and boundary conditions in a simplified form as

$$[B \otimes A - N \otimes M - \Lambda (N \otimes A)]\,C = P_1,$$

$$[I_1 \otimes A]\,C = 0,$$

$$[R \otimes B_1]\,C = 0,$$

$$[R \otimes B_2]\,C = \theta_w,$$

which can also be written in the form

$$A_1 C = P_1. \tag{13}$$

$$\begin{aligned} A_2 C &= 0, \\ A_3 C &= 0, \\ A_4 C &= \theta_w. \end{aligned} \tag{14}$$

Dimension of A_1 is $(m-1)n, (m+1)(n+1)$, dimension of C is $((m+1)(n+1),1)$ and it is defined as $C = [c_{00},\ c_{01},...,c_{0n},\ c_{10},\ c_{11},\ c_{12},...,c_{1n},\ c_{21},\ c_{22},...,c_{2n},\ c_{20},\ c_{21},....,\ c_{2n},\ c_{m0},\ c_{m1},....,\ c_{mn},\ ]^T$, dimension of P_1 is $((m-1)n,1)$. The dimension of A_2, A_3 and A_4 is $(m-1,(m+1)(n+1)), ((n+1,(m+1)(n+1)))$ and $((n+1),(m+1)(n+1))$ respectively. Now a complete matrix can be obtained by collectively writing the equations as follows

$$EC = Z,$$

where $E = [A_1, A_2, A_3, A_4]^T$ is of dimension $((m+1)(n+1),(m+1)(n+1))$ and $Z = [P_1, 0, 0, \theta_w]^T$ is of dimension $((m+1)(n+1),1)$. This is a system of linear equations in which the unknowns coefficients C can be obtained by solving it. The required solution will be then extracted from Eq. (7).

5   Numerical Results and Discussion

For numerical analysis, we use the following parametric values [34] for living biological tissues $R = 0.05\,\mathrm{m}^{-1}, K_t = 0.5\,\mathrm{m}^{-1}{}^\circ\mathrm{C}^{-1}, S = 12.5\,\mathrm{kg}^{-1}, a = -127\,\mathrm{m}^{-1}, \rho = 1000\,\mathrm{kg\,m}^{-3}, C_b = 3344\,\mathrm{J\,kg}^{-1}{}^\circ\mathrm{C}^{-1}, Q_{mo} = 1091\,\mathrm{W\,m}^{-3}, C = 4180\,\mathrm{J\,kg}^{-1}{}^\circ\mathrm{C}^{-1}, T_0 = T_a = T_w = T_f = 37\,{}^\circ\mathrm{C}^{-1}, b_0 = -6.35, a_0 = -5.08, W_b = 8\,\mathrm{kg\,m}^{-3}\mathrm{s}^{-1}$. The value of antenna power can be vary and depend upon requirement, in this analysis we consider $P = 10\,\mathrm{W}$.

On applying the proposed Chebyshev collocation method, we examine the behavior of the model with initial and boundary conditions as defined in equation (5). We consider following cases to study the anomalous diffusion to this model for various values of fractional-order α and β including the variable-order space fractional derivative.

Case 1. The temperature profile for integer-order case, i.e., for $\alpha = 2$ and $\beta = 1$ along x for different time is shown in the Fig. 1. We utilize the proposed method with m = n = 20. In hyperthermia treatment, the non-dimensional temperature (θ) for heating the cancer cells ranges from 0.102 to 0.216, and to achieve this, the non-dimensional time (τ) normally varies from 0.04 to 0.08. It should be clear from Fig. 1 that 0.04 is the maximum non-dimensional temperature which, occurs at $\tau = 5 \times 10^{-3}$ whereas, at $\tau = 0.04$ it reaches to the hyperthermia position.

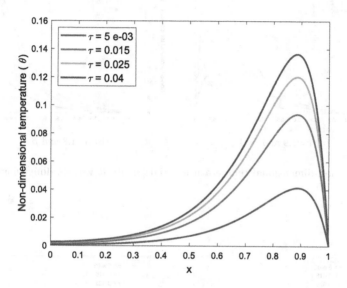

Fig. 1. The non-dimensional temperature variation at various non-dimensional time for $\alpha = 2$ and $\beta = 1$

Case 2. By decreasing the order of space fractional derivative from 1.9 to 1.8, the maximum non-dimensional temperature raises from 0.155 to 0.165 at $\tau = 0.04$ which is shown in Fig. 2. Similarly, Fig. 3 shows the graph of θ with respect to x for dissimilar values of β. There are minor changes in the maximum temperature by changing the non-integer time derivative from 0.9 to 0.8. It should be noted from Fig. 2 and 3 that by changing any of the fractional-order either space or time, there is a different time to achieve the hyperthermia position. It can also be concluded that space fractional derivative has a more considerable impact on temperature variation.

Case 3. Figure 4 represents the temperature variation for various space fractional variable-order derivative α which depend on x and τ. Figure 5 shows the temperature profile at different time. The graph follows the same pattern of elevation and decrease in temperature as constant fractional-order. It reaches to the peak at x = 0:9.

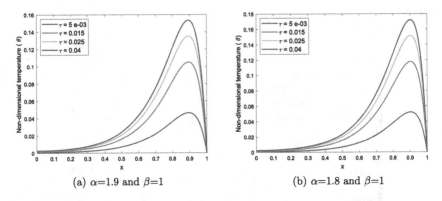

(a) α=1.9 and β=1   (b) α=1.8 and β=1

Fig. 2. The non-dimensional temperature variation (θ) at various dimensionless time (τ), m = n = 20.

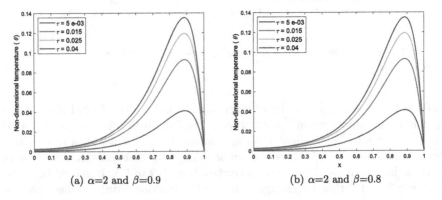

(a) α=2 and β=0.9   (b) α=2 and β=0.8

Fig. 3. The non-dimensional temperature variation (θ) at various dimensionless time (τ), m = n = 20.

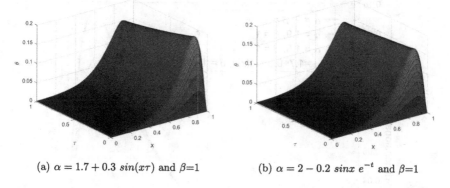

(a) $\alpha = 1.7 + 0.3 \, sin(x\tau)$ and $\beta = 1$ (b) $\alpha = 2 - 0.2 \, sinx \, e^{-t}$ and $\beta = 1$

Fig. 4. The non-dimensional temperature variation (θ) for different space fractional variable-order, m = n = 20.

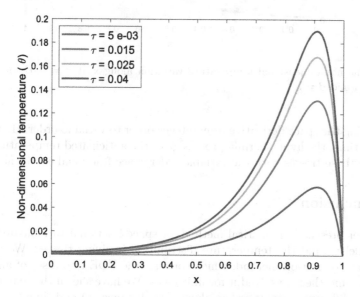

Fig. 5. The non-dimensional temperature variation along x for $\alpha = 1.7 + 0.3 \, sin(x\tau)$ and $\beta = 1$

5.1 Comparison of Variable-Order and Constant Fractional-Order Derivative

We applied the proposed Chebyshev collocation method to variable-order bioheat model for m = 20, n = 20. Figure 6 depicts the non-dimensional temperature field within the living tissues along x and at a fixed time. It can be observed from Fig. 6 that up to a certain length, the increase in temperature is lump sum similar for every choices of α. The variation in temperature profile can be seen when the length of the tissue (x) is greater than 0.7. From the achieved results, the maximum temperature increases from 0.105 and 0.145, which is about 38%,

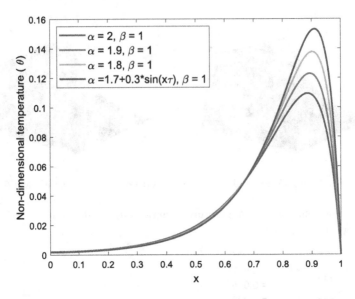

Fig. 6. The non-dimensional temperature variation along x at $\tau = 0.02$ for various values of α and $\beta = 1$.

by changing the space derivative from integer-order to variable-order. Hence at a specified time, the hyperthermia process gives the anticipated temperature level faster for those tissues with the variable-order space fractional exponent.

6   Conclusion

This paper presents a numerical method for space fractional variable-order bioheat model to find the temperature distribution in living tissues. We use the Chebyshev collocation method, which converts the PDE into a set of algebraic equations, and then we solved it for unknowns. We have shown the temperature profile for different variable-orders along with distance (x) and time (τ).

We examine the results of anomalous diffusion with variable-order fractional derivative and compare it with constant fractional diffusion. We observed that the extreme temperature profile in the variable-order case is higher than the later one.

Temperature prediction is significant in many clinical practices like cryosurgery, cancer hyperthermia, etc. The obtained result may be useful for predicting the temperature in space fractional variable-order bioheat model. The proposed method may also be helpful for the researchers working on a variable-order fractional derivative field.

Acknowledgement. One of the authors acknowledges the Council of Scientific & Industrial Research (CSIR), New Delhi, India for JRF fellowship (File no. 09/1007(0005)/2019 EMR-1) during the preparation of this paper.

References

1. Adolfsson, K., Enelund, M., Olsson, P.: On the fractional order model of viscoelasticity. Mech. Time-Depend. Mater. **9**(1), 15–34 (2005)
2. Arafa, A., Khalil, M., Sayed, A.: A non-integer variable order mathematical model of human immunodeficiency virus and malaria coinfection with time delay. Complexity **2019** (2019)
3. Bagaria, H., Johnson, D.: Transient solution to the bioheat equation and optimization for magnetic fluid hyperthermia treatment. Int. J. Hyperth. **21**(1), 57–75 (2005)
4. Baleanu, D., Asad, J.H., Petras, I.: Numerical solution of the fractional Euler-Lagrange's equations of a thin elastica model. Nonlinear Dyn. **81**(1–2), 97–102 (2015)
5. Bhrawy, A.H., Zaky, M.A.: Numerical simulation for two-dimensional variable-order fractional nonlinear cable equation. Nonlinear Dyn. **80**, 101–116 (2014). https://doi.org/10.1007/s11071-014-1854-7
6. Bhrawy, A., Zaky, M.: Numerical algorithm for the variable-order Caputo fractional functional differential equation. Nonlinear Dyn. **85**(3), 1815–1823 (2016)
7. Coccarelli, A., Boileau, E., Parthimos, D., Nithiarasu, P.: An advanced computational bioheat transfer model for a human body with an embedded systemic circulation. Biomech. Model. Mechanobiol. **15**(5), 1173–1190 (2015). https://doi.org/10.1007/s10237-015-0751-4
8. Coimbra, C.F.: Mechanics with variable-order differential operators. Ann. Phys. **12**(11–12), 692–703 (2003)
9. Damor, R., Kumar, S., Shukla, A.: Numerical simulation of fractional bioheat equation in hyperthermia treatment. J. Mech. Med. Biol. **14**(02), 1450018 (2014)
10. Deka, K., Bhanja, D., Nath, S.: Fundamental solution of steady and transient bio heat transfer equations especially for skin burn and hyperthermia treatments. Heat Transf. Asian Res. **48**(1), 361–378 (2019)
11. Durkee, J., Jr., Antich, P.: Exact solutions to the multi-region time-dependent bioheat equation with transient heat sources and boundary conditions. Phys. Med. Biol. **36**(3), 345 (1991)
12. Durkee, J., Jr., Antich, P., Lee, C.: Exact solutions to the multiregion time-dependent bioheat equation. I: solution development. Phys. Med. Biol. **35**(7), 847 (1990)
13. Ezzat, M.A., AlSowayan, N.S., Al-Muhiameed, Z.I., Ezzat, S.M.: Fractional modelling of Pennes' bioheat transfer equation. Heat Mass Transf. **50**(7), 907–914 (2014)
14. Fahmy, M.A.: A new LRBFCM-GBEM modeling algorithm for general solution of time fractional-order dual phase lag bioheat transfer problems in functionally graded tissues. Numer. Heat Transf. Part A Appl. **75**(9), 616–626 (2019)
15. Gage, A.A., Baust, J.G.: Cryosurgery - a review of recent advances and current issues. CryoLetters **23**(2), 69–78 (2002)
16. Giordano, M.A., Gutierrez, G., Rinaldi, C.: Fundamental solutions to the bioheat equation and their application to magnetic fluid hyperthermia. Int. J. Hyperth. **26**(5), 475–484 (2010)
17. Goodwine, B.: Modeling a multi-robot system with fractional-order differential equations. In: 2014 IEEE International Conference on Robotics and Automation (ICRA), pp. 1763–1768. IEEE (2014)

18. Habash, R., Krewski, D., Bansal, R., Alhafid, H.T.: Principles, applications, risks and benefits of therapeutic hyperthermia. Front. Biosci. (Elite Ed.) **3**, 1169–1181 (2011)
19. Hesameddini, E., Rahimi, A., Asadollahifard, E.: On the convergence of a new reliable algorithm for solving multi-order fractional differential equations. Commun. Nonlinear Sci. Numer. Simul. **34**, 154–164 (2016)
20. Heydari, M.H.: A new approach of the Chebyshev wavelets for the variable-order time fractional mobile-immobile advection-dispersion model. arXiv preprint arXiv:1605.06332 (2016)
21. Heydari, M.H., Avazzadeh, Z.: Legendre wavelets optimization method for variable-order fractional Poisson equation. Chaos Solitons Fractals **112**, 180–190 (2018)
22. Heydari, M.H., Avazzadeh, Z., Haromi, M.F.: A wavelet approach for solving multi-term variable-order time fractional diffusion-wave equation. Appl. Math. Comput. **341**, 215–228 (2019)
23. Hosseininia, M., Heydari, M., Roohi, R., Avazzadeh, Z.: A computational wavelet method for variable-order fractional model of dual phase lag bioheat equation. J. Comput. Phys. **395**, 1–18 (2019)
24. Jamil, M., Ng, E.Y.K.: Ranking of parameters in bioheat transfer using Taguchi analysis. Int. J. Therm. Sci. **63**, 15–21 (2013)
25. Khader, M.: On the numerical solutions for the fractional diffusion equation. Commun. Nonlinear Sci. Numer. Simul. **16**(6), 2535–2542 (2011)
26. Khan, N.A., Razzaq, O.A., Mondal, S.P., Rubbab, Q.: Fractional order ecological system for complexities of interacting species with harvesting threshold in imprecise environment. Adv. Differ. Equ. **2019**(1), 1–34 (2019). https://doi.org/10.1186/s13662-019-2331-x
27. Li, X., et al.: A physics-based fractional order model and state of energy estimation for lithium ion batteries. Part I: model development and observability analysis. J. Power Sources **367**, 187–201 (2017)
28. Magin, R., Sagher, Y., Boregowda, S.: Application of fractional calculus in modeling and solving the bioheat equation. WIT Trans. Ecol. Environ. **73** (2004)
29. Martínez-Salgado, B.F., Rosas-Sampayo, R., Torres-Hernández, A., Fuentes, C.: Application of fractional calculus to oil industry. In: Fractal Analysis: Applications in Physics, Engineering and Technology, p. 21 (2017)
30. Pennes, H.H.: Analysis of tissue and arterial blood temperatures in the resting human forearm. J. Appl. Physiol. **1**(2), 93–122 (1948)
31. Roohi, R., Heydari, M., Aslami, M., Mahmoudi, M.: A comprehensive numerical study of space-time fractional bioheat equation using fractional-order Legendre functions. Eur. Phys. J. Plus **133**(10), 412 (2018)
32. Samko, S.G., Ross, B.: Integration and differentiation to a variable fractional order. Integral Transform. Spec. Funct. **1**(4), 277–300 (1993)
33. Shekari, Y., Tayebi, A., Heydari, M.H.: A meshfree approach for solving 2D variable-order fractional nonlinear diffusion-wave equation. Comput. Methods Appl. Mech. Eng. **350**, 154–168 (2019)
34. Singh, J., Gupta, P.K., Rai, K.: Solution of fractional bioheat equations by finite difference method and HPM. Math. Comput. Model. **54**(9–10), 2316–2325 (2011)
35. Sun, H., Chen, W., Wei, H., Chen, Y.: A comparative study of constant-order and variable-order fractional models in characterizing memory property of systems. Eur. Phys. J. Spec. Top. **193**(1), 185 (2011)
36. Wang, Y., Zhu, L., Rosengart, A.J.: Targeted brain hypothermia induced by an interstitial cooling device in the rat neck: experimental study and model validation. Int. J. Heat Mass Transf. **51**(23–24), 5662–5670 (2008)

An Efficient Numerical Method for Singularly Perturbed Parabolic Problems with Non-smooth Data

Narendra Singh Yadav$^{(\boxtimes)}$ and Kaushik Mukherjee

Department of Mathematics, Indian Institute of Space Science and Technology,
Thiruvanthapuram 695547, India
narendrasingh.16@res.iist.ac.in, kaushik@iist.ac.in

Abstract. This article deals with a class of singularly perturbed convection-diffusion parabolic problems with discontinuous source term exhibiting both boundary and weak interior layers. In order to solve this class of problems, we discretize the time derivative by the backward-Euler method on the uniform mesh; and for the spatial discretization, a new finite difference scheme is proposed utilizing a layer resolving piecewise-uniform Shishkin mesh. We discuss the monotonocity of the proposed method. Further, we demonstrate through the numerical experiment that the proposed method converges uniformly with respect to the perturbation parameter ε; and is almost second-order accurate in space, regardless of the larger and smaller values of ε. In addition to this, we compare the present numerical method with the classical implicit upwind finite difference scheme.

Keywords: Singularly perturbed parabolic problem · Boundary layer · Interior layer · Numerical scheme · Piecewise-uniform Shishkin mesh · Uniform convergence

1 Introduction

At first, for defining the domain associated with the model problem, the following notations are introduced :

$$\mathtt{I} = (0,1), \mathtt{I}^- = (0,\xi), \mathtt{I}^+ = (\xi,1); \mathtt{Q}^- = \mathtt{I}^- \times (0,\mathtt{T}], \mathtt{Q}^+ = \mathtt{I}^+ \times (0,\mathtt{T}], \mathtt{Q} = \mathtt{I} \times (0,\mathtt{T}].$$

In this article, we deal with singularly perturbed parabolic initial-boundary-value problems (IBVPs) of the form:

$$\begin{cases} \mathtt{L}_\varepsilon y(x,t) \equiv \varepsilon \dfrac{\partial^2 y}{\partial x^2} + \mathtt{a}(x)\dfrac{\partial y}{\partial x} - \mathtt{b}(x)y - \dfrac{\partial y}{\partial t} = \mathtt{f}(x,t), \quad (x,t) \in \mathtt{Q}^- \cup \mathtt{Q}^+, \\ y(x,0) = \mathtt{g}_0(x), \quad x \in \overline{\mathtt{I}} = [0,1], \\ y(0,t) = \mathtt{g}_l(t), \, y(1,t) = \mathtt{g}_r(t), \quad t \in (0,\mathtt{T}], \end{cases} \qquad (1)$$

© Springer Nature Singapore Pte Ltd. 2021
A. Awasthi et al. (Eds.): CSMCS 2020, CCIS 1345, pp. 159–171, 2021.
https://doi.org/10.1007/978-981-16-4772-7_12

together with the interface conditions

$$[u](\xi, t) = 0, \quad \left[\frac{\partial u}{\partial x}\right](\xi, t) = 0, \quad t \in (0, \mathrm{T}], \tag{2}$$

Here, ε is a small parameter such that $\varepsilon \in (0, 1]$; and it is assumed that the convection coefficient $a(x)$, the reaction term $b(x)$ and the source term $f(x, t)$ are sufficiently smooth functions in their respective domains such that

$$\begin{cases} a(x) \geq \alpha > 0, \text{ on } I^- \cup I^+, \quad b(x) \geq 0, \text{ on } \overline{I}, \\ |[a](\xi)| \leq C, \ |[f](\xi, t)| \leq C. \end{cases} \tag{3}$$

Since the source term is considered to be discontinuous at $x = \xi$ and the convection coefficient $a(x)$ satisfies the conditions given in (3); the solution of the IBVP (1)–(3) generally possesses a weak interior layer to the right side of $x = \xi$, in addition to the boundary layer at $x = 0$. This type of model problem appears in the semiconductor device modeling (see, *e.g.*, [5]).

We assume that the given data g_0, g_l and g_r are sufficiently smooth functions and also satisfy the necessary compatibility conditions at the points $(0, 0), (1, 0)$ and $(\xi, 0)$; and under these hypothesis, the IBVP (1)–(3) possesses a unique solution $y(x, t) \in \mathcal{C}^{1+\lambda}(Q) \cap \mathcal{C}^{2+\lambda}(Q^- \cup Q^+)$ (for detailed discussion, one can refer [4]).

Finding the robust numerical solution of singularly perturbed problems having non-smooth data has been an interesting subject in the last few years. To cite a few, Farrell et al. discuss about the first-order classical numerical schemes for singularly perturbed convection-diffusion BVPs possessing strong interior layers in [2], and boundary and interior layers in [3]. Further, Cen in [1] develop a second-order numerical scheme to solve singularly perturbed BVPs having discontinuous convection coefficient . Moreover, Shanthi et al. in [11] also analyze numerical aspects of singularly perturbed reaction-diffusion problems with discontinuous source term. Apart from the above mentioned literature, O'Riordan and Shishkin in [9] and Mukherjee and Natesan in [7, 8], respectively develop and analyze the first-order and the second-order implicit numerical schemes for time-dependent singularly perturbed parabolic IBVPs possessing strong interior layers.

In this article, our primary objective is to develop an ε-uniformly convergent numerical method with better numerical accuracy for the IBVP (1)–(3) whose solution exhibits both boundary and weak interior layers. To achieve this purpose, the domain is discretized using a spacial rectangular mesh which consists of a layer-resolving piecewise-uniform Shishkin mesh in the spatial direction, and an equidistant mesh in the temporal direction. The proposed numerical method is comprised of the backward-Euler method for the time derivative and a new hybrid finite difference scheme for the spatial discretization. In the recent past, similar scheme has been proposed and analyzed for singularly perturbed BVPs in [6] and singularly perturbed IBVPs in [12], having discontinuous convection coefficient.

The rest of this paper is arranged as follows: In Sect. 2, the proposed numerical method is described using the spacial rectangular mesh. Further, the monotonocity of the proposed method is discussed at the end of this section. In Sect. 3, the numerical results are presented for some test example and also compared with the implicit upwind scheme in order to verify the efficiency and the accuracy of the present numerical method. Moreover, a brief conclusion has been presented in Sect. 4.

In this paper, the notation $[\phi](\xi,t)$ denotes the jump of ϕ across $x = \xi$ and is defined by $[\phi](\xi,t) = \phi(\xi^+,t) - \phi(\xi^-,t)$, where $\phi(\xi^\pm,t) = \lim_{x\to\xi\pm0}\phi(x,t)$. Throughout the paper, we use $\|\phi\|_{\infty,D}$ as the supremum norm, where D corresponds to the domain of the function ϕ and for simplicity, we sometimes omit D whenever the domain is obvious.

2 Numerical Approximation

Here, we give description of the spacial rectangular mesh for the discretization of the domain. We also provide the detailed construction of the proposed numerical method and discuss the monotonocity of the method.

2.1 Discretization of the Domain

Here, we choose $N_x(\geq 8)$ as an even positive integer. Now, to discretize the domain $\overline{Q} = \overline{I} \times [0,T]$, we construct a rectangular mesh $\overline{q}^{N_x,N_t} = \overline{I}^{N_x} \times S^{N_t}$, where S^{N_t} denotes the equidistant mesh on the temporal domain $[0,T]$ such that $S^{N_t} = \{t_n = n\Delta t, n = 0,\ldots,N_t, \Delta t = T/N_t\}$, whereas $\overline{I}^{N_x} = \{x_k\}_{k=0}^{N_x}$ denotes the piecewise-uniform Shishkin mesh on the spatial domain $\overline{I}$. The Shishkin mesh is condensed near $x = 0$ and in the vicinity of the right side of $x = \xi$. To obtain $\overline{I}^{N_x}$, we divide $\overline{I}$ into four subintervals as $\overline{I} = [0,\varsigma_1] \cup [\varsigma_1,\xi] \cup [\xi,\xi+\varsigma_2] \cup [\xi+\varsigma_2,1]$, where

$$\varsigma_1 = \min\left\{\frac{\xi}{2},\varsigma_0\varepsilon\ln N_x\right\}, \quad \varsigma_2 = \min\left\{\frac{1-\xi}{2},\varsigma_0\varepsilon\ln N_x\right\}, \quad \varsigma_0 \text{ is a positive constant;}$$

and we introduce equidistant mesh with $N_x/4$ mesh intervals on each sub-interval. Let $h_k = x_k - x_{k-1}, k = 1,\ldots,N_x$, with $\widehat{h}_k = h_k + h_{k+1}, k = 1,\ldots,N_x - 1$. We further denote the mesh width h_k as follows: $h_k = h_1 = \dfrac{4\varsigma_1}{N_x}$, for $k = 1,\ldots,N_x/4$; $h_k = H_1 = \dfrac{4(\xi-\varsigma_1)}{N_x}$, for $k = N_x/4+1,\ldots,N_x/2$; $h_k = h_2 = \dfrac{4\varsigma_2}{N_x}$, for $k = N_x/2+1,\ldots,3N_x/4$; and $h_k = H_2 = \dfrac{4(1-\xi-\varsigma_2)}{N_x}$, for $k = 3N_x/4+1,\ldots,N_x$.

2.2 Proposed Numerical Method

We define the difference operators $D_x^+, D_x^-, D_x^+D_x^-, D_x^*,$ and D_t^-, respectively for a given mesh function $Y(x_k,t_n) = Y_k^n$ as follows:

$$\begin{cases} D_x^+Y_k^n = \dfrac{Y_{k+1}^n - Y_k^n}{h_k}, & D_x^-Y_k^n = \dfrac{Y_k^n - Y_{k-1}^n}{h_k}, & D_x^+D_x^-Y_k^n = \dfrac{2}{\widehat{h}_k}(D_x^+Y_k^n - D_x^-Y_k^n), \\ D_x^*Y_k^n = \dfrac{h_k}{\widehat{h}_k}D_x^+Y_k^n + \dfrac{h_{k+1}}{\widehat{h}_k}D_x^-Y_k^n, & D_t^-Y_k^n = \dfrac{Y_k^n - Y_k^{n-1}}{\Delta t}. \end{cases}$$

(4)

Further, we define $Y^n_{k+\frac{1}{2}} = (Y^n_{k+1} + Y^n_k)/2$, $a_{k+\frac{1}{2}} = (a_k + a_{k+1})/2$, $b^n_{k+\frac{1}{2}} = (b^n_k + b^n_{k+1})/2$, $f^n_{k+\frac{1}{2}} = (f^n_k + f^n_{k+1})/2$.

In order to constitute the numerical method, we use the backward-Euler method for discretizing the time derivative and for the spatial discretization, we propose a new hybrid finite difference scheme, which is comprised of a modified central difference scheme whenever $\varepsilon > 2\|a\|N_x^{-1}$; and a suitable combination of the midpoint upwind scheme and the modified central difference scheme whenever $\varepsilon \leq 2\|a\|N_x^{-1}$. We impose second-order one-sided difference approximation at the point of discontinuity. The proposed numerical method is now described in the following form on the mesh $\overline{Q}^{N_x,N_t}$:

$$
\begin{cases}
Y(x_k,0) = g_0(x_k), \quad \text{for } k = 0,\dots,N_x, \\[4pt]
L_{mc}^{N_x,N_t} Y_k^{n+1} = f_k^{n+1}, \quad \text{for } k = 1,\dots,N_x/4 - 1 \text{ and } k = N_x/2+1,\dots,3N_x/4-1, \\[4pt]
L_{mc}^{N_x,N_t} Y_k^{n+1} = f_k^{n+1}, \quad \text{for } k = N_x/4,\dots,N_x/2-1 \text{ and } k = 3N_x/4,\dots,N_x-1, \\
\qquad\qquad\qquad\qquad \text{and when } \varepsilon > 2\|a\|N_x^{-1}, \\[4pt]
L_{mu}^{N_x,N_t} Y_k^{n+1} = f_{k+1/2}^{n+1}, \quad \text{for } k = N_x/4,\dots,N_x/2-1, \\
\qquad\qquad\qquad\qquad \text{and when } \varepsilon \leq 2\|a\|N_x^{-1}, \\[4pt]
L_{mu}^{N_x,N_t} Y_k^{n+1} = f_{k+1/2}^{n+1}, \quad \text{for } k = 3N_x/4,\dots,N_x-1, \\
\qquad\qquad\qquad\qquad \text{and when } \varepsilon \leq 2\|a\|N_x^{-1}, \\[4pt]
D_x^F Y_k^{n+1} - D_x^B Y_k^{n+1} = 0, \quad \text{for } k = N_x/2, \\[4pt]
Y(0,t_{n+1}) = g_l(t_{n+1}), \; Y(1,t_{n+1}) = g_r(t_{n+1}), \quad \text{for } n = 0,\dots,N_t - 1,
\end{cases}
\tag{5}
$$

where the midpoint upwind operator $L_{mu}^{N_x,N_t}$ and the modified central difference operator $L_{mc}^{N_x,N_t}$ are respectively defined as

$$
\begin{cases}
L_{mc}^{N_x,N_t} Y_k^{n+1} = \varepsilon D_x^+ D_x^- Y_k^{n+1} + a_k D_x^* Y_k^{n+1} - b_k Y_k^{n+1} - D_t^- Y_k^{n+1}, \\[4pt]
L_{mu}^{N_x,N_t} Y_k^{n+1} = \varepsilon D_x^+ D_x^- Y_k^{n+1} + a_{k+1/2} D_x^+ Y_k^{n+1} - b_{k+1/2} Y_{k+1/2}^{n+1} - D_t^- Y_{k+1/2}^{n+1},
\end{cases}
\tag{6}
$$

and

$$
D_x^F Y_{N_x/2}^{n+1} = (-Y_{N_x/2+2}^{n+1} + 4Y_{N_x/2+1}^{n+1} - 3Y_{N_x/2}^{n+1})/2h_2, \quad D_x^B Y_{N_x/2}^{n+1} = (Y_{N_x/2-2}^{n+1} - 4Y_{N_x/2-1}^{n+1} + 3Y_{N_x/2}^{n+1})/2H_1. \tag{7}
$$

Now, we rewrite (5) in the following form:

$$
\begin{cases}
Y(x_k,0) = g_0(x_k), \quad \text{for } k = 0,\dots,N_x, \\[4pt]
L_\varepsilon^{N_x,N_t} Y_k^{n+1} = F_k^{n+1}, \quad \text{for } k = 1,\dots,N_x - 1, \\[4pt]
Y(0,t_{n+1}) = g_l(t_{n+1}), \quad Y(1,t_{n+1}) = g_r(t_{n+1}), \quad \text{for } n = 0,\dots,N_t - 1,
\end{cases}
\tag{8}
$$

where the difference operator $L_\varepsilon^{N_x,N_t}$ is defined as

$$L_\varepsilon^{N_x,N_t} Y_k^{n+1} = \begin{cases} [\mu_k^- Y_{k-1}^{n+1} + \mu_k^c Y_k^{n+1} + \mu_k^+ Y_{k+1}^{n+1}] + [\lambda_k^- Y_{k-1}^n + \lambda_k^c Y_k^n + \lambda_k^+ Y_{k+1}^n], \\ \qquad \text{for } k = 1,\ldots,N_x/2-1, N_x/2+1,\ldots,N_x-1, \\ [v_k^{-,2} Y_{k-2}^{n+1} + v_k^{-,1} Y_{k-1}^{n+1} + v_k^c Y_k^{n+1} + v_k^{+,1} Y_{k+1}^{n+1} + v_k^{+,2} Y_{k+2}^{n+1}], \\ \qquad \text{for } k = N_x/2, \end{cases} \tag{9}$$

and the term F_k^{n+1} as

$$F_k^{n+1} \equiv \begin{cases} f_k^{n+1}, & \text{for } k = 1,\ldots,N_x/4-1, \text{ and } k = N_x/2+1,\ldots,3N_x/4-1, \\ f_k^{n+1}, & \text{for } k = N_x/4,\ldots,N_x/2-1, \text{ and } k = 3N_x/4,\ldots,N_x-1, \text{ and when } \varepsilon > 2\|a\|N_x^{-1} \\ f_{k+1/2}^{n+1}, & \text{for } k = N_x/4,\ldots,N_x/2-1, \text{ and when } \varepsilon \leq 2\|a\|N_x^{-1}, \\ f_{k+1/2}^{n+1}, & \text{for } k = 3N_x/4,\ldots,N_x-1, \text{ and when } \varepsilon \leq 2\|a\|N_x^{-1}, \\ 0, & \text{for } k = N_x/2. \end{cases} \tag{10}$$

The coefficients $\mu_k^-, \mu_k^c, \mu_k^+; \lambda_k^-, \lambda_k^c, \lambda_k^+; v_k^{-,2} v_k^{-,1}, v_k^c, v_k^{+,1}, v_k^{+,2}$ associated with the difference operator $L_\varepsilon^{N_x,N_t}$ in (9) can be obtained from (5)–(7).

2.3 Monotonocity of the Proposed Method

We convert the difference scheme (8)–(10) into the following system of equations:

$$\begin{cases} Y(x_k,0) = g_0(x_k), & \text{for } 0 \leq k \leq N_x, \\ \begin{cases} L_{hyb}^{N_x,N_t} Y_k^{n+1} = \widetilde{F}_k^{n+1}, & \text{for } 1 \leq k \leq N_x-1, \\ Y(0,t_{n+1}) = g_l(t_{n+1}), \quad Y(1,t_{n+1}) = g_r(t_{n+1}), & \text{for } n = 0,\ldots,N_t-1, \end{cases} \end{cases} \tag{11}$$

where the difference operator $L_{hyb}^{N_x,N_t}$ and the term $\widetilde{F}_k^{n+1}$ are respectively defined as

$$L_{hyb}^{N_x,N_t} Y_k^{n+1} = \begin{cases} [\widetilde{\mu}_k^- Y_{k-1}^{n+1} + \widetilde{\mu}_k^c Y_k^{n+1} + \widetilde{\mu}_k^+ Y_{k+1}^{n+1}] + [\widetilde{\lambda}_k^- Y_{k-1}^n + \widetilde{\lambda}_k^c Y_k^n + \widetilde{\lambda}_k^+ Y_{k+1}^n], \\ \qquad \text{for } k = N_x/2, \\ L_\varepsilon^{N_x,N_t} Y_k^{n+1}, \qquad \text{for } k \neq N_x/2, \end{cases} \tag{12}$$

and

$$\widetilde{F}_k^{n+1} = \begin{cases} \widetilde{\gamma}_k^- f_{k-1}^{n+1} + \widetilde{\gamma}_k^c f_k^{n+1} + \widetilde{\gamma}_k^+ f_{k+1}^{n+1}, & \text{for } k = N_x/2, \\ F_k^{n+1}, & \text{for } k \neq N_x/2. \end{cases} \tag{13}$$

Here, one can derive the coefficients $\widetilde{\mu}_k^-, \widetilde{\mu}_k^c, \widetilde{\mu}_k^+; \widetilde{\lambda}_k^-, \widetilde{\lambda}_k^c, \widetilde{\lambda}_k^+; \widetilde{\gamma}_k^-, \widetilde{\gamma}_k^c, \widetilde{\gamma}_k^+$, for the scheme (5)–(7). Now, we set

$$-L_{hyb}^{N_x,N_t} Y_k^{n+1} = \left[A_{k,k-1} Y_{k-1}^{n+1} + A_{k,k} Y_k^{n+1} + A_{k,k+1} Y_{k+1}^{n+1} \right] - \left[B_{k,k-1} Y_{k-1}^n + B_{k,k} Y_k^n + B_{k,k+1} Y_{k+1}^n \right],$$

where for $k \neq N_x/2$,

$$
\begin{cases}
A_{k,k} = -\mu_k^c, \ A_{k,k+1} = -\mu_k^+, \ A_{k,k-1} = -\mu_k^-, \\
B_{k,k} = \lambda_k^c, \ B_{k,k+1} = \lambda_k^+, \ B_{k,k-1} = \lambda_k^-,
\end{cases}
$$

and for $k = N_x/2$,

$$
\begin{cases}
A_{k,k} = -\widetilde{\mu}_k^c, \ A_{k,k+1} = -\widetilde{\mu}_k^+, \ A_{k,k-1} = -\widetilde{\mu}_k^-, \\
B_{k,k} = \widetilde{\lambda}_k^c, \ B_{k,k+1} = \widetilde{\lambda}_k^+, \ B_{k,k-1} = \widetilde{\lambda}_k^-.
\end{cases}
$$

It is obvious that the matrix $B = (B_{k,j}) \geq 0$. Now, by considering the case $\varepsilon \leq 2\|\mathsf{a}\|N_x^{-1}$, one can derive that

$$
\begin{cases}
A_{N_x/2,N_x/2} = -\left[\dfrac{2\varepsilon - \mathsf{a}_{N_x/2+1}h_2}{2h_2(2\varepsilon + \mathsf{a}_{N_x/2+1}h_2)} - \dfrac{3}{2h_2} - \dfrac{3}{2H_1} - \dfrac{H_1}{2\varepsilon}\left(\dfrac{1}{2\Delta t} + \dfrac{\mathsf{b}_{(N_x/2-1)+\frac{1}{2}}}{2} - \dfrac{\mathsf{a}_{(N_x/2-1)+\frac{1}{2}}}{H_1} - \dfrac{\varepsilon}{H_1^2} \right) \right] > 0, \\[3mm]
A_{N_x/2,N_x/2+1} = -\dfrac{1}{2h_2}\left[4 - \dfrac{4\varepsilon + 2h_2^2(\mathsf{b}_{N_x/2+1} + \frac{1}{\Delta t})}{2\varepsilon + \mathsf{a}_{N_x/2+1}h_2} \right] \leq 0,
\end{cases}
$$

under the assumptions $N_x/\ln N_x > 2\varsigma_0\|\mathsf{a}\|$ and $\left(\frac{1}{\Delta t} + \|\mathsf{b}\| \right) \leq \alpha N_x/2$. Also, we have

$$
A_{N_x/2,N_x/2-1} = -\dfrac{1}{2H_1}\left[4 - \dfrac{2\varepsilon + \mathsf{a}_{(N_x/2-1)+\frac{1}{2}}H_1 + H_1^2(\mathsf{b}_{(N_x/2-1)+\frac{1}{2}} + \frac{1}{\Delta t})/2}{\varepsilon} \right],
$$

$$
= \dfrac{1}{2H_1}\left[-4 + \dfrac{2\varepsilon + \mathsf{a}_{(N_x/2-1)+\frac{1}{2}}H_1 + H_1^2(\mathsf{b}_{(N_x/2-1)+\frac{1}{2}} + \frac{1}{\Delta t})/2}{\varepsilon} \right],
$$

$$
\leq \dfrac{1}{2H_1}\left[-2 + \dfrac{\|\mathsf{a}\|H_1}{\varepsilon} + \dfrac{H_1^2(\|\mathsf{b}\| + \frac{1}{\Delta t})}{2\varepsilon} \right].
$$

Now, using $H_1 \leq 4N_x^{-1}$ and $\left(\frac{1}{\Delta t} + \|\mathsf{b}\| \right) \leq \alpha N_x/2$, we have

$$
A_{N_x/2,N_x/2-1} \leq \dfrac{1}{\varepsilon H_1}\left[-\varepsilon + 4\|\mathsf{a}\|N_x^{-1} \right] \nleq 0,
$$

since $\varepsilon \leq 2\|\mathsf{a}\|N_x^{-1} \leq 4\|\mathsf{a}\|N_x^{-1}$. This shows that the matrix $A := (A_{k,j})$ does not satisfy the M-matrix criterion; and hence according to [Lemma 3.12, Part II] given in the book of Roos et al. [10], the discrete operator $L_{hyb}^{N_x,N_t}$ does not satisfy the discrete maximum principle, i.e., if the conditions $Y_k^n \leq 0$ on $\partial \mathsf{Q}^{N_x,N_t}$, and $L_{hyb}^{N_x,N_t} Y_k^n \geq 0$ in Q^{N_x,N_t} are satisfied by a mesh function Y_k^n, it implies that $Y_k^n \leq 0$ at each point $(x_k, t_n) \in \overline{\mathsf{Q}}^{N_x,N_t}$, where $\mathsf{Q}^{N_x,N_t} = \overline{\mathsf{Q}}^{N_x,N_t} \cap \mathsf{Q}$ and $\partial \mathsf{Q}^{N_x,N_t} = \overline{\mathsf{Q}}^{N_x,N_t} \setminus \mathsf{Q}^{N_x,N_t}$.

3 Numerical Experiment

Here, we present the numerical results computed using the newly proposed method (5)–(8) and also compare the numerical results of the proposed method with the classical implicit upwind scheme (14)–(16) .

3.1 The Classical Implicit Upwind Scheme

We introduce the classical implicit upwind scheme for the problem (1)–(3), which takes the following form on the mesh $\overline{Q}^{N_x,N_t}$:

$$
\begin{cases}
Y(x_k,0) = g_0(x_k), \quad \text{for } k = 0,\ldots,N_x, \\
L_{up}^{N_x,N_t} Y_k^{n+1} = f_k^{n+1}, \quad \text{for } k = 1,\ldots,N_x/2-1, \text{ and } k = N_x/2+1,\ldots N_x-1, \\
D_x^+ Y_k^{n+1} - D_x^- Y_k^{n+1} = 0, \text{ for } k = N_x/2, \\
Y(0,t_{n+1}) = g_l(t_{n+1}), \ Y(1,t_{n+1}) = g_r(t_{n+1}), \quad \text{for } n = 0,\ldots,N_t-1,
\end{cases}
\tag{14}
$$

where the difference operator $L_{up}^{N_x,N_t}$ is defined as

$$
L_{up}^{N_x,N_t} Y_k^{n+1} = \varepsilon D_x^+ D_x^- Y_k^{n+1} + a_k D_x^+ Y_k^{n+1} - b_k Y_k^{n+1} - D_t^- Y_k^{n+1},
\tag{15}
$$

and

$$
D_x^+ Y_{N_x/2}^{n+1} = (Y_{N_x/2+1}^{n+1} - Y_{N_x/2}^{n+1})/h_2, \quad D_x^- Y_{N_x/2}^{n+1} = (Y_{N_x/2}^{n+1} - Y_{N_x/2-1}^{n+1})/H_1.
\tag{16}
$$

3.2 Numerical Results

We carry out numerical experiments for the following test example . In all the numerical experiments, we choose the constant $\varsigma_0 = 2.2$.

Example 1. *Consider the parabolic IBVP of the form* (1)–(3), *where* $a(x) = 1 + x(1 - x)$, $b(x) = 1 + x$ *and the term* $f(x,t)$ *is given by*

$$f(x,t) = -9, \quad for \ (x,t) \in (0,1/2) \times (0,1], \quad f(x,t) = 9(x-1)^2, \quad for \ (x,t) \in (1/2,1) \times (0,1].$$

We set $g_0(x) = 0, \quad for \ x \in [0,1] \ and \ g_l(t) = -1, \ g_r(t) = 0, \quad for \ t \in [0,1].$

In Fig. 1, we plot numerical solutions with $N_x = 128$ and $\Delta t = 1.6/N_x$ using the proposed numerical method for $\varepsilon = 2^{-6}$ and $\varepsilon = 2^{-14}$. Since, the exact solution of Example 1 is not known, we use the following technique in order to demonstrate the ε-uniform convergence and the accuracy of the proposed method.

For each ε, we compute the maximum point-wise errors by

$$
\tilde{e}_\varepsilon^{N_x,N_t} = \max_{0 \le k \le N_x, \, n=N_t} \left| Y^{N_x,N_t}(x_k,t_n) - \hat{Y}^{2N_x,2N_t}(x_k,t_n) \right|,
$$

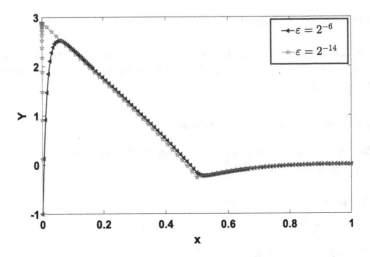

Fig. 1. Numerical solutions obtained using the proposed method.

and the corresponding order of convergence by

$$\widehat{r}_\varepsilon^{N_x,N_t} = \log_2 \left(\frac{\widehat{e}_\varepsilon^{N_x,N_t}}{\widehat{e}_\varepsilon^{2N_x,2N_t}} \right).$$

Here, $Y^{N_x,N_t}(x_k,t_n)$ and $\widehat{Y}^{2N_x,2N_t}(x_k,t_n)$, respectively denote the numerical solutions computed on $\overline{\mathbb{Q}}^{N_x,N_t}$ and $\widehat{\mathbb{Q}}^{2N_x,2N_t}$, where $\widehat{\mathbb{Q}}^{2N_x,2N_t}$ is the fine mesh with $2N_x$ mesh-intervals in the spatial direction and $2N_t$ mesh-intervals in the temporal direction such that the transition parameters ς_1, ς_2 remain unaltered after doubling the mesh-intervals.

Further, for each N_x and N_t, we calculate the ε-uniform maximum point-wise errors by $\widehat{e}^{N_x,N_t} = \max_\varepsilon \widehat{e}_\varepsilon^{N_x,N_t}$ and the corresponding order of convergence by $\widehat{r}^{N_x,N_t} = \log_2 \left(\frac{\widehat{e}^{N_x,N_t}}{\widehat{e}^{2N_x,2N_t}} \right).$

We display the computed ε-uniform maximum point-wise errors and corresponding order of convergence for Example 1 using the proposed method and the implicit upwind scheme, respectively in Tables 1 and 2. Moreover, those ε-uniform errors are depicted in Fig. 2. It indicates that both the proposed numerical method as well as the implicit upwind scheme are ε-uniformly convergent.

Table 1. ε-uniform errors and order of convergence for Example 1 computed using the proposed method, using $\Delta t = 1.6/N_x$.

$\varepsilon \in \{2^0, 2^{-2}, \ldots, 2^{-20}\}$	Number of mesh intervals N_x/time step size Δt				
	$64/\frac{1}{40}$	$128/\frac{1}{80}$	$256/\frac{1}{160}$	$512/\frac{1}{320}$	$1024/\frac{1}{640}$
$\widehat{e}^{N_x,N_t}$	3.4355e−02	1.1767e−02	3.7684e−03	1.2055e−03	4.4002e−04
$\widehat{r}^{N_x,N_t}$	1.5457	1.6428	1.6443	1.4540	

Table 2. ε-uniform errors and order of convergence for Example 1 computed using the implicit upwind scheme, using $\Delta t = 1.6/N_x$.

$\varepsilon \in \{2^0, 2^{-2}, \ldots, 2^{-20}\}$	Number of mesh intervals N_x/time step size Δt				
	$64/\frac{1}{40}$	$128/\frac{1}{80}$	$256/\frac{1}{160}$	$512/\frac{1}{320}$	$1024/\frac{1}{640}$
$\widehat{e}^{N_x,N_t}$	1.5889e−01	1.0211e−01	6.3016e−02	3.7045e−02	2.1099e−02
$\widehat{r}^{N_x,N_t}$	0.63794	0.69634	0.76644	0.81212	

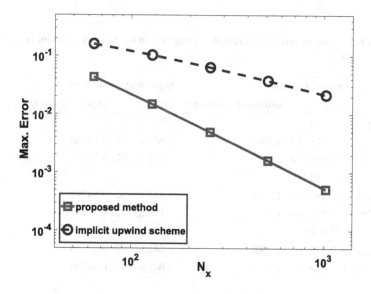

Fig. 2. Loglog plot of ε-uniform errors

Apart from the above, we compare region-wise errors and order of convergence of the proposed method and the implicit upwind scheme in Tables 3, 4, 5 and 6. It confirms that proposed method is almost second-order accurate in space, whereas the implicit upwind scheme is almost first-order accurate is space.

Finally, to demonstrate the computational efficiency, we compare the computational time of both the methods in Table 7. It shows that the proposed method takes approximately same compuational time in comparison with the implicit upwind scheme to produce more accuarte results.

Table 3. Comparison of errors for Example 1 computed in the boundary layer region, *i.e.*, in $[0, \varsigma_1)$, using $\Delta t = 1/N_x^2$.

N	Proposed method		Implicit upwind scheme	
	Error	Order of convergence	Error	Order of convergence
$\varepsilon = 2^{-4} \approx 10^{-1}$				
128	2.0966e−03	2.1601	3.2950e−02	0.93205
256	4.6909e−04	2.4473	1.7269e−02	0.96712
$\varepsilon = 2^{-6} \approx 10^{-2}$				
128	1.1626e−02	1.6466	9.3721e−02	0.69956
256	3.7131e−03	1.6542	5.7709e−02	0.76086
$\varepsilon = 2^{-14} \approx 10^{-4}$				
128	1.0142e−02	1.6307	1.0225e−01	0.69814
256	3.2752e−03	1.6642	6.3021e−02	0.76695

Table 4. Comparison of errors for Example 1 computed in the left outer region, *i.e.*, in $[\varsigma_1, \xi)$, using $\Delta t = 1/N_x^2$.

N	Proposed method		Implicit upwind scheme	
	Error	Order of convergence	Error	Order of convergence
$\varepsilon = 2^{-4} \approx 10^{-1}$				
128	1.1314e−03	2.7203	2.5086e−02	0.97540
256	1.7169e−04	2.6860	1.2759e−02	0.98778
$\varepsilon = 2^{-6} \approx 10^{-2}$				
128	2.0038e−03	1.3156	1.8673e−02	0.74241
256	8.0507e−04	1.5815	1.1162e−02	0.79448
$\varepsilon = 2^{-14} \approx 10^{-4}$				
128	5.4657e−05	2.0008	8.0483e−03	0.99238
256	1.3657e−05	1.9879	4.0455e−03	0.99738

Table 5. Comparison of errors for Example 1 computed in the interior layer region, *i.e.*, in $[\xi, \xi + \varsigma_2)$, using $\Delta t = 1/N_x^2$.

N	Proposed method		Implicit upwind scheme	
	Error	Order of convergence	Error	Order of convergence
$\varepsilon = 2^{-4} \approx 10^{-1}$				
128	1.1423e−03	2.7266	2.5396e−02	0.98431
256	1.7258e−04	2.6490	1.2837e−02	0.99219
$\varepsilon = 2^{-6} \approx 10^{-2}$				
128	2.0049e−03	1.3074	1.9109e−02	0.76166
256	8.1008e−04	1.5865	1.1271e−02	0.80253
$\varepsilon = 2^{-14} \approx 10^{-4}$				
128	6.7940e−05	1.9834	4.6467e−03	1.0023
256	1.7182e−05	1.9645	2.3197e−03	1.0015

Table 6. Comparison of errors for Example 1 computed in the right outer region, *i.e.*, in $[\xi + \varsigma_2, 1]$, using $\Delta t = 1/N_x^2$.

N	Proposed method		Implicit upwind scheme	
	Error	Order of convergence	Error	Order of convergence
$\varepsilon = 2^{-4} \approx 10^{-1}$				
128	8.4070e−06	2.0983	4.3284e−04	0.88021
256	1.9634e−06	2.1945	2.3516e−04	0.94147
$\varepsilon = 2^{-6} \approx 10^{-2}$				
128	2.2584e−04	4.8935	1.6133e−03	1.2596
256	7.5984e−06	2.3127	6.7384e−04	1.2808
$\varepsilon = 2^{-14} \approx 10^{-4}$				
128	6.7601e−05	1.9835	4.6468e−03	1.0023
256	1.7094e−05	1.9668	2.3197e−03	1.0015

Table 7. Comparison of computational time (in seconds) for Example 1 taking $\Delta t = 1/N_x$.

N	Implicit upwind scheme		Proposed method	
	Time	Error	Time	Error
$\varepsilon = 2^{-6} \approx 10^{-2}$				
64	0.028958	1.4376e−01	0.065921	3.4440e−02
128	0.237022	9.3638e−02	0.296579	1.1808e−02
256	1.629150	5.7675e−02	1.712070	3.7911e−03
512	12.395647	3.4041e−02	12.624179	1.2180e−03
1024	93.223280	1.9445e−02	95.346525	3.8100e−04
$\varepsilon = 2^{-14} \approx 10^{-4}$				
64	0.038513	1.5909e−01	0.079077	3.0435e−02
128	0.226236	1.0216e−01	0.451556	1.0151e−02
256	1.771846	6.3008e−02	2.142324	3.2759e−03
512	12.419061	3.7034e−02	13.709683	1.0335e−03
1024	94.296317	2.1092e-02	98.567949	3.1886e-04

4 Conclusions

In this article, we propose an efficient numerical method for solving a class of singularly perturbed parabolic IBVPs with non-smooth data utilizing a layer-resolving piecewise-uniform Shishkin mesh. The current investigation reveals the following important features of the newly proposed method and also addresses the future challenges.

It is shown that the discrete maximum principle can not be established by converting the system (8) associated with the proposed method into a new system (11). However, it has been computationally demonstrated that the proposed method is uniformly convergent, and is at least almost second-order spatially accurate throughout the spatial domain $\overline{I} = [0, 1]$, regardless of the larger and the smaller values of the parameter ε. It is also observed that the proposed method exhibits notable improvement over the implicit upwind scheme. Moreover, the comparison of the computational time shows that the present method is compuatationally more efficent than the implicit upwind scheme.

Henceforth, by considering the performance of the newly proposed method from the computational perspective, we are currently working to resolve the theoretical difficulty in establishing the discrete maximum principle, and also interested to obtain the parameter-uniform error estimate of the proposed method.

Acknowledgements. The authors would like to express their sincere thanks to the reviewers for their valuable comments and suggestions; and also gratefully acknowledge the financial support received from Indian Institute of Space Science and Technology (IIST), Thiruvananthapuram.

References

1. Cen, Z.: A hybrid difference scheme for a singularly perturbed convection-diffusion problem with discontinuous convection coefficient. Appl. Math. Comput. **169**, 689–699 (2005)

2. Farrell, P.A., Hegarty, A.F., Miller, J.J.H., O'Riordan, E., Shishkin, G.I.: Global maximum norm parameter-uniform numerical method for a singularly perturbed convection-diffusion problem with discontinuous convection coefficient. Math. Comput. Model. **40**, 1375–1392 (2004)
3. Farrell, P., Hegarty, A., Miller, J., O'Riordan, E., Shishkin, G.: Singularly perturbed convection diffusion problems with boundary and weak interior layers. J. Comput. Appl. Math. **166**(1), 133–151 (2004)
4. Ladyzenskaja, O.A., Solonnikov, V.A., Ural'ceva, N.N.: Linear and Quasi-Linear Equations of Parabolic Type, Translations of Mathematical Monographs, vol. 23. American Mathematical Society, Providence (1968)
5. Markowich, P., Ringhofer, C., Schmeiser, C.: Semiconductor Equations. Springer-Verlag, New York (2012)
6. Mukherjee, K.: Parameter-uniform improved hybrid numerical scheme for singularly perturbed problems with interior layers. Math. Model. Anal. **23**(2), 167–189 (2018)
7. Mukherjee, K., Natesan, S.: An efficient numerical scheme for singularly perturbed parabolic problems with interior layers. Neural Parallel Sci. Comput. **16**, 405–418 (2008)
8. Mukherjee, K., Natesan, S.: ε-Uniform error estimate of hybrid numerical scheme for singularly perturbed parabolic problems with interior layers. Numer. Algor. **58**(1), 103–141 (2011)
9. O'Riordan, E., Shishkin, G.: Singularly perturbed parabolic problems with non-smooth data. J. Comput. Appl. Math. **166**, 233–245 (2004)
10. Roos, H.G., Stynes, M., Tobiska, L.: Robust Numerical Methods for Singularly Perturbed Differential Equations, 2nd edn. Springer, Berlin (2008)
11. Shanthi, V., Ramanujam, N., Natesan, S.: Fitted mesh method for singularly perturbed reaction convection-diffusion problems with boundary and interior layers. J. Appl. Math. Comput. **22**(1–2), 49–65 (2006)
12. Yadav, N.S., Mukherjee, K.: Uniformly convergent new hybrid numerical method for singularly perturbed parabolic problems with interior layers. Int. J. Appl. Comput. Math. **6**, 53 (2020)

Soft Computing

Scheduling of Jobs on Computational Grids by Fuzzy Particle Swarm Optimization Algorithm Using Trapezoidal and Pentagonal Fuzzy Numbers

Debashis Dutta and Subhabrata Rath[✉]

Department of Mathematics, National Institute of Technology, Warangal 506004,
Telangana, India
ddutta@nitw.ac.in

Abstract. Grid computing is a computational technique used to meet growing computational demands. Scheduling of independent tasks in Computational Grids commonly arises in many Grid-enabled large scale applications. This paper introduces fuzzy particle swarm optimization (FPSO) using trapezoidal and pentagonal fuzzy numbers for job scheduling problems. It is to be noted that the Job scheduling on computational grid is an N-P Complete problem, this requires a lot of attention because of its practical importance and complexity. The position and velocity of particles in conventional particle swarm optimization (PSO) is extended from real vectors to fuzzy matrices where trapezoidal and pentagonal fuzzy numbers play a key role. We evaluated the performances of FPSO with trapezoidal fuzzy numbers and FPSO with pentagonal fuzzy numbers. Results are compared for scheduling.

Keywords: Fuzzy particle swarm optimization · Trapezoidal fuzzy number · Pentagonal fuzzy number

1 Introduction

A computational grid is a large scale and heterogeneous collection of autonomous systems [1]. The sharing of computational job among the grid is one of the major applications of the grids. It's resources may be distributed among different owners, who may have some constraints and various access policy. Several meta-heuristic methods are developed for minimizing the average completion time of jobs on each Grid node through optimal Job allocation [2]. A more complete analysis of the scheduling on the grid was provided by Dong and Akl [3], which is known as a N-P complete problem [4]. Every grid node has a processing speed of its own and requirements of its own. So here we are using fuzzy PSO a job

Supported by organization NIT W.

A. Awasthi et al. (Eds.): CSMCS 2020, CCIS 1345, pp. 175–185, 2021.
https://doi.org/10.1007/978-981-16-4772-7_13

scheduling problem on computational grids. Then we will compare the results obtained using trapezoidal fuzzy number with the result obtained using pentagonal fuzzy number. The objective in this problem is to minimize the time complexity and efficient use of grid nodes. PSO is a particle swarm optimization method developed by Kennedy and Eberhart in [5]. The success in a PSO problem depends on the mapping between PSO particle and possible solution.

The rest of the paper is organized as follows. In Sect. 2, it explains about the problem we have tackled in this paper. It explains some basic concepts required for proper understanding of this paper and our objective in this paper. In Sect. 3, it explains about the process used in this paper. In Sect. 3.1, it explains about the fuzziness used for solving this paper. In Sect. 3.1, it explains the basic PSO method which will extended fuzzy PSO. In Sect. 3.3, it explains the fuzzy PSO algorithm that is used to solve this problem. In Sect. 4, Numerical experiment is given. In Sect. 5, the conclusion of the above approach is given.

2 Problem Formulation

Here in this problem on computational Grids, there is generally a framework focusing on the interaction between grid information server, resource broker of the grid and the manager of the domain resource [7]. In this section the problem of this paper is explained. In the computational grid environment, scheduling of jobs on the grids using fuzzy PSO is explained. For our proper understanding some important terms and concepts are defined. They are as follows-

Scheduling Problem. Schedule is a function from jobs to the specific intervals of time of the grid node. Scheduling problem is defined as the jobs allocated to the machines with an optimal criteria. In this paper, scheduling problem is defined as allocation of jobs to specific computational grid with an optimal criteria. Here optimal criteria is the max number of iterations allowed.

Grid Nodes. A gird node is a computational resource whose capacity is limited. It can be a computer lab, workstation, personal computer or a collection of computers at a specific location. The computational capacity of the grid node depends on amount of memory, number of the Central processing unit, basic storage space and other types of specifications. Every Grid node has a processing speed of its own which is expressed as number of cycles per unit time.

Jobs. Job is a collection of the operations or a single operation allocated to the computational grid.

Now we are going to explain the concerned problem. Now J(j) means Job on machine j and G(i) means Grid at node i. Now let us consider jobs J(j), $j \in (1, 2, \ldots, b)$ that are independent on Grid nodes G(i), $i \in (1, 2, \ldots, a)$. The speed of each grid node is expressed as number of CPUT. The objective of this problem is to minimize the time complexity and efficient use of grid nodes.

Now we define d(i,j) as the completion time, In other words time taken by grid node G(i) to finish the job J(j). The time taken by the grid node to execute all the jobs allocated to that grid node only is represented by $(\sum d(i))$. Now max $(\sum d(i))$ is called makespan. $(\sum_{i=1}^{a}(\sum d(i)))$ is called as the flow time. These concepts are used while applying fuzzy PSO algorithm.

The Objective in this paper is to minimize the makespan value. We have to optimize a job scheduling that minimizes the makespan value. That is to minimize the maximum time taken by all Grids to complete all the jobs assigned to them. And then to see the effects of trapezoidal fuzzy number and pentagonal fuzzy number on it.

3   Fuzzy PSO

In this section fuzzy particle swarm optimization is explained, that has been used in solving the above job scheduling problem in a computational grid atmosphere. This paper compares fuzzy PSO with trapezoidal fuzzy number and pentagonal fuzzy number and then compares them.

In Sect. 3.1 fuzziness that has been used in this paper is explained. In Sect. 3.2 particle swarm optimization is explained. In Sect. 3.3 fuzzy particle swarm optimization that has been used in this paper has been explained in detail.

3.1   Fuzziness

Fuzzy Number. A fuzzy set is called a fuzzy number when the following properties are satisfied

1) A must be a normal fuzzy set.
2) All alpha cut of A must be in a closed interval.
3) The support of A must be bounded.
   If it satisfies all the conditions then it is called a fuzzy number.

Trapezoidal Fuzzy Number. A Trapezoidal fuzzy number denoted by A is defined as (a, b, c, u) where the membership function is given by

$$(x) = \begin{cases} 0, & x \leq a \\ (x-a)/(b-a), & a \leq x \leq b \\ 1, & b \leq x \leq c \\ (u-x)/(u-c), & c \leq x \leq u \\ 0, & x \geq u \end{cases}$$

Pentagonal Fuzzy Number. A fuzzy number is called pentagonal fuzzy number if the following conditions are satisfied-

1) Let the pentagonal fuzzy number denoted by (a, b, c, d, e) with membership function (x)

2) (x) must be a continuous membership function whose interval is [0, 1].
3) (x) must be a strictly non-decreasing function that is continues on the intervals [a, b] and [b, c].
4) (x) must be a strictly non-increasing function that is continues on the intervals
   [c, d] and [d, e].

3.2   Particle Swarm Optimization (PSO)

To successfully apply PSO, one of the factor is to find the map between solution
of the problem and PSO particle. PSO is originally designed by Kennedy and
Eberhart. It is naturally inspired by fish schooling and bird flocking [8,9]. In
PSO each particle represents a possible solution. The main strength of this algorithm is it has faster convergence when compared with other global optimization
algorithms [10–14]. Each particle moves with some velocity, keeping it's personal
best position and global best position. Mathematically the particle is guided in
the domain by the following formula -

$$x(t+1) = q * x(t) + (q1 * a1) * (y'(t) - y(t)) + (q2 * a2) * (y''(t) - y(t)) \quad (1)$$

$$y(t+1) = y(t) + x(t+1) \quad (2)$$

Here $x(t+1)$ represents velocity of a particle at $(t+1)$ iteration. $x(t)$ represents velocity of a particle at (t) iteration. Here a1 and a2 are random numbers
taken initially. $y(t)$ represents the position of the particle at (t) iteration. $y'(t)$
represents the personal best position of a particle at (t) iteration. $y''(t)$ represents the global best position among all the particles at (t) iteration. Also $y'(t)$
and $y''(t)$ are known as personal best of a particle and global best respectively.

Here q1 and q2 are constants that are positive, these are called acceleration
coefficients. q represents the inertia weight. Now, a1 and a2 are two random
numbers in the range [0, 1]. When inertia weight is very large it implies a global
exploration and small inertia weight implies local exploitation. q plays an important role, as it affects the convergence behaviour of PSO.

3.3   Fuzzy PSO Algorithm

In this section, fuzzy PSO algorithm is explained in detail. Here in the scheduling
of jobs on the computational Grids environment using PSO, the position and
velocities of particles are taken in the form of fuzzy matrices [6]. In this section
it is explained how fuzzy PSO is used for solving problems on scheduling of jobs
on the computational grid nodes. Then their results are compared for fuzzy PSO
with trapezoidal fuzzy number and fuzzy PSO with pentagonal fuzzy number.
To successfully apply PSO, one of the factor is to find the map between problem
solution and PSO particle. The performance and feasibility are directly affected
by it. Suppose G = (G(1), G(2), ...,G(a)) , J = (J(1), J(2), ..., J(b)) are the
grid nodes and jobs respectively. The number of Grids and Jobs are a and b
respectively. Let the position of the particle is defined as

$$Y = [(y(1, 1) \quad y(1, b)$$
$$y(a, 1) \quad y(a, b))]$$

The elements of Y must satisfy the following criteria-

$$y(i,j) \in [0, 1], \ i \in (1, 2, ..., a) \text{ and } j \in (1, 2, ..., b)$$
$$\sum_{i=1}^{a}(y(i,j)) = 1, \ i \in (1,2,...,a) \text{ and } j \in (1, 2, ..., b)$$

Similarly the velocity of the particle is defined as

$$x = [(x(1, 1) \quad x(1, b)$$
$$x(a, 1) \quad x(a, b))]$$

Now in order to update the velocities and positions of the particles, we have to follow the following mathematical expression-

$$x(t+1) = q * x(t) + (q1 * a1) * (y'(t) - y(t)) + (q2 * a2) * (y''(t) - y(t)) \quad (3)$$

$$y(t+1) = y(t) + x(t+1) \quad (4)$$

Here x(t + 1) represents velocity of a particle at (t + 1) iteration. x(t) represents velocity of a particle at (t) iteration. Here a1 and a2 are random numbers taken initially. y(t) represents the position of the particle at (t) iteration. y'(t) represents the personal best position of a particle at (t) iteration. y''(t) represents the global best position among all the particles at (t) iteration. Also y'(t) and y''(t) are known as personal best of a particle and global best respectively. Here q1 and q2 are constants that are positive, these are called acceleration coefficients. q represents the inertia weight. Now, a1 and a2 are two random numbers in the range [0, 1].

Before going into detail of the fuzzy PSO algorithm, first let us see some of the notations required on the way, they are as follows-

$\alpha1 =$ Collection of all the jobs to be processed.
$\alpha2 =$ Collection of all the jobs that are being scheduled
$\alpha3 =$ Collection of all the jobs after job allocation is already completed.
$\alpha4 =$ Collection of all the available grid nodes.
$\alpha5 =$ Collection of all the grids nodes that has already been allocated to the jobs.
$\alpha6 =$ Collection of all the grid nodes that are available or free.

The algorithm of fuzzy PSO is as follows
STEP 1 When the nodes are active and no new jobs are available, then we have to wait for the jobs that are new or update $\alpha4$ and $\alpha1$.
STEP 2
At t = 0, If $\alpha4 = 0$, wait for new grids to be available. If $\alpha1 > 0$, update JL2. If $\alpha2 < \alpha4$, then jobs are allocated on the principle called first come first serve basis. If $\alpha2 > \alpha4$, job allocation is done by the following -
STEP 3

3.0) Now we have to initialize all the parameters of the particle swarm. The size of the particle swarm (N) depends on the experiment and its value is given before the start of the algorithm. The values of the parameters are as taken as q = 0.8, q1 = 2, q2 = 1.3.

3.1) Now we have to initialize the position for each particle. So we have taken random matrices which will be treated as position of the particles. Then the matrices are normalized.

3.2.0) t = t + 1 (Here we will start the iteration process from t = 1 to the maximum iteration, which can be changed in the programming code depending on the requirement of the coder)

3.2.1) Then the makespan value is calculated for each particle.

3.2.2) The latest best solution is calculated as follows-
$$y'' = argmin(f(y''(t-1)), f(y(1)(t)), f(y(2)(t)),...f(y(b)(t)))$$

3.2.3) For each particle personal best solution is computed as follows
$$y'(t) = argmin(f(y'(t-1)), f(y(t)))$$

3.2.4.a) Take random velocity as a trapezoidal matrix, i.e. every element of the matrix is a trapezoidal fuzzy number for the first Case.

3.2.4.b) Take random velocity as a pentagonal matrix, i.e. every element of the matrix is a pentagonal fuzzy number for the second Case.

3.2.5) Now update each particle using Eq. (1) and (2).

3.2.6) Now for each particle the position matrix is normalized.

3.3) The iteration process is continued until the optimality criteria is fulfilled.

STEP 4

Repeat the process as long as the grid is active.

4 Experiment

Now we have taken some parameters required to solve the problem. They are Inertia weight (q) = 0.8. Acceleration coefficient q1 and q2 are as follows 2 and 1.3 respectively. The two random numbers are generated automatically.

4.1 Experiment 1

Number of Grid Nodes = 2 Number of Jobs = 3 Now we will explain this by taking a small example. In this example we are taking two Grid nodes and three Jobs.

Total number of particles (N) = 3. Inertia weight (q) = 0.8. Acceleration coefficient q1 and q2 are as follows 2 and 1.3 respectively. Two random numbers are generated automatically. The speed of two grid nodes are 4 and 4.1 respectively. The time required for each job are as follows 1119, 1112 and 1811 respectively.

For first case we have taken velocity matrix with each element as trapezoidal fuzzy number. Here we have observed that job1 is scheduled on grid 2, job2 is scheduled on grid 1 and job3 is scheduled on grid 2. With increase in number of iterations makespan value decreases and after some iterations it remains more or less constant.

For Second case we have taken velocity matrix with each element as pentagonal fuzzy number. Here we have observed that job1 is scheduled on grid 2, job2 is scheduled on grid 2 and job3 is scheduled on grid 1. With increase in number of iterations makespan value decreases and after some iterations it remains more or less constant.

4.2   Experiment 2

Number of Grid Nodes = 3 Number of Jobs = 7 Here '1' represents job assigned to the Grid and '0' represents no job assigned to the Grid. Total number of particles (N) = 20.

Optimal Schedule with Trapezoidal Fuzzy Number

Table 1. Optimal schedule with trapezoidal fuzzy number.

	J1	J2	J3	J4	J5	J6	J7
G1	0	0	0	0	0	1	1
G2	0	0	1	0	0	0	0
G3	1	1	0	1	1	0	0

Here the Grid Speed are as following – 17, 34, 13 and the time required for each job is as following-45, 103, 80, 62, 91, 113, 88 respectively. Here '1' represents job assigned to the Grid and '0' represents no job assigned to the Grid.

The above Table 1 is the Optimal Schedule. Here job 1 is scheduled on grid 3, job 2 is scheduled on grid 3, job 3 is scheduled on grid 2, job 4 is scheduled on grid 3, job 5 is scheduled on grid 3, job 6 is scheduled on grid 1, job 7 is scheduled on grid 1. With increase in number of iterations makespan value decreases and after some iterations it remains more or less constant.

Optimal Schedule with Pentagonal Fuzzy Number

Table 2. Optimal schedule with Pentagonal fuzzy number.

	J1	J2	J3	J4	J5	J6	J7
G1	0	1	0	1	0	0	1
G2	0	0	0	0	1	1	0
G3	1	0	1	0	0	0	0

Here the Grid Speed are as following – 28, 9, 23 and the time required for each job is as following-124, 71, 132, 99, 83, 78, 64 respectively. Here '1' represents job assigned to the Grid and '0' represents no job assigned to the Grid. The above Table 2 is the Optimal Schedule. Here job 1 is scheduled on grid 3, job 2 is scheduled on grid 1, job 3 is scheduled on grid 3, job 4 is scheduled on grid 1, job 5 is scheduled on grid 2, job 6 is scheduled on grid 2, job is scheduled on grid 1. With increase in number of iterations makespan value decreases and after some iterations it remains more or less constant.

4.3   Experiment 3

Number of Grid Nodes = 4 Number of Jobs = 12 Here '1' represents job assigned to the Grid and '0' represents no job assigned to the Grid. Total number of particles (N) = 20.

Optimal Schedule with Trapezoidal Fuzzy Number

Table 3. Optimal schedule with Trapezoidal fuzzy number.

	J1	J2	J3	J4	J5	J6	J7	J8	J9	J10	J11	J12
G1	0	0	0	0	1	0	0	1	0	0	0	0
G2	0	0	0	0	0	1	0	0	0	0	1	0
G3	0	0	1	1	0	0	0	0	0	1	0	0
G4	1	1	0	0	0	0	1	0	1	0	0	1

Here the Grid Speed are as following – 44, 3, 11, 23 and the time required for each job is as following-144, 119, 68, 51, 9, 112, 77, 30, 65, 26, 113, 56 respectively. Here '1' represents job assigned to the Grid and '0' represents no job assigned to the Grid.

The above Table 3 is the Optimal Schedule. Here job 1 is scheduled on grid 4, job 2 is scheduled on grid 4, job 3 is scheduled on grid 3, job 4 is scheduled on grid 3, job 5 is scheduled on grid 1, job 6 is scheduled on grid 2, job 7 is scheduled on grid 4, job 8 is scheduled on grid 1, job 9 is scheduled on grid 4, job 10 is scheduled on grid 3, job 11 is scheduled on grid 2, job 12 is scheduled on grid 4. With increase in number of iterations makespan value decreases and after some iterations it remains more or less constant.

Optimal Schedule with Pentagonal Fuzzy Number

Here the Grid Speed are as following – 50, 21, 45, 10 and the time required for each job is as following-101, 74, 71, 58, 23, 123, 17, 123, 124, 41, 15, 84 respectively. Here '1' represents job assigned to the Grid and '0' represents no job assigned to the Grid.

The above Table 4 is the Optimal Schedule. Here job 1 is scheduled on grid 3, job 2 is scheduled on grid 4, job 3 is scheduled on grid 4, job 4 is scheduled on grid 3, job 5 is scheduled on grid 4, job 6 is scheduled on grid 1, job 7 is scheduled on grid 3, job 8 is scheduled on grid 1, job 9 is scheduled on grid 2,

Table 4. Optimal schedule with Pentagonal fuzzy number.

	J1	J2	J3	J4	J5	J6	J7	J8	J9	J10	J11	J12
G1	0	0	0	0	0	1	0	1	0	0	0	1
G2	0	0	0	0	0	0	0	0	1	0	1	0
G3	1	0	0	0	0	0	0	0	0	1	0	0
G4	0	1	1	0	1	0	0	0	0	0	0	0

job 10 is scheduled on grid 3, job 11 is scheduled on grid 2, job 12 is scheduled on grid 1. With increase in number of iterations makespan value decreases and after some iterations it remains more or less constant.

4.4 Experiment 4

Number of Grid Nodes = 5 Number of Jobs = 20 Here '1' represents job assigned to the Grid and '0' represents no job assigned to the Grid. Total number of particles (N) = 20.

Optimal Schedule with Trapezoidal Fuzzy Number

Table 5. Optimal schedule with Trapezoidal fuzzy number.

	J1	J2	J3	J4	J5	J6	J7	J8	J9	J10	J11	J12	J13	J14	J15	J16	J17	J18	J19	J20
G1	0	0	0	1	0	0	0	0	0	1	0	0	1	0	0	0	1	0	0	1
G2	0	0	0	0	0	0	0	1	1	0	0	0	0	0	0	0	0	0	0	0
G3	0	1	0	0	0	1	0	0	0	0	1	0	0	1	0	0	0	1	0	0
G4	1	0	0	0	1	0	1	0	1	0	0	0	0	0	0	0	0	0	1	0
G5	0	0	1	0	0	0	0	0	1	0	0	1	0	0	1	1	0	0	0	0

Here the Grid Speed are as following – 4, 34, 47, 24, 19 and the time required for each job is as following-13, 82, 140, 105, 124, 129, 17, 106, 83, 79, 48, 25, 48, 101, 79, 92, 115, 25, 75, 115 respectively. Here '1' represents job assigned to the Grid and '0' represents no job assigned to the Grid.

The above Table 5 is the Optimal Schedule. Here job 1 is scheduled on grid 4, job 2 is scheduled on grid 3, job 3 is scheduled on grid 5, job 4 is scheduled on grid 1, job 5 is scheduled on grid 4, job 6 is scheduled on grid 3, job 7 is scheduled on grid 4, job 8 is scheduled on grid 2, job 9 is scheduled on grid 2, job 10 is scheduled on grid 1, job 11 is scheduled on grid 3, job 12 is scheduled on grid 5, job 13 is scheduled on grid 1, job 14 is scheduled on grid 3, job 15 is scheduled on grid 5, job 16 is scheduled on grid 5, job 17 is scheduled on grid 1, job 18 is scheduled on grid 3, job 19 is scheduled on grid 4, job 20 is scheduled on grid 1. With increase in number of iterations makespan value decreases and after some iterations it remains more or less constant.

Table 6. Optimal schedule with Pentagonal fuzzy number.

	J1	J2	J3	J4	J5	J6	J7	J8	J9	J10	J11	J12	J13	J14	J15	J16	J17	J18	J19	J20
G1	0	1	0	0	0	0	0	0	0	0	0	1	0	0	0	0	0	0	1	0
G2	0	0	1	0	0	0	0	0	0	1	0	0	0	0	0	0	0	0	0	0
G3	0	0	0	0	1	0	1	1	0	0	0	0	0	1	1	0	0	0	0	0
G4	1	0	0	0	0	0	0	0	1	0	1	0	0	0	0	0	1	0	0	0
G5	0	0	0	1	0	1	0	0	0	0	0	0	1	0	0	1	0	1	0	1

Optimal Schedule with Pentagonal Fuzzy Number

Here the Grid Speed are as following – 19, 12, 8, 36, 15 and the time required for each job is as following-86, 16, 142, 55, 96, 145, 81, 5, 51, 13, 119, 61, 20, 107, 100, 42, 75, 112, 71, 134 respectively. Here '1' represents job assigned to the Grid and '0' represents no job assigned to the Grid.

The above Table 6 is the Optimal Schedule. Here job 1 is scheduled on grid 4, job 2 is scheduled on grid 1, job 3 is scheduled on grid 2, job 4 is scheduled on grid 5, job 5 is scheduled on grid 3, job 6 is scheduled on grid 5, job 7 is scheduled on grid 3, job 8 is scheduled on grid 3, job 9 is scheduled on grid 4, job 10 is scheduled on grid 2, job 11 is scheduled on grid 4, job 12 is scheduled on grid 1, job 13 is scheduled on grid 5, job 14 is scheduled on grid 3, job 15 is scheduled on grid 3, job 16 is scheduled on grid 5, job 17 is scheduled on grid 4, job 18 is scheduled on grid 5, job 19 is scheduled on grid 1, job 20 is scheduled on grid 5. With increase in number of iterations makespan value decreases and after some iterations it remains more or less constant.

Then we have taken some more examples with more Grids Nodes and Jobs, i.e. 10 Grid Node and 50 Jobs, 40 Grid Node and 100 Jobs. We are getting similar result.

Here the termination criteria is Maximum iteration. Here the optimal Criteria is the makespan value. We have to optimize a job scheduling that minimizes the makespan value. That is to minimize the maximum time taken by all Grids to complete all the jobs assigned to them. Here we can see the makespan value of fuzzy PSO using trapezoidal fuzzy number and the makespan value of fuzzy PSO using pentagonal fuzzy number remains the same. Here we can also see that the global best position is same for both the approaches the approaches.

5   Conclusion

Here a job scheduling problem on a computational grid is solved by fuzzy particle swarm optimization with trapezoidal fuzzy number and pentagonal fuzzy number. In this paper fuzzy PSO using trapezoidal fuzzy number is calculated and then compared with fuzzy PSO using pentagonal fuzzy number. Here we see that fuzzy PSO with pentagonal fuzzy number gives the same result as compared with fuzzy PSO with trapezoidal fuzzy number. For future work we can take other fuzzy numbers in this process and compare the results.

References

1. Foster, I., Kesselman, C. (eds.): The Grid 2: Blueprint for a New Computing Infrastructure, 2nd edn. Morgan Kaufmann, Boston (2003)
2. Abraham, A., Liu, H., Zhao, M.: Particle swarm scheduling for work-flow applications in distributed computing environments. In: Xhafa, F., Abraham, A. (eds.) Metaheuristics for Scheduling in Industrial and Manufacturing Applications. Studies in Computational Intelligence, vol. 128, pp. 327–342. Springer, Heidelberg (2008). https://doi.org/10.1007/978-3-540-78985-7_13
3. Dong, F., Akl, S.G.: Scheduling algorithms for grid computing: state of the art and open problems. Queen's University, Technical report (2006)
4. Garey, M.R., Johnson, D.S.: Computers and Intractability: A Guide to the Theory Of NP Completeness. Freeman, San Diego (1979)
5. Kennedy, J., Eberhart, R.C.: Particle swarm optimization. In: Proceedings of the IEEE International Conference on Neural Networks, Recovery effects in binary aluminum alloys
6. Liu, H., Abraham, A.: An hybrid fuzzy variable neighborhood particle swarm optimization algorithm for solving quadratic assignment problems. J. Univ. Comput. Sci. **13**, 1309–1331 (2007)
7. Abraham, A., Buyya, R., Nath, B.: Nature's heuristics for scheduling jobs oncomputational Grids. In: Brown, M.P., Austin, K. (eds.) 8th International Conference on Advanced Computing and Communications. Tata McGraw-Hill, India (2000)
8. Kennedy, J., Eberhart, R.: Swarm Intelligence. In: Furnish, M.D. (ed.) AIP Conference Proceedings, San Francisco, CA, Morgan Kaufmann Publishers (2001)
9. Clerc, M.: Particle Swarm Optimization. ISTE Publishing Company, London (2006)
10. Goldberg, D.E.: Genetic Algorithms in Search, Optimization, and Machine Learning. Addison-Wesley Publishing Corporation Inc., Boston (1989)
11. Kirkpatrick, C.S., Gelatt, D., Vecchi, M.P.: Optimization by simulated annealing. Science
12. Parsopoulos, K.E., Vrahatis, M.N.: Recent approaches to global optimization problems through particle swarm optimization. Nat. Comput. **1**, 235–306 (2002)
13. Grosan, C., Abraham, A., Nicoara, M.: Search optimization using hybrid particle sub-Swarms and evolutionary algorithms. Int. J. Simul. Syst. Sci. Technol. **6**, 60–79 (2005)
14. Liu, H., Abraham, A., Li, Y.: Nature inspired population-based heuristics for rough set reduction. In: Abraham, A., Falcón, R., Bello, R. (eds.) Rough Set Theory: A True Landmark in Data Analysis. Studies in Computational Intelligence, vol. 174, pp. 261–278. Springer, Heidelberg (2009). https://doi.org/10.1007/978-3-540-89921-1_10

S-Metacompact Spaces

Baiju Thankachan$^{(\boxtimes)}$ [ID]

Manipal Institute of Technology, Manipal Academy of Higher Education,
Manipal, India
baiju.t@manipal.edu

Abstract. Semi-open sets in topological spaces was introduced by
Levine in 1963. In 2006, Al-Zoubi introduced the concept of S-
paracompact spaces using locally finite semi-open refinement and studied
some characterizations and basic properties of S-paracompact spaces. In
this paper we introduce the class of S-metacompact spaces as a generalization of metacompact spaces using point finite semi-open refine-
ments. A topological space (X, τ) is said to be S-metacompact if every
open cover of X has a point finite semi-open refinement. Moreover we
obtain a characterization of S-metacompact spaces and study some basic
properties of S-metacompact spaces. Also we investigate the relationship
between S-metacompact spaces and metacompact spaces.

Keywords: Point finite · Metacompactness · S-metacompactness

1 Introduction

A number of generalizations of open sets have been considered by many authors
in the past few years and some of these notions were defined similarly using
the closure and interior operators; viz semi-open set by Levine [1], preopen set
Corson and Michael [2], semi-preopen set by Abd El-Monsef et al. [3]. After
the introduction of semi-open sets, several authors studied its properties and
characteristics in topological spaces and also introduced a number of spaces in
terms of semi-open sets such as S-expandable spaces by Al-Zoubi in 2004 [4],
H-closed topological spaces by Velicko [5], countably S-closed spaces by dlaka
et al. [6] etc.

Al-Zoubi [7] introduced the concept of S-paracompact spaces using locally
finite semi-open refinement and studied its basic properties and characterizations. The purpose of this paper is to introduce the class of S-metacompact
spaces as a generalization of metacompact spaces using point finite semi-open
refinement. Moreover we obtain a characterization of S-metacompact spaces and
study some basic properties of it. Also we investigate the relationship between
S-metacompact spaces and metacompact spaces.

2 Preliminaries

Let (X, τ) be a topological spaces and A be a subset of X. We shall denote the
closure of A, interior of A by clA, $intA$ respectively.

© Springer Nature Singapore Pte Ltd. 2021
A. Awasthi et al. (Eds.): CSMCS 2020, CCIS 1345, pp. 186–191, 2021.
https://doi.org/10.1007/978-981-16-4772-7_14

Definition 1. [1] *A subset A is called a semi-open subset of (X, τ) if there exists an open set U of X such that $U \subseteq A \subseteq cl(U)$. i.e. $A \subseteq cl(intA)$.*

Semi-closed sets are defined as the complement of semi-open sets. i.e.A is semi-closed if and only if $int(cl(A)) \subseteq A$. The concepts of preclosed and semi-preclosed subsets are similarly defined.

Definition 2. [8] *The smallest semi-closed set containing the subset A is defined as the semi closure of A and is denoted by $scl(A)$.*

$SO(X, \tau)$ denots the family of all semi-open subsets of a space(X, τ).

Definition 3. [9] *A collection $\mathbf{B} = \{B_t : t \in T\} \subseteq X$ is a closure preserving collection if for every subfamily $\mathbf{B_0} = \{B_t : t \in T_0 \subset T\}$ of $\mathbf{B}$, $cl(\cup B_t : t \in T_0) = \cup(cl(B_t : t \in T_0)$*

Definition 4. [10] *A collection $\mathbf{U}$ of subsets of a space (X, τ) is said to be interior-preserving if $int(\cap \mathbf{W}) = \cap \{intW : W \in \mathbf{W}\}$ for every $\mathbf{W} \subset \mathbf{U}$.*

Definition 5. [9] *A partially ordered set P is called a well-ordered set, if for every nonempty subset A of P , there exists $x_0 \in A$ such that $x_0 \leq x$ for every $x \in A$. Then the partial order is called a well order.*

Definition 6. [9] *For any subset A of X and collection $\mathbf{U}$, the star of $\mathbf{U}$ about A, denoted by $st(A, \mathbf{U})$,is defined as the set $\cup \{U \in \mathbf{U} : U \cap A \neq \phi\}$. For any $x \in X$, $st(\{x\}, \mathbf{U})$ is denoted as $st(x, \mathbf{U})$.*

Definition 7. [9] *Let $\mathbf{A}, \mathbf{B}$ be covers of a space (X, τ). $\mathbf{B}$ is said to be a pointwise W-refinement of $\mathbf{A}$ if for any $x \in X$ there is a finite $\mathbf{C} \subset \mathbf{A}$ such that if $x \in B \in \mathbf{B}$, then $B \subset A$ for some $A \in \mathbf{C}$.*

Definition 8. [9] *A collection $\mathbf{V}$ of subsets of a space (X, τ) is said to be well-monotone if the subset relation $\subset$ is a well-order on $\mathbf{V}$*

Definition 9. [9] *A collection $\mathbf{V}$ is said to be directed if $U, V \in \mathbf{V} \Rightarrow U \cup V \subset W$ for some $W \in \mathbf{V}$*

3 S-Metacompact Spaces

Definition 10. [9] *A ccollection $\mathbf{B}$ of subsets of a space (X, τ) is said to be point finite if each point $x \in X$is an element of atmost finitely many members of $\mathbf{B}$.*

Definition 11. [11] *A space (X, τ) is metacompact if every open cover of X has a point finite open refinement.*

Definition 12. *A space (X, τ) is said to be S-metacompact if every open cover of X has a point finite semi-open refinement.*

The proof of the following proposition follows immediately from the definition.

Proposition 1. *Every locally finite family is point finite.*

Remark 1. From the above proposition it follows that
S-paracompactness ⇒ S-metacompactness.

Remark 2. We have the following obvious implication
metacompactness ⇒ S-metacompactness.

Corollary 1. *The unit interval $[0,1]$ is S-metacompact.*

Example 1. The Space (X, τ), where X is the set of real numbers and the topology $\tau = \{\phi, X, \{1\}\}$, is a S-metacompact space.

Proposition 2. *If $\mathbf{B}$ is a closure preserving collection of closed sets in X and A a closed subset of X, then $\{B \cap A : B \in \mathbf{B}\}$ is closure preserving.*

Proposition 3. *A point finite closure preserving closed collection is always locally finite.*

Remark 3. A collection $\mathbf{U} = \{U : U \in \mathbf{U}\}$ is locally finite implies that so is $\{cl(U) : U \in \mathbf{U}\}$

Proposition 4. *Let (X, τ) be a topological space, A be a subset of X and $B \in \tau'$. Then
A is S-metacompact ⇒ $A \cap B$ is S-metacompact.*

Proof. Let $\mathbf{U}$ be an open cover of $A \cap B$ and take $\mathbf{V} = \mathbf{U} \cup \{B'\}$. It is clear that $\mathbf{V}$ is an open cover A. Since A is S-metacompact, it has point finite semi-open refinement $\mathbf{W}$ such that $\mathbf{W}$ is an open cover of A. It is also obvious that $\mathbf{W}' = \{W \in \mathbf{W} : W \subseteq U \text{ for some } U \in \mathbf{U}\}$ is point finite in $A \cap B$. Next we will show that $\mathbf{W}'$ is an open cover of $A \cap B$.
Let $D \subseteq A \cap B \subseteq A$. Then there exist $W \in \mathbf{W}$ such that $D \subseteq W$. Now since $D \subseteq B$, we have $B \supseteq D \not\subseteq W'$ and hence $W \not\subseteq B'$. Since $\mathbf{W}$ refines $\mathbf{V} = \mathbf{U} \cup \{B'\}$ there exist $U \in \mathbf{U}$ such that $W \subseteq U$. Therefore $D \subseteq W \in \mathbf{W}'$. Thus $\mathbf{W}'$ is an open cover of $A \cap B$ and it follows that $A \cap B$ is S-metacompact.

Lemma 1. *Let $\mathbf{U} = \{U_\alpha : \alpha \in \Lambda, \Lambda \text{ is well order }\}$ is an open cover of (X, τ) and $\mathbf{V}_\alpha = \cup\{U_\beta : \beta \leq \alpha, \alpha \in \Lambda\}$. If $\{\mathbf{V}_\alpha : \alpha \in \Lambda\}$ has a precise point finite semi-open refinement $\{\mathbf{W}_\alpha : \alpha \in \Lambda\}$ and each $X - \cup\{W_\gamma : \gamma > \alpha\}$ has a point finite semi-open cover which is a partial refinement of $\{U_\beta : \beta \leq \alpha\}$, then $\mathbf{U}$ has a point finite semi-open refinement.*

Proof. Let $W_\alpha \neq \Phi$ and $W_\alpha \neq W_\beta$ if $\alpha \neq \beta$. For each $\alpha \in \Lambda$, suppose $\mathbf{H}_\alpha$ is a point finite semi-open cover of $X - \cup\{W_\gamma : \gamma > \alpha\}$ such that $H \in \mathbf{H}_\alpha \Rightarrow H \subset U_\beta$ for some β.
Consider $\mathbf{K}_\alpha = \{W_\alpha \cap H : H \in \mathbf{H}_\alpha, H \subset U_\beta \text{ for some } \beta \leq \alpha\}$. Then the family $\mathbf{M} = \bigcup_{\alpha \in \Lambda} \mathbf{K}_\alpha$ is a point finite semi-open collection such that $M \in \mathbf{M}$

implies $M \subset U_\beta$ for some β. Now to show that $\mathbf{M}$ covers (X, τ), let $x \in X$. Then the set $\{\alpha \in \Lambda : x \in W_\alpha$ is finite and δ be the largest element. Then $x \in X - \cup\{W_\gamma : \gamma > \alpha\}$ and so $x \in P$ for some $P \in H_\delta$. It follows that $x \in W \cap P \in \mathbf{K}_\alpha$ and this completes the proof.

Theorem 1. *Let (X, τ) be a topological space. Then the following are equivalent:*

(i) (X, τ) is S-metacompact

(ii) Every well monotone open cover of X has a point finite semi-open refinement which is also an open cover of X.

Proof. (i)$\Rightarrow$(ii) is obvious.

(ii)$\Rightarrow$(i) Assume that (ii) holds. If (X, τ) is not S-metacompact there is a smallest cardinal number α such that there exist an open cover $\mathbf{U}$ of (X, τ) with no point finite semi-open refinement and $\mid \mathbf{U} \mid = \alpha$. i.e. whenever $\mathbf{V}$ is an open cover of (X, τ) with $\mid \mathbf{V} \mid < \mid \mathbf{U} \mid$, then $\mathbf{V}$ has a point finite semi-open refinement. Represent $\mathbf{U}$ as $\mathbf{U} = \{\mathbf{U}_\beta : \beta < \alpha\}$ and for each $\beta < \alpha$, let $\mathbf{W}_\beta = \bigcup_{\gamma \leq \beta} \mathbf{U}_\gamma$. The collection $\mathbf{W} = \{W_\beta : \beta < \alpha\}$ is a well monotone open cover of X so there is a point finite semi-open refinement $\{V_\beta : \beta < \alpha\}$ of $\mathbf{W}$. For $\beta < \alpha$, let $G_\beta = X - \cup\{V_\gamma : \gamma > \beta\}$. Then $\{X - G_\beta\} \cup \{U_\gamma : \gamma \geq \beta\}$ is an open cover of X and the cardinality is less than α and so by the minimal condition on α it must have a point finite semi-open refinement $\mathbf{H}_\beta$. If we take $\mathbf{K}_\beta = \{P \in \mathbf{H}_\beta : P \cap G_\beta \neq \phi\}$, then by the above lemma $\mathbf{U}$ has a point finite semi-open refinement which is a contradiction and it completes the proof.

Lemma 2. *Let (X, τ) be a topological space. If the open cover $\mathbf{C}$ has a point finite semi-open refinement $\mathbf{V}$ such that $x \in int(st(x, \mathbf{V}))$ for all $x \in X$, then $\mathbf{C}$ has a semi-open pointwise W-refinement.*

Proof. For each $V \in \mathbf{V}$ take $C_V \in \mathbf{C}$ such that $V \subset C_V$. Let $U_x = [int(st(x, \mathbf{V}))] \cap [\cap\{C_V : x \in V \in \mathbf{V}\}], x \in X$. Then the collection $\mathbf{U} = \{U_x : x \in X\}$ is the required semi-open pointwise W-refinement of $\mathbf{C}$.

Theorem 2. *Let (X, τ) be a topological space. If $\mathbf{C}$ is an interior-preserving open cover of X, then $\mathbf{C}^F$ has a closure preserving semi-closed refinement if and only if $\mathbf{C}$ has an interior-preserving semi-open pointwise W-refinement. $\mathbf{C}^F$ denotes the collection of all unions of finite subcollections from $\mathbf{C}$.*

Proof. Let $U_x = [\cap\{C \in \mathbf{C} : x \in \mathbf{C}\}] - \cup\{G \in \mathbf{G} : x \notin G\}$, where $\mathbf{G}$ is a closure preserving semi-closed refinement of $\mathbf{C}^F$. Then $\mathbf{U} = \{U_x : x \in X\}$ is an interior-preserving semi-open pointwise W-refinement of $\mathbf{C}$.

Conversely assume that $\mathbf{U}$ is an interior-preserving semi-open pointwise W-refinement of $\mathbf{C}$. For $H \in \mathbf{C}^F$, take $M_H = \{x \in X : st(x, \mathbf{U}) \subset H\}$ and the collection $\mathbf{M} = \{M_H : H \in \mathbf{C}^F\}$ is a closure preserving semi-closed refinement of $\mathbf{C}^F$.

The proof of the following lemma follows from the above Theorem 2 since a point finite open cover of X is an interior-preserving semi-open pointwise W-refinement of itself.

Lemma 3. *If* **C** *is a point finite open cover of* X, *then* $\mathbf{C}^F$ *has a closure preserving semi-closed refinement.*

Theorem 3. *Let* (X, τ) *be a topological space. Then every directed open cover of* X *has a closure preserving semi-closed refinement if and only if for every open cover* **C** *of* X, $\mathbf{C}^F$ *has a closure preserving semi-closed refinement.*

Proof. Necessary part: Clearly $\mathbf{C}^F$ is directed and hence it has a closure preserving semi-closed refinement.

Sufficient part: Let **C** be a directed open cover of (X, τ). Since **C** is directed, $\mathbf{C}^F$ is a refinement of **C**. Then by our assumption, $\mathbf{C}^F$ has a closure preserving semi-closed refinement say **V** which is an open cover of (X, τ). Now $\mathbf{V} \subset \mathbf{C}^F \subset \mathbf{C}$ and this follows that **V** is the required closure preserving semi-closed refinement of **C**.

Theorem 4. *Let* (X, τ) *be a topological space. Then* (X, τ) *is S-metacompact if and only if for every open cover of* (X, τ), *there is a point finite semi-open refinement* **B** *such that* $x \in int(st(x, \mathbf{B}))$ *for every* $x \in X$.

Proof. Necessary part is obvious.

Sufficient part: Let **U** be an open cover of X. Then there exists a point finite semi-open refinement **B** such that $x \in int(st(x, \mathbf{B}))$. i.e. $x \in int(\cup\{B \in \mathbf{B} : B \cap \{x\} \neq \phi\}$. Then **B** is an open cover of (X, τ) and hence (X, τ) is S-metacompact.

Theorem 5. *Let* (X, τ) *be a topological space. If for every open cover* **U** *of* X, $\mathbf{U}^F$ *has a closure preserving semi-closed refinement then every well-monotone open cover of* X *has a point finite semi-open refinement which covers* X.

Proof. Let **U** be a well-monotone open cover of (X, τ). It is obvious that **U** is an interior-preserving open cover of X. By the repeated application of Lemma 3.4(ii) there is a sequence $\{\mathbf{U}_n\}_1^\infty$ of open covers X such that $\mathbf{U} = \mathbf{U}_1$ and $\mathbf{U}_{n+1}$ is an interior-preserving semi-open pointwise W-refinement of $\mathbf{U}_n$. Then **U** has a semi-open refinement $\mathbf{W} = \bigcup_{n=1}^{\infty} \mathbf{W}_n$, $\mathbf{W}_n$ is point finite. Take $\mathbf{H}_n = \cup\{W : W \in \mathbf{W}_k, k \leq n\}$. Then $\{\mathbf{H}_n : n \in \mathbb{N}\}$ is a directed open cover of X and it has a closure preserving semi-closed refinement **D** expressed as $\mathbf{D} = \{D_n : D_n \subset H_n, n \in \mathbb{N}\}$. Let $\mathbf{E}_n = \{W - \bigcup_{k<n} D_n : W \in \mathbf{W}_n\}$. It follows that $\bigcup_{n=1}^{\infty} \mathbf{E}_n$ is a point finite semi-open refinement of **U**.

Combining Theorem 1, Theorem 3, Theorem 4, Theorem 5 and Lemma 3 we obtain the following characterization of S-metacompact spaces.

Theorem 6. *Let* (X, τ) *be a topological space. Then the following are equivalent:*

(i) (X, τ) *is S-metacompact.*
(ii) *Every well-monotone open cover of* (X, τ) *has a point finite semi-open refinement.*

(iii) Every directed open cover of (X, τ) *has a closure preserving semi-closed refinement.*

(iv) For every open cover **V** *of* (X, τ), $\mathbf{V}^F$ *has a closure preserving semi-closed refinement.*

(v) Every open cover **V** *of* (X, τ) *has a point finite semi-open refinement* **B** *such that* $x \in int(st(x, \mathbf{B}))$ *for every* $x \in X$.

References

1. Levine, N.: Semi-open sets and semi-continuity in topological spaces. Amer. Math. Monthly **70**, 36–41 (1963)
2. Corson, H.H., Michael, E.: Metrizability of certain countable unions. Illinois J. Math. **8**, 351–360 (1964)
3. Abd El-Monsef, M.E., El-Deeb, S.N., Mahmoud, R.A.: β-open sets and β-continuous mappings. Bull. Fac. Sci. Assiut. Univ. **12**, 77–90 (1964)
4. Al-Zoubi, K.Y.: s-expandable spaces. Acta Math. Hungar. **102**, 203–212 (2004)
5. Veličko, K.Y.: H-closed toplogical spaces. Amer. Math. Soc. Transl. **78**, 103–118 (1968)
6. Dlaka, K., Ergun, N., Ganster, M.: Countably S-closed spaces. Math. Slovaca **44**, 337–348 (1944)
7. Al-Zoubi, K.Y.: s-paracompact spaces. Acta Math. Hungar. **110**(1–2), 165–174 (2006)
8. Crossely, S.G.: Semi-closed and semi-continuity in topological spaces. Texas J. Sci. **22**, 123–126 (1971)
9. Burke, D.K.: Covering properties. In: Kunen, K., Vaughan, J.E. (eds.) Hand Book of Set Theoretic Topology, pp. 349–422. Elsevier Science Publishers (1984)
10. Junnila, H.J.K.: Metacompactness, paracompactness and interior-preserving open covers. Trans. Amer. Math. Soc. **249**, 373–385 (1979)
11. Junnila, H.J.K.: Paracompactness, metacompactness, and semi-open covers. Proc. Amer. Math. Soc. **73**, 244–248 (1979)
12. Engelking, R.: General Topology. Heldermann, Berlin (1989)
13. Williard, S.: General Topology. Addison-Wesley, London (1970)
14. Ganster, M.: On covering properties and generalized open sets in topological spaces. Math. Chron. **19**, 27–33 (1990)

Interval Valued Fuzzy Graph and Complement Number

Deepthi Mary Tresa Souriar[1] ⓘ, Divya Mary Daise Souriar[1] ⓘ,
and Shery Fernandez[2](✉) ⓘ

[1] Department of Mathematics, St. Albert's College (Autonomous), Kochi, Kerala, India
[2] Department of Mathematics, Cochin University of Science and Technology, Kochi, Kerala, India

Abstract. In this paper, we present an example to show that many non-isomorphic IVFGs may have a same complement. To overcome this limitation, we introduce the notion of complement number of an edge and prove that given a complement IVFG $\bar{G}$ along with its complement numbers, the IVFG for which $\bar{G}$ acts as the complement can be uniquely determined. We also study the range of variation of complement number and some of its properties.

Keywords: Classic interval valued fuzzy graph · Complement · Complement number

1 Introduction

George Cantor defined set (crisp set) as a well-defined collection of objects. That is, after defining a set we must be able to say whether a given object belongs to the set or not. But in real life situation we cannot always say yes or no. In 1965, L. A. Zadeh [27] introduced the notion of fuzzy set, in which the partial membership of objects is also possible. Ever since, the theory of fuzzy sets has become a vigorous area of research in different disciplines including medical and life sciences, management sciences, engineering, statistics, graph theory, artificial intelligence, signal processing, etc. In 1975, Rosenfeld [15] introduced the notion of fuzzy graph, which allows partial membership of vertices and edges of a graph. The complement of a fuzzy graph was defined by Mordeson and Nair [10]. Complement of fuzzy graphs and the complement of the operations of union, join and composition of fuzzy graphs that were introduced in [27]. Also, Sunitha and Vijayakumar [24] studied complement of fuzzy graphs. Hawary [2] defined complete fuzzy graphs and presented new operations on it. Mathew and Sunitha [11, 12] studied different types of connectivity of fuzzy graphs. Mordeson [9], defined several new operations on fuzzy graphs. As an extension of fuzzy sets, Zadeh [27] introduced the notion of interval valued fuzzy sets, in which the values of the membership degree are intervals of numbers instead of fixed numbers. In 2009, Hongmei and Lianhua [8] defined interval valued fuzzy graphs (IVFG's) and in 2011, Akram and Dudek [1] defined some operations on them. Talebi and Rashmanlou [24] studied properties of isomorphism and complement of interval valued fuzzy graphs. Samanta et al. studied strong edge, weak edge of an interval-valued fuzzy

A. Awasthi et al. (Eds.): CSMCS 2020, CCIS 1345, pp. 192–202, 2021.
https://doi.org/10.1007/978-981-16-4772-7_15

graph, degree of a vertex in bipolar fuzzy graph, regular, irregular bipolar fuzzy graphs, bipolar fuzzy hyper graphs, fuzzy k-competition, p-competition of fuzzy graphs, m-step fuzzy competition graphs, Fuzzy planar graphs, tolerance, threshold fuzzy graphs [13, 16–23]. In 2018, were defined the definition of complement of IVFGs [4] as we come across with an example for which the definition of complement given in [24] was not valid. In the light of new definition, we introduced two classes of Interval Valued Fuzzy Graphs (IVFGs), Classic IVFGs and Non-classic IVFGs, and classify edges as Perfect and Imperfect edges of an IVFG. In [3], we studied some of their properties regarding degree, isomorphism, weak – isomorphism, union, join etc. and in [4], we derived a necessary and sufficient condition for an IVFG to be classic. In this paper we present a drawback of the new definition of complement IVFG [4] and to overcome it we introduce the concept of complement number of an edge of an IVFG. We study the range of variation of complement number and we proved some results regarding it. Also, using this idea of complement numbers we constructed a lattice of IVFGs in [6]. Many researches are still going on in this area of IVFGs and it has numerous applications in various fields since it provides a more adequate description of uncertainty than fuzzy graph [14, 26] are some recent works in this area.

2 Some Basic Concepts

A *fuzzy set* [27]. A on a set X is characterized by a mapping $\mathfrak{M} : X \rightarrow [0, 1]$ which is called the membership function and fuzzy set A on X is denoted by $A = \{(x, \mathfrak{M}(x)) : x \in X\}$.

An *interval valued fuzzy set(IVFS)* [28]. A on X is characterized by an interval-valued function $i : X \rightarrow \mathbb{P}[0, 1]$, where $\mathbb{P}[0,1]$ denotes the *power set* of $[0,1]$, such that $i(x) = \left[a_x^-, a_x^+\right]$ where $0 \le a_x^- \le a_x^+ \le 1$. For each $x \in X$, $i(x)$ is called the *interval number* of x. An IVFS A on X is denoted by $A = \{(x, i(x)) : x \in X\}$.

A *graph* (or a *crisp graph*) [7] is defined as a pair, $G^* = (V, E)$ consisting of a nonempty finite set V of elements called *vertices* and a finite set E of pairs of vertices called *edges*.

An *interval valued fuzzy graph(IVFG)* [8]. $G = (V, \sigma, \mu)$ consists of a nonempty set V together with a pair of interval valued functions $\sigma : V \rightarrow \mathbb{P}[0, 1]$ and $\mu : V \times V \rightarrow \mathbb{P}[0, 1]$ where

$$\sigma(A) = \left[\sigma_A^-, \sigma_A^+\right], 0 \le \sigma_A^-, \le \sigma_A^+ \le 1 \text{ and}$$

$$\mu(AB) = \left[\mu_{AB}^-, \mu_{AB}^+\right], 0 \le \mu_{AB}^- \le \mu_{AB}^+ \le 1$$

represent the interval numbers of the vertex A and edge (A, B) respectively in G satisfying

$$\mu_{AB}^- \le \min\{\sigma_A^-, \sigma_B^-\} \text{ and } \mu_{AB}^+ \le \min\{\sigma_A^+, \sigma_B^+\}$$

for all $A, B \in V$.

Let $G = (V, \sigma, \mu)$ and $G' = (V', \sigma', \mu')$ be two IVFG's. Then G and G' are said to be *isomorphic* [1], written as $G \cong G'$, if there exist a bijection $h : V \rightarrow V'$ such that

1. $\sigma_A^- = \sigma'^-_{h(A)}, \sigma_A^+ = \sigma'^+_{h(A)}$ for every vertex $A \in V$.
2. $\mu_{AB}^- = \mu'^-_{h(A)h(B)}, \mu_{AB}^+ = \mu'^+_{h(A)h(B)}$ for every edge AB in G.

Otherwise we say that G and G' are *non – isomorphic*, and denote it as $G \not\cong G'$.

2.1   Classic and Non-classic IVFG

An IVFG $G = (V, \sigma, \mu)$ is called a ***classic IVFG*** [4] if all its edges satisfy the condition

$$\min\{\sigma_A^-, \sigma_B^-\} - \mu_{AB}^- \leq \min\{\sigma_A^+, \sigma_B^+\} - \mu_{AB}^+.$$

Otherwise we call it as a ***non-classic IVFG*** [4].
We shall refer to the above condition as the *classic condition for an edge AB*.

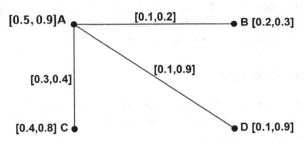

Fig. 1. An example for a classic IVFG.

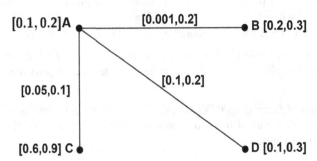

Fig. 2. An example for a non-classic IVFG.

Example
In Fig. 1 and (2) edges BC, CD, BD are not drawn. It means that the interval number those edges are [0,0].

In the graph in Fig. 2, edge AB does not satisfy the classic condition. So it is a non-classic IVFG. Also, in a non-classic IVFG, there may be some edges satisfying the classic condition. For example, edge AC in Fig. 2. This observation leads to the next definition.

Let $G = (V, \sigma, \mu)$ be an IVFG. Then edges AB in G satisfying

$$\min\{\sigma_A^-, \sigma_B^-\} - \mu_{AB}^- \leq \min\{\sigma_A^+, \sigma_B^+\} - \mu_{AB}^+$$

are called *perfect edges* [4] and all other edges AB for which

$$\min\{\sigma_A^-, \sigma_B^-\} - \mu_{AB}^- > \min\{\sigma_A^+, \sigma_B^+\} - \mu_{AB}^+$$

are called *imperfect edges* [4].

That is, an edge satisfying the classic condition is perfect and others imperfect. It may be observed that, in a classic IVFG, all edges are perfect; and a non-classic IVFG must contain at least one imperfect edge.

Now, seeing that the definition of complement given by Talebi and Rashmanlou [25] does not apply to all IVFGs, we redefined complement as follows (Fig. 3).

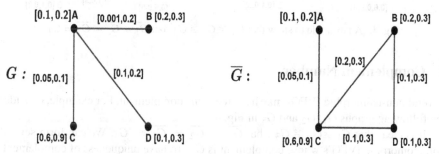

Fig. 3. An example of complement of IVFG.

2.2   Complement of an Interval Valued Fuzzy Graph

The *complement of IVFG* $G = (V, \sigma, \mu)$ is an IVFG $\overline{G} = (V, \sigma, \overline{\mu})$ where

$$\overline{\mu}(AB) = \left[\overline{\mu}_{AB}^-, \overline{\mu}_{AB}^+\right] = \begin{cases} [\min\{\sigma_A^-, \sigma_B^-\} - \mu_{AB}^-, \ \min\{\sigma_A^+, \sigma_B^+\} - \mu_{AB}^+], & \text{if } AB \text{ is a perfect edge of } G \\ [\min\{\sigma_A^+, \sigma_B^+\} - \mu_{AB}^+, \ \min\{\sigma_A^+, \sigma_B^+\} - \mu_{AB}^+], & \text{if } AB \text{ is an imperfect edge of } G \end{cases}$$

for all $A, B \in V$.

Example

Observe that, the IVFG G given above is the non-classic graph in Fig. 2.

Remark 1. If edge AB is an imperfect edge, then $\overline{\mu}(AB)$ is always a real number in $[0,1)$.

Theorem 2 [4]. For any IVFG $G = (V, \sigma, \mu)$, $\overline{G}$ is always classic.

Theorem 3 [5]. If $G \cong \overline{G}$ then, G is a classic IVFG, but the converse is not true.

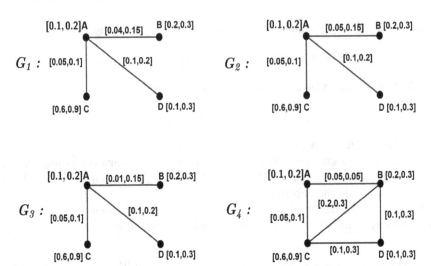

Fig. 4. An example to show that $G_1 \ncong G_2 \ncong G_3$, but $\overline{G_1} = \overline{G_2} = \overline{G_3} = G_4$

3 Complement Number

Several non-isomorphic IVFGs may have the same complement. For example, consider the following graphs G_1, G_2 and G_3 in Fig. 4.

It is clear that $G_1 \ncong G_2 \ncong G_3$, but $\overline{G_1} = \overline{G_2} = \overline{G_3} = G_4$. We can form several non-isomorphic IVFG's whose complement is G_4. To have uniqueness of complement, we need another notion called *complement number of an edge*. This is defined below (Fig. 5).

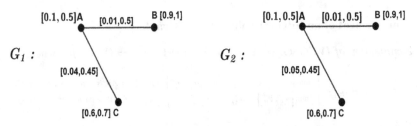

Fig. 5. Two IVFGs used to illustrate complement number.

3.1 Complement Number of an Edge

Let $G = (V, \sigma, \mu)$ be any IVFG, then *in $\overline{G}$ complement number of an edge AB w.r.t.* G (denoted as c_{AB}^G or simply c_{AB}) is defined as

$$c_{AB} = \begin{cases} \min\{\sigma_A^-, \sigma_B^-\} - \mu_{AB}^-, & \textit{if AB is an imperfect edge of } G \\ 0, & \textit{otherwise} \end{cases}$$

Remark 4. In $\bar{G}$, if c_{AB} is not given we will consider that *complement number of edge AB* as zero.

Example. Consider the two IVFGs G_1, and G_2 given in Fig. 4.

Observe that, in G_1 both the edges AB and AC are imperfect, and BC is a perfect edge with membership [0,0]. Since BC is perfect its complement number is zero and need not be specified. So, the collection of complement numbers of $\overline{G_1}$ is $\{C_{AB} = 0.09, C_{AC} = 0.06\}$. Similarly, since the only imperfect edge of G_2 is AB, all other edges in $\overline{G_2}$ have complement number zero and hence, the collection of complement numbers of $\overline{G_2}$ is $\{C_{AB} = 0.09\}$.

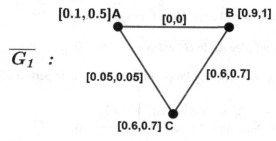

Fig. 6. Complement of G_1 with $\{c_{AB} = 0.09, c_{AC} = 0.06\}$

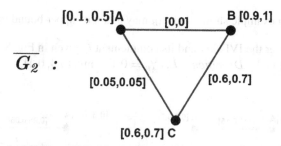

Fig. 7. Complement of G_2 with $c_{AB} = 0.09$.

From Fig. 6 and Fig. 7 it can be noted that, the visualization of both $\overline{G_1}$ and $\overline{G_2}$ are same but their complement numbers are different. Also, G_1 is the only IVFG having $\overline{G_1}$ as complement with $c_{AB} = 0.09$ and $c_{AC} = 0.06$. Similarly, observation holds good for G_2 also. Thus, specification of complement numbers, ensures unique of IVFG while going back from complement to IVFG.

Remark 5.

1. $c_{AB} = 0$ for all edges AB in $\overline{\overline{G}}$ (by Theorem 2).
2. If G is self-complementary IVFG, then $c_{AB}^G = 0$, for all edge AB in $\overline{G}$ (by Theorem 3).

In $\overline{G} = c_{AB} \neq 0 \Rightarrow \overline{\mu}(AB) = \left[\overline{\mu}_{AB}^{+}, \overline{\mu}_{AB}^{+},\right]$, a real number in $[0,1)$.

Proposition 6. Let $G = (V, \sigma, \mu)$ be any IVFG. Then in $\overline{G}$, $c_{AB} = 0 \Leftrightarrow AB$ is a perfect edge of G.

Proof. Suppose AB is a perfect edge of G. Then by definition of complement number, $c_{AB} = 0$.

    Conversely, suppose $c_{AB} = 0$. If possible assume that, AB is an imperfect edge of G

$$\min\{\sigma_A^-, \sigma_B^-\} - \mu_{AB}^- > \min\{\sigma_A^+, \sigma_B^+\} - \mu_{AB}^+ \Rightarrow c_{AB} > \min\{\sigma_A^+, \sigma_B^+\} - \mu_{AB}^+$$

$$\Rightarrow 0 > \min\{\sigma_A^+, \sigma_B^+\} - \mu_{AB}^+, \text{ which is not possible by definition of IVFG.} \qquad \blacksquare$$

Corollary 7. $c_{AB}^G = 0$, for all edge AB in $\overline{G} \Leftrightarrow G$ is classic.

Proposition 8. For an edge AB in $\overline{G}$, either $c_{AB} = 0$ or $\overline{\mu}_{AB}^+ < c_{AB} \leq \min\{\sigma_A^-, \sigma_B^-\}$

Proof. Suppose $c_{AB} \neq 0$. Then by Proposition 6, AB is an imperfect edge of G.

$$\Rightarrow \min\{\sigma_A^-, \sigma_B^-\} - \mu_{AB}^- > \min\{\sigma_A^+, \sigma_B^+\} - \mu_{AB}^+$$
$$\Rightarrow c_{AB} > \overline{\mu}_{AB}^+ \text{ (by Definition (3.1) and Definition (2.2))}$$

    Also by Definition (3.1), $c_{AB} \leq \min\{\sigma_A^-, \sigma_B^-\}$. Hence proved. $\qquad \blacksquare$

    The following example shows that c_{AB} may attain its upper bound $\min\{\sigma_A^-, \sigma_B^-\}$.

Example. Consider the IVFG G and its complement $\overline{G}$ given in Fig. 8. Since AB is an imperfect edge of G, by *Definition 3.1.*, $c_{AB}^G = 0.1 = \min\{0.3, 0.1\}$.

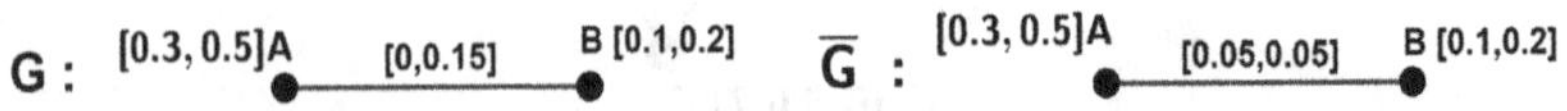

$$G: \quad \overset{[0.3,0.5]A}{\bullet} \xrightarrow{\quad [0,0.15] \quad} \overset{B\ [0.1,0.2]}{\bullet} \qquad \overline{G}: \quad \overset{[0.3,0.5]A}{\bullet} \xrightarrow{\quad [0.05,0.05] \quad} \overset{B\ [0.1,0.2]}{\bullet}$$

Fig. 8. An example to show that c_{AB} attains its upper bound $\min\{\sigma_A^-, \sigma_B^-\}$

    The next theorem gives information regarding the range of μ_{AB}^- and μ_{AB}^+, if c_{AB} attains its upper bound $\min\{\sigma_A^-, \sigma_B^-\}$.

    We have observed from examples in Fig. 4 that many non-isomorphic IVFGs may have same complement. So, if we are given any complement graph $\overline{G}$, we may not uniquely determine the IVFG G from which the $\overline{G}$ is made. The proof of following theorem gives us a method to construct an unique IVFG, if its complement IVFG with complement numbers are given.

Theorem 9. Given any complement graph $\overline{G} = (V, \sigma, \overline{\mu})$ along with its complement numbers, the IVFG G can be uniquely determined.

Proof. Let $G_1 = (V, \sigma, \mu_1)$ where

$$\mu_1(AB) = \left[\mu_1\bar{}_{AB}, \mu_1{}^+_{AB}\right]$$

$$= \begin{cases} \left[\min\{\sigma_A^-, \sigma_B^-\} - c_{AB}, \min\{\sigma_A^+, \sigma_B^+\} - \bar{\mu}{}^+_{AB}\right], & if \ c_{AB} \neq 0 \\ \left[\min\{\sigma_A^-, \sigma_B^-\} - \bar{\mu}{}^-_{AB}, \min\{\sigma_A^+, \sigma_B^+\} - \bar{\mu}{}^+_{AB}\right], & if \ c_{AB} = 0. \end{cases} \quad (1)$$

First, we are going to show that $\overline{G_1} = \overline{G}$. It is enough to show that $\overline{\mu_1} = \overline{\mu}$.

Case 1. $c_{AB} = 0.$

$\Rightarrow AB$ is a perfect edge of G_1, by Proposition 6.

Now, $\overline{\mu_1}(AB) = [\min\{\sigma_A^-, \sigma_B^-\} - \mu_1\bar{}_{AB}, \ \min\{\sigma_A^+, \sigma_B^+\} - \mu_1{}^+_{AB}]$, by Definition (2.2).

$= [\min\{\sigma_A^-, \sigma_B^-\} - (\ \min\{\sigma_A^-, \sigma_B^-\} - \bar{\mu}{}^-_{AB})\ , \min\{\sigma_A^+, \sigma_B^+\} - (\min\{\sigma_A^+, \sigma_B^+\} - \bar{\mu}{}^+_{AB})] = [\bar{\mu}{}^-_{AB}, \bar{\mu}{}^+_{AB}] = \overline{\mu}(AB).$

Case 2. $c_{AB} \neq 0.$

By Proposition 8,

$$c_{AB} > \bar{\mu}{}^+_{AB}. \quad (2)$$

Now,

$$\min\{\sigma_A^-, \sigma_B^-\} - \mu_1\bar{}_{AB} = \min\{\sigma_A^-, \sigma_B^-\} - (\min\{\sigma_A^-, \sigma_B^-\} - c_{AB}) = c_{AB}$$

and

$$\min\{\sigma_A^+, \sigma_B^+\} - \mu_1{}^+_{AB} = \min\{\sigma_A^+, \sigma_B^+\} - (\min\{\sigma_A^+, \sigma_B^+\} - \bar{\mu}{}^+_{AB} = \bar{\mu}{}^+_{AB}) \quad (3)$$

$\min\{\sigma_A^-, \sigma_B^-\} - \mu_1\bar{}_{AB} > \min\{\sigma_A^+, \sigma_B^+\} - \mu_1{}^+_{AB}$, by (3)

$\Rightarrow AB$ is an imperfect edge of G_1.

Hence by Definition (2.2),

$\overline{\mu_1}(AB) = \left[\min\{\sigma_A^+, \sigma_B^+\} - \mu_1{}^+_{AB}, \ \min\{\sigma_A^+, \sigma_B^+\} - \mu_1{}^+_{AB}\right] = [\bar{\mu}{}^+_{AB}, \bar{\mu}{}^+_{AB}]$, by (2)

$= \overline{\mu}(AB)$, by Remark 5 (3).

Hence $\overline{G_1} = \overline{G}$.

Now to prove uniqueness let $G_2 = (V, \sigma, \mu_2)$ be an IVFG such that $\overline{G_2} = \overline{G}$ and let the complement numbers of $\overline{G}$ w.r.t. both G_1 and G_2 are same. $\quad (4)$

Case 1: $c_{AB} \neq 0.$

Then by Definition (3.1) and Proposition 6,

$c_{AB} = \min\{\sigma_A^-, \sigma_B^-\} - \mu_1\bar{}_{AB} = $ w.r.t. G_1 and $c_{AB} = \min\{\sigma_A^-, \sigma_B^-\} - \mu_2\bar{}_{AB}$ w.r.t. G_2.

$\Rightarrow \mu_1\bar{}_{AB} = \mu_2\bar{}_{AB}$, by (4)

Case 2: $c_{AB} = 0.$ Then by Proposition 6, AB is a perfect edge of both G_1 and G_2. Hence by Definition (2.2),

$C_{AB} = \min\{\sigma_A^-, \sigma_B^-\} - \mu_{1\bar{A}B}$ w.r.t. G_1 and $C_{AB} = \min\{\sigma_A^-, \sigma_B^-\} - \mu_{2\bar{A}B}$ w.r.t G_2
$\Rightarrow \mu_{1\bar{A}B} = \mu_{2\bar{A}B}$

In either case, $\overline{\mu}_{\overset{+}{AB}} = \min\{\sigma_A^+, \sigma_B^+\} - \mu_{1\overset{+}{AB}}$ w.r.t. G_1 and $C_{AB} = \min\{\sigma_A^+, \sigma_B^+\} - \mu_{2\overset{+}{AB}}$ w.r.t G_2
$\Rightarrow \mu_{1\overset{+}{AB}} = \mu_{2\overset{+}{AB}}$

Hence $\left[\mu_{1\bar{A}B}, \mu_{1\overset{+}{AB}}\right] = \left[\mu_{2\bar{A}B}, \mu_{2\overset{+}{AB}}\right] \Rightarrow G_1 G_2$.
Taking $G = G_1$ will complete the proof. ∎

Remark 10. Let any complement graph $\overline{G} = (V, \sigma, \overline{\mu}))$ along with its complement numbers be given. Then the unique IVFG from which $\overline{G}$ made is, $G = (V, \sigma, \mu)$ where

$$\mu(AB) = \left[\mu_{\bar{AB}}, \mu_{\overset{+}{AB}}\right] = \begin{cases} \left[\min\{\sigma_A^-, \sigma_B^-\} - c_{AB}, \min\{\sigma_A^+, \sigma_B^+\} - \overline{\mu}_{\overset{+}{AB}}\right], & \text{if } c_{AB} \neq 0 \\ \left[\min\{\sigma_A^-, \sigma_B^-\} - \overline{\mu}_{\bar{AB}}, \min\{\sigma_A^+, \sigma_B^+\} - \overline{\mu}_{\overset{+}{AB}}\right], & \text{if } c_{AB} = 0. \end{cases}$$

Example. Consider the complement IVFG $G_4 = (V, \sigma, \overline{\mu})$ in Fig. 4 (without complement numbers). Since $\overline{\mu}(BC) = [0.2, 0.3]$ is not a real number, by Definition (2.2) and by Proposition 6 the only possible complement number for edge BC is zero.ie. $c_{BC} = 0$. By Proposition 8, for edge AB, $c_{AB} = 0$ or $0.05 < c_{AB} \leq \min\{0.1, 0.2\} \Rightarrow c_{AB} = 0$ or $0.05 < c_{ab} \leq 0.1$. In similar way, we can find the range of variation of complement number of all other edges and hence we get,

- $C_{AB} = 0$ or $0.05 < c_{ab} \leq 0.1$
- $c_{AC} = 0$
- $0 \leq c_{AD} \leq 0.1$
- $C_{BC} = 0$
- $c_{BD} = 0$
- $C_{CD} = 0$

Now, consider the IVFG G_4 along with the collection $\{c_{AB} = 0.06, c_{AC} = 0, c_{AD} = 0, c_{BC} = 0, c_{BD} = 0, c_{CD} = 0\}$ of complement number. To determine the IVFG $G = (V, \sigma, \mu)$ from which the complement G_4 is made, we will use Remark 10. Here, $\overline{\mu}(AB) = [0.05, 0.05]$ with $c_{AB} = 0.06$. So, $\mu(AB) = [0.1 - 0.06, 0.2 - 0.05] = [0.04, 0.15]$. And $\overline{\mu}(AC) = [0.05, 0.1]$ with $c_{AC} = 0$. So, $\mu(AC) = [0.1 - 0.05, 0.2 - 0.1] = [0.05, 0.1]$. Similarly, $\mu(AD) = [0.1, 0.2]$, $\mu(BC) = [0,0]$, $\mu(BD) = [0,0]$, $\mu(CD) = [0,0]$. It is now clear that the IVFG $G = G_1$ in Fig. 4.

In G_4, corresponding to the collection $\{c_{AB} = 0.09, c_{AC} = 0, c_{AD} = 0, c_{BC} = 0, c_{BD} = 0, c_{CD} = 0\}$ of complement number, we get G_3 in Fig. 4 which is nonisomorphic to G_1 and having complement G_4.

Similarly, in the given range of complement numbers of each edge, we can vary complement numbers and corresponding to each collection of complement numbers we will get an IVFG and all those IVFGs will have the same complement G_4 in Fig. 4.

Proposition 11. If $\overline{\mu}(AB)$ is not a real number for all edge AB in $\overline{G}$, then there exist unique graph whose complement is $\overline{G}$ and the determined graph will be classic IVFG.

Proof. Let $\overline{\mu}(AB) \neq r$, $r \in [0, 1)$ for all edge AB in $\overline{G}$. Then by Remark 1, AB is not an imperfect edge of G.

$\Rightarrow G$ is classic IVFG and also by Definition (3.1), $c_{AB} = 0$ for all edge AB in $\overline{G}$. Hence by Theorem 9, we can conclude the result. ∎

Proposition 12. If $\min\{\sigma_A^-, \sigma_B^-\} = 0$, for all edge AB in $\overline{G}$ with $\overline{\mu}(AB) = r$, $r \in [0, 1)$, then there exist unique graph whose complement is $\overline{G}$ and the determined graph will be classic IVFG.

Proof. If $\overline{\mu}(AB)$ is not a real number, then $c_{AB} = 0$, in $\overline{G}$. Now for those edges AB in $\overline{G}$ with $\overline{\mu}(AB) = r$, $r \in [0, 1)$, it is given that $\min\{\sigma_A^-, \sigma_B^-\} = 0$. Also, by Proposition 8, either $c_{AB} = 0$ or $\bar{\mu}_{AB}^+ < c_{AB} \le \min\{\sigma_A^-, \sigma_B^-\}$. So, $\min\{\sigma_A^-, \sigma_B^-\} = 0 \Rightarrow c_{AB} = 0$.

Hence, $c_{AB} = 0$, for all edge AB in $\overline{G}$. Then by Theorem 9, we can uniquely determine G and G will be classic IVFG, by Proposition 6. ∎

4 Conclusion

In our previous work we redefined complement of IVFG as its definition in [24] is invalid for some IVFGs. In this paper we confirmed existence of infinitely many non-isomorphic IVFGs having same complement by providing an example. To overcome this problem, we assigned a number to each edge of a complement IVFG, which is termed as complement number of that edge. Also, we proved that given a complement IVFG $\bar{G}$ along with its complement numbers, the IVFG for which $\bar{G}$ acts as the complement can be uniquely determined. sWe studied some relevant properties and derived the range of variation of complement number of each edge of a complement IVFG. Finally, we derived two cases in which the complement numbers to all edges of a complement IVFG are unique and hence has a unique IVFG whose complement is the given one.

References

1. Akram, M., Dudek, W.A.: Interval valued fuzzy graphs. Comput. Math. Appl. **61**(2), 289–299 (2011)
2. AL-Hawary, T.: Complete fuzzy graphs. Int. J. Math. Comb. **4**, 26–34 (2011)
3. Tresa, S.D.M., Daise, S.D.M., Fernandez, S.: A study onclassic and non-classic interval valued fuzzy graphs. In: Proceedings of the International Conference on Mathematics 2018, pp. 72–77 (2018)
4. Tresa, S.D.M., Daise, S.D.M., Fernandez, S.: Classic and nonclassic interval valued fuzzy graphs. Int. J. Appl. Eng. Res. **13**(3), 1–4 (2018)
5. Tresa, S.D.M., Daise, S.D.M., Fernandez, S.: On complement of interval valued fuzzy graphs. Int. J. Current Adv. Res. **7**(5), 12954–56 (2018)
6. Tresa, S.D.M., Daise, S.D.M., Fernandez, S.: The lattice of pre-complements of a classic interval valued fuzzy graph. Malaya J. Matematik, **8**, 1311–1320 (2020)
7. Deo, N.: Graph Theory with Applications to Engineering and Computer Science. PHI Learning Pvt. Ltd., Delhi (2014)

8. Hongmei, J., Lianhua, W.: Interval-valued fuzzy subsemigroups and subgroups associated by interval-valued fuzzy graphs. In: Global Congress on Intelligent Systems, pp. 484–487 (2009)

9. Mordeson, J.N., Peng, C.S.: Operation on fuzzy graphs. Inf. Sci. **19**, 159–170 (1994)

10. Mordeson, J.N., Nair, P.S.: Fuzzy Graphs and Fuzzy Hypergraphs. Physica-verlay, Heidelberg (2000)

11. Mathew, S., Sunitha, M.S.: Node connectivity and arc connectivity of a fuzzy graph. Inf. Sci. **180**(4), 519–531 (2010)

12. Mathew, S., Sunitha, M.S.: Cycle connectivity in fuzzy graphs. J. Intell. Fuzzy Syst. **24**(3), 549–554 (2013)

13. Pramanik, T., Samanta, S., Pal, M.: Interval-valued fuzzy planar graphs. Int. J. Mach. Learn. Cybern. **7**(4), 653–664 (2014). https://doi.org/10.1007/s13042-014-0284-7

14. Rashmanlou, H., Borzooei, R.A., Samanta, S., Pal, M.: Properties of interval valued intuitionistic (S, T)–fuzzy graphs. Pacific Sci. Rev. A Nat. Sci. Eng. **18**(1), 30–37 (2016)

15. Rosenfeld, A.: Fuzzy graphs, Fuzzy sets and their applications to cognitive and decision processes, pp. 77–95 (1975)

16. Samanta, S., Pal, M.: Fuzzy tolerance graphs. Int. J. Latest Trend Math. **1**, 57–67 (2011)

17. Samanta, S., Pal, M.: Fuzzy threshold graphs, CiiT. Int. J. Fuzzy Syst. **3**, 360–364 (2011)

18. Samanta, S., Pal, M.: Irregular bipolar fuzzy graphs. Int. J. Appl. Fuzzy Sets **2**, 91–102 (2012)

19. Samanta, S., Pal, M.: Bipolar fuzzy hyper graphs. Int. J. Fuzzy Logic Syst. **2**, 17–28 (2012)

20. Samanta, S., Pal, M.: Fuzzy K-Competition graphs and P-Competition fuzzy graphs. Fuzzy Eng. Inf. **5**, 191–204 (2013)

21. Samanta, S., Pal, M.: New concepts of fuzzy planar graph. Int. J. Adv. Res. Artif. Intell. **3**, 52–59 (2014)

22. Samanta, S., Pal, M., Akram, M.: m-step fuzzy competition graphs. J. Appl. Math. Comput. **47**, 461–472 (2015)

23. Samanta, S., Pal, M.: Fuzzy Planar graphs. IEEE Trans. Fuzzy Syst. **23**, 1936–1942 (2015)

24. Sunitha, M.S., Vijayakumar, A.: Complement of fuzzy graphs. Indian J. Pure and Appl. Math. **33**, 1451–1464 (2002)

25. Talebi, A. A., and Rashmanlou. H., Isomorphism on interval-valued fuzzy graphs. Ann. Fuzzy Math. Inform. **6**(1), 47–58 (2013)

26. Talebi, A.A., Rashmanlou, H., Sadati, S.H.: Interval valued intuitionistic fuzzy competition graph. J. Mult. Valued Logic Soft Comput. **34**, 5 (2020)

27. Zadeh, L.A.: Fuzzy sets. Inf. Control **8**, 338–353 (1965)

28. Zadeh, L.A.: The concept of a linguistic and application to approximate reasoning. Inf. Sci. **8**(3), 199–249 (1975)

Fermatean Fuzzy Soft Sets and Its Applications

Aparna Sivadas[✉] and Sunil Jacob John

Department of Mathematics, National Institute of Technology,
Calicut 673601, India
{aparna_p190067ma,sunil}@nitc.ac.in

Abstract. A hybrid structure-Fermatean fuzzy soft set is proposed, which includes the characteristics of Fermatean fuzzy set along with the parameterization of soft set. A few basic operations are specified such as union, intersection and complement. Further an algorithm to solve decision-making problem is developed using the aggregation operators defined for this structure.

Keywords: Fermatean fuzzy sets · Soft sets · Fermatean fuzzy soft sets · Decision-making problems

1 Introduction

Orthopair fuzzy set is a potential tool to express uncertain information.The membership grade of this non standard fuzzy set contains a pair of numbers in $[0,1]$. The first value depicts the degree of membership and the other the degree of non-membership. Clearly the classical intuitionistic fuzzy set [1,2], can be viewed as an orthopair fuzzy set with the condition that the sum of degree of membership and degree of non-membership is bounded by one. Several researchers have developed various algorithms to deal with problems in MCDM using aggregation operators (AOs) on IFSs. Some weighted averaging operators were introduced by Xu [3]. Garg [4,5] presented AOs which made use of Einstein t-norms and t-conorms in operational laws on intuitionistic fuzzy numbers (IFNs). AOs using Hamacher operators were developed by Huang [6] and Garg [7]. Cheng and Chang [8] suggested a technique to convert intuitionistic fuzzy values into right-angled triangular fuzzy numbers and vice-versa which was then used to define geometric AOs.

Another orthopair fuzzy set is the Pythagorean fuzzy set [9] whose sum of squares of membership value and non-membership value is confined to the value one. Clearly it is a generalization of IFSs and can handle more information than IFSs. Moreover Yager [10] suggested a variety of AOs on this set. Later Yager and Abbasov [11] suggested that Pythagorean membership grades can be expressed as complex numbers and used this relation in geometric aggregation operations. Along with some properties of already defined averaging operators Peng and

© Springer Nature Singapore Pte Ltd. 2021
A. Awasthi et al. (Eds.): CSMCS 2020, CCIS 1345, pp. 203–216, 2021.
https://doi.org/10.1007/978-981-16-4772-7_16

Yuan [12] introduced some generalized averaging operators for the Pythagorean fuzzy information. Garg [13] suggested the notion of using Einstein t-norm and t-conorm operations to aggregate Pythagorean fuzzy numbers (PFNs). Another Pythagorean fuzzy AO was put forward by Zeng *et al.* [14] and they illustrated its application in MAGDM problem. Unfamiliarity of decision makers about the objects evaluated is a concern in decision making problems. Thus Garg [15] developed AOs which included the notion of confidence levels for each evaluation of decision maker in the form of PFN.

The concept of q-rung orthopair fuzzy set proposed by Yager [16] has a pair of values as membership grades with the property that sum of the qth power of the membership value and the qth power of non-membership value is bounded by one. Thus IFS and PFS can be considered as its particular cases. Since an increase in the rung q increases the number of pairs satisfying the mentioned condition, it is clear that q-rung orthopair fuzzy set allows us to capture uncertain information more effectively. Along with various set operations Yager [16] proposed the aggregation of q-rung orthopair fuzzy sets. Liu and Wang [17] suggested two new aggregation operators for this fuzzy set. They developed some methods using these operators to approach decision-making problems.

Senapati and Yager [18] suggested an extension of IFSs called the Fermatean fuzzy set. When $q = 3$ q-rung orthopair fuzzy set is called a Fermatean fuzzy set. They listed some set operations like union, intersection complement etc. on this set. Two values specifically score function and accuracy function were defined to rank FFSs. Moreover they applied TOPSIS technique to tackle MCDM problems with Fermatean fuzzy information. Recently they developed several AOs for FFSs [19] and also applied the WPM method, a frequently applied MCDM method, to Fermatean fuzzy data.

Molodtsov [20] proposed the notion of soft set, a family of subsets of universal set where the subsets are obtained based on a parameter set. Maji *et al.* [21] deliberated on this new concept of soft set and provided several operations on soft set. In most of the practical situations the parameter set involved is fuzzy in nature and this resulted in hybrid structures like Fuzzy Soft Set (FSS) [22], Intuitionistic Fuzzy Soft Set (IFSS) [23] and Pythagorean Fuzzy Soft Set (PFSS) [24]. R.Arora and H.Garg [25] presented averaging and geometric operators for IFSNs and discussed a MCDM method using these operators. Along with various binary operations Peng [24] suggested a decision making algorithm based on this set. Abhishek Guleria and Rakesh Kumar Bajaj [26] represented PFSS as Pythagorean Fuzzy Soft Matrix and formulated decision making algorithm using these matrices. This matrix formulation is further utilised by Naeem *et al.* [27] to employ PFSS in MCGDM. In addition to this they developed a PFS TOPSIS and PFS VIKOR techniques and illustrated its application in real life decision making problems.

Within this work we have introduced Fermatean fuzzy soft set (FFSS) and defined a few basic operations on it. Two Fermatean fuzzy soft aggregation operators are developed for the Fermatean fuzzy soft numbers (FFSNs) and a decision making approach is discussed using these two AOs.

2   Preliminaries

In this section, some basic concepts of soft set and Fermatean fuzzy set required for further discussions are enlisted.

Definition 1. [21] *Let U be the universe set and E be a set of parameters. Let $P(U)$ be the power set of U and $A \subseteq E$. A pair (F, A) is called a soft set over U where F is a mapping $F : A \rightarrow P(U)$. In other words a soft set is a parameterized family of subsets of U.*

Definition 2. [21] *Let (F, A) and (G, B) be two soft sets over a common universe U and $A, B \subseteq E$ then*

1. *(F, A) is a soft subset of (G, B), if $A \subseteq B$ and $F(a) \subseteq G(a)$ for all $a \in A$, denoted as $(F, A) \preccurlyeq (G, B)$.*
2. *(F, A) and (G, B) are soft equal, if $(F, A) \preceq (G, B)$ and $(G, B) \preceq (F, A)$.*
3. *The union of (F, A) and (G, B) is the soft set (H, C), where $C = A \cup B$ and for all $e \in C$,*

$$H(e) = \begin{cases} F(e) & \text{if } e \in A \setminus B \\ G(e) & \text{if } e \in B \setminus A \\ F(e) \cup G(e) & \text{if } e \in A \cap B \end{cases}$$

4. *The intersection of (F, A) and (G, B) is the soft set (H, C), where $C = A \cap B \neq \emptyset$ and for all $e \in C$, $H(e) = F(e) \cap G(e)$.*
5. *The complement of a soft set (F, A) is the soft set (F^c, A) where $F^c(a) = U \setminus F(a)$ for all $a \in A$.*

Definition 3. [18] *Let U be the universe set, a Fermatean fuzzy set F in U is an object having the form $F = \{(u, \alpha_F(u), \beta_F(u)) : u \in U\}$ where $\alpha_F : U \rightarrow [0, 1]$ and $\beta_F : U \rightarrow [0, 1]$, with the condition $0 \leq (\alpha_F(u))^3 + (\beta_F(u))^3 \leq 1$ for all $u \in U$.*

The numbers $\alpha_F(u)$ and $\beta_F(u)$ indicate respectively, the degree of membership and the degree of non-membership of the element u to the set F. The value $\sqrt[3]{1 - ((\alpha_F(u))^3 + (\beta_F(u))^3)}$ is defined as the degree of indeterminacy of u to F. Each $(\alpha_F(u), \beta_F(u))$ is called a Fermatean fuzzy number (FFN) and for simplicity it is denoted as (α_F, β_F).

Definition 4. [18] *For two Fermatean fuzzy sets F and G in the universe U, the union, intersection and complement are defined as follows:*

1. *$F \subseteq G$, if for all $u \in U, \alpha_F(u) \leq \alpha_G(u)$ and $\beta_F(u) \geq \beta_G(u)$.*
2. *$F = G$ if $F \subseteq G$ and $G \subseteq F$.*
3. *$F \cup G = \{(u, max\{\alpha_F(u), \alpha_G(u)\}, min\{\beta_F(u), \beta_G(u)\}) : u \in U\}$.*
4. *$F \cap G = \{(u, min\{\alpha_F(u), \alpha_G(u)\}, max\{\beta_F(u), \beta_G(u)\}) : u \in U\}$.*
5. *$F^c = \{(u, \beta_F(u), \alpha_F(u)) : u \in U\}$.*

Definition 5. [18] *Let $F = (\alpha_F, \beta_F), F_1 = (\alpha_{F_1}, \beta_{F_1})$ and $F_2 = (\alpha_{F_2}, \beta_{F_2})$ be three FFNs and $\lambda > 0$, then their operations are defined as follows:*

1. $F_1 \boxplus F_2 = (\sqrt[3]{\alpha_{F_1}^3 + \alpha_{F_2}^3 - \alpha_{F_1}^3 \alpha_{F_2}^3}, \beta_{F_1} \beta_{F_2})$

2. $F_1 \boxtimes F_2 = (\alpha_{F_1} \alpha_{F_2}, \sqrt[3]{\beta_{F_1}^3 + \beta_{F_2}^3 - \beta_{F_1}^3 \beta_{F_2}^3})$

3. $\lambda F = (\sqrt[3]{1 - (1 - \alpha_F^3)^\lambda}, \beta_F^\lambda)$

4. $F^\lambda = (\alpha_F^\lambda, \sqrt[3]{1 - (1 - \beta_F^3)^\lambda})$.

In order to rank FFNs Senapati and Yager defined the score function for FFNs.

Definition 6. [18] *Let $F = (\alpha_F, \beta_F)$ be a FFN, then the score function of F can be defined as $score(F) = \alpha_F^3 - \beta_F^3$*
Clearly $score(F)$ lies in $[-1, 1]$.

For any two FFNs F_1 and F_2 if $score(F_1) < score(F_2)$ then $F_1 < F_2$, if $score(F_1) > score(F_2)$, then $F_1 > F_2$ and if $score(F_1) = score(F_2)$, then $F_1 \sim F_2$. But in some situations this score function is not suitable to compare two FFNs, for instance if $F_1 = (0.7, 0.7)$ and $F_2 = (0.4, 0.4)$ then $score(F_1) = score(F_2) = 0$ but these FFNs are not identical. For this reason accuracy function for FFN was defined.

Definition 7. [18] *Let $F = (\alpha_F, \beta_F)$ be a FFN, then the accuracy function of F can be defined as $acc(F) = \alpha_F^3 + \beta_F^3$.*

Evidently $acc(F) \in [0, 1]$. Greater the value of $acc(F)$, higher is the degree of accuracy of the FFN F. In accordance with the values of the score and accuracy functions of FFNs, the ordering for any two FFNs is explained as:

Definition 8. [18] *Let $F_1 = (\alpha_{F_1}, \beta_{F_1})$ and $F_2 = (\alpha_{F_2}, \beta_{F_2})$ be two FFNs, then*

1. *if $score(F_1) < score(F_2)$, then $F_1 < F_2$;*
2. *if $score(F_1) > score(F_2)$, then $F_1 > F_2$;*
3. *if $score(F_1) = score(F_2)$, then*
   (a) if $acc(F_1) < acc(F_2)$, then $F_1 < F_2$.
   (b) if $acc(F_1) > acc(F_2)$, then $F_1 > F_2$.
   (c) if $acc(F_1) = acc(F_2)$, then $F_1 = F_2$.

3   Fermatean Fuzzy Soft Sets

Definition 9. *Let U be the universe set and E be a set of parameters, $A \subseteq E$ a Fermatean fuzzy soft set (FFSS) on U is defined as the pair (F, A) where where F is mapping given by $F : A \to FFS(U)$, where $FFS(U)$ is the set of all Fermatean fuzzy sets over U. Here for any parameter $e \in A$, $F(e)$ is the Fermatean fuzzy set given as $F(e) = \{(u, \alpha_{F(e)}(u), \beta_{F(e)}(u)) : u \in U\}$ where $\alpha_{F(e)}(u)$ and $\beta_{F(e)}(u)$ are the degree of membership and the degree of non-membership respectively with the condition $0 \leqslant (\alpha_{F(e)}(u))^3 + (\beta_{F(e)}(u))^3 \leqslant 1$.*
Hence $(F, A) = \{(e, \{(u, \alpha_{F(e)}(u), \beta_{F(e)}(u)\}) : e \in A, u \in U\}$.

Example: Let $U = \{p_1, p_2, p_3, p_4\}$ be a set of jobs available for an individual. Let $E = \{$ high salary (e_1), interesting job (e_2), close driving distance $(e_3)\}$ be the set of parameters under consideration. Then we can dsecribe the "feasibility of the job" using a Fermatean fuzzy soft set. Let

$F(e_1) = \{(p_1, 0.5, 0.9), (p_2, 0.8, 0.6), (p_3, 0.4, 0.6), (p_4, 0.2, 0.9\}$

$F(e_2) = \{(p_1, 0.6, 0.6), (p_2, 0.8, 0.76), (p_3, 0.3, 0.6), (p_4, 0.5, 0.2)\}$

$F(e_3) = \{(p_1, 0.35, 0.7), (p_2, 0.8, 0.9), (p_3, 0.1, 0.5), (p_4, 0.5, 0.7)\}$

Then the Fermatean fuzzy soft set $(F, E) = \{F(e_1), F(e_2), (F(e_3)\}$ is a parameterized collection of Fermatean fuzzy sets over U. Each pair in the set $F_{e_j}(u_i) = \{(u_i, \alpha_{F(e_j)}(u_i), \beta_{F(e_j)}(u_i)) : u_i \in U\}$ is called a Fermatean fuzzy soft number (FFSN) and is denoted as $F_{e_{ij}} = (\alpha_{ij}, \beta_{ij})$.

Definition 10. *Let (F, A) and (G, B) be two FFSSs over a common universe U and $A, B \subseteq E$ then (F, A) is a soft subset of (G, B) denoted as $(F, A) \widehat{\subset} (G, B)$ if*

1. $A \subseteq B$
2. for all $e \in A$, $F(e)$ is a Fermatean fuzzy subset of $G(e)$.

If $(F, A) \widehat{\subset} (G, B)$ and $(G, B) \widehat{\subset} (F, A)$ then (F, A) and (G, B) are said to be equal.

Definition 11. *Let (F, A) and (G, B) be two FFSSs over a common universe U, intersection of (F, A) and (G, B) represented as $(F, A) \widehat{\cap} (G, B)$ is the Fermatean fuzzy soft set (H, C) where $C = A \cap B \neq \emptyset$ and $H(e) = F(e) \cap G(e)$ for all $e \in C$.*

Definition 12. *Let (F, A) and (G, B) FFSSs over a common universe U, union of (F, A) and (G, B) represented as $(F, A) \widehat{\cup} (G, B)$ is the Fermatean fuzzy soft set (H, C) where $C = A \cup B$ and*

$$H(e) = \begin{cases} F(e) & \text{if } e \in A \setminus B \\ G(e) & \text{if } e \in B \setminus A \\ F(e) \cup G(e) & \text{if } e \in A \cap B \end{cases}$$

Definition 13. *Let (F, A) is a FFSS over U, the complement of (F, A) is denoted by (F^c, A) where $F^c : A \to FFS(U)$ is the mapping given by $F^c(e) = (F(e))^c$ for all $e \in A$.*

Definition 14. *A FFSS (F, E) over U is known as a null FFSS represented as $F_\emptyset$ if for all $e \in E$, $F_\emptyset(e) = 0$ where 0 denote the null Fermatean fuzzy set, $F_\emptyset = \{(e, \{u, 0, 1\}) : e \in E, u \in U\}$.*

Definition 15. *A FFSS (F, E) over U is known as an absolute FFSS represented as $\widetilde{U}$ if $F(e) = \widetilde{1}$ for all $e \in E, \widetilde{1}$ denote absolute Fermatean fuzzy set, hence $\widetilde{U} = \{(e, \{u, 1, 0\}) : e \in E, u \in U\}$.*

Proposition 1. *Let (F, A) and (G, B) are FFSS over U, then*

1. $(F,A) \mathbin{\widehat{\cap}} (F,A) = (F,A)$
2. $(F,A) \mathbin{\widehat{\cup}} (F,A) = (F,A)$
3. $(F,A) \mathbin{\widehat{\cap}} (G,B) = (G,B) \mathbin{\widehat{\cap}} (F,A)$
4. $(F,A) \mathbin{\widehat{\cup}} (G,B) = (G,B) \mathbin{\widehat{\cup}} (F,A)$
5. $(F,A) \mathbin{\widehat{\cap}} ((G,B) \mathbin{\widehat{\cap}} (H,C)) = ((F,A) \mathbin{\widehat{\cap}} (G,B)) \mathbin{\widehat{\cap}} (H,C)$
6. $(F,A) \mathbin{\widehat{\cup}} ((G,B) \mathbin{\widehat{\cup}} (H,C)) = ((F,A) \mathbin{\widehat{\cup}} (G,B)) \mathbin{\widehat{\cup}} (H,C)$.

Proposition 2. *Let (F,A) and (G,B) are FFSS over U, then*

1. $((F,A) \mathbin{\widehat{\cap}} (G,B)) \mathbin{\widehat{\subset}} (F,A) \mathbin{\widehat{\subset}} ((F,A) \mathbin{\widehat{\cup}} (G,B))$.
2. $((F,A) \mathbin{\widehat{\cap}} (G,B)) \mathbin{\widehat{\subset}} (G,B) \mathbin{\widehat{\subset}} ((F,A) \mathbin{\widehat{\cup}} (G,B))$.

Proof: $min\{\alpha_{F(e)}(u), \alpha_{G(e)}(u)\} \leq \alpha_{F(e)}(u) \leq max\{\alpha_{F(e)}(u), \alpha_{G(e)}(u)\}$ and $max\{\beta_{F(e)}(u), \beta_{G(e)}(u)\} \geq \beta_{F(e)}(u) \geq min\{\beta_{F(e)}(u), \beta_{G(e)}(u)\}$ for all $u \in U$, hence (1) is true. Similarly (2) is true.

Remark 1. Let $U = \{u_1, u_2, u_3, u_4, u_5, u_6\}$, $E = \{e_1, e_2, e_3, e_4, e_5\}$, $A_1 = \{e_1, e_3\}$, $A_2 = \{e_3, e_5\}$.
Consider the FFSSs (F, A_1) and (G, A_2) given as
$(F, A_1) = \{(e_1, \{u_1, 0.3, 0.9\}, \{u_3, 0.6, 0.9\}, \{u_6, 0.8, 0.78\}), (e_3, \{u_2, 0.5, 0.9\}, \{u_6, 0.3, 0.8\})\}$
$(G, A_2) = \{(e_3, \{u_1, 0.7, 0.4\}, \{u_6, 0.6, 0.6\}), (e_5, \{u_2, 0.7, 0.2\}, \{u_3, 0.6, 0.5\}, \{u_6, 0.4, 0.2\})\}$.
Then
$(F, A_1)^c = \{(e_1, \{u_1, 0.9, 0.3\}, \{u_3, 0.9, 0.6\}, \{u_6, 0.78, 0.8\}), (e_3, \{u_2, 0.9, 0.5\}, \{u_6, 0.8, 0.3\})\}$
$(G, A_2)^c = \{(e_3, \{u_1, 0.4, 0.7\}, \{u_6, 0.6, 0.6\}), (e_5, \{u_2, 0.2, 0.7\}, \{u_3, 0.5, 0.6\}, \{u_6, 0.2, 0.4\})\}$
$(F, A_1) \mathbin{\widehat{\cup}} (G, A_2) = \{(e_1, \{u_1, 0.3, 0.9\}, \{u_3, 0.6, 0.9\}\{u_6, 0.8, 0.78\}), (e_3, \{u_1, 0.7, 0.4\}, \{u_2, 0.5, 0.9\}, \{u_6, 0.6, 0.6\}), (e_5, \{u_2, 0.7, 0.2\}, \{u_3, 0.6, 0.5\}, \{u_6, 0.4, 0.2\})\}$
$((F, A_1) \mathbin{\widehat{\cup}} (G, A_2))^c = \{(e_1, \{u_1, 0.9, 0.3\}, \{u_3, 0.9, 0.6\}, \{u_6, 0.78, 0.8\}), (e_3, \{u_1, 0.4, 0.7\}, \{u_2, 0.9, 0.5\}, \{u_6, 0.6, 0.6\}), (e_5, \{u_2, 0.2, 0.7\}, \{u_3, 0.5, 0.6\}, \{u_6, 0.2, 0.4\})\}$
$(F, A_1)^c \mathbin{\widehat{\cap}} (G, A_2)^c = \{(e_3, \{u_6, 0.6, 0.6\})\}$
Here $((F, A_1) \mathbin{\widehat{\cup}} (G, A_2))^c \neq (F, A_1)^c \mathbin{\widehat{\cap}} (G, A_2)^c$
$(F, A_1) \mathbin{\widehat{\cap}} (G, A_2) = \{(e_3, \{u_6, 0.3, 0.8\})\}$
$((F, A_1) \mathbin{\widehat{\cap}} (G, A_2))^c = \{(e_3, \{u_6, 0.8, 0.3\})\}$
$(F, A_1)^c \mathbin{\widehat{\cup}} (G, A_2)^c = \{(e_1, \{u_1, 0.9, 0.3\}, \{u_3, 0.9, 0.6\}, \{u_6, 0.78, 0.8\}), (e_3, \{u_1, 0.4, 0.7\}, \{u_2, 0.9, 0.5\}, \{u_6, 0.8, 0.3\}), (e_5, \{u_2, 0.4, 0.7\}, \{u_3, 0.5, 0.6\}, \{u_6, 0.2, 0.4\})\}$.
Here $((F, A_1) \mathbin{\widehat{\cap}} (G, A_2))^c \neq (F, A_1)^c \mathbin{\widehat{\cup}} (G, A_2)^c$.
Thus FFSSs do not satisfy the De Morgan's laws.

Remark 2. Let $U = \{u_1, u_2, u_3, u_4, u_5, u_6\}$, $E = \{e_1, e_2, e_3, e_4, e_5\}$, $A = \{e_1, e_3\}$
Consider the FFSS $(F, A) = \{(e_1, \{u_1, 0.3, 0.9\}, \{u_3, 0.6, 0.9\}, \{u_6, 0.8, 0.78\})$,
$(e_3, \{u_2, 0.5, 0.9\}, \{u_6, 0.3, 0.8\})\}$.
Then $(F, A)^c = \{(e_1, \{u_1, 0.9, 0.3\}, \{u_3, 0.9, 0.6\}, \{u_6, 0.78, 0.8\}), (e_3, \{u_2, 0.9,$
$0.5\}, \{u_6, 0.8, 0.3\})\}$
$(F, A) \,\widehat{\cup}\, (F, A)^c = \{(e_1, \{u_1, 0.9, 0.3\}, \{u_3, 0.9, 0.6\}, \{u_6, 0.8, 0.78\}), (e_3, \{u_2, 0.9,$
$0.5\}, \{u_6, 0.8, 0.3\})\}$
Here $(F, A) \,\widehat{\cup}\, (F, A)^c \neq \widetilde{U}$.
$(F, A) \,\widehat{\cap}\, (F, A)^c = \{(e_1, \{u_1, 0.3, 0.9\}, \{u_3, 0.6, 0.9\}, \{u_6, 0.78, 0.8\}), (e_3, \{u_2, 0.5,$
$0.9\}, \{u_6, 0.3, 0.8\})\}$.
Here $(F, A) \,\widehat{\cap}\, (F, A)^c \neq F_\emptyset$.

3.1 Aggregation Operators for Fermatean Fuzzy Soft Numbers

In this section, two aggregation operators namely Fermatean Fuzzy Soft
Weighted Averaging operator (FFSWA) and Fermatean Fuzzy Soft Weighted
Geometric operator (FFSWG) are discussed. Let Λ denote a collection of FFSNs.

Definition 16. *Let $F_{e_{ij}} = (\alpha_{ij}, \beta_{ij})(i = 1, 2, ...n; j = 1, 2, ...m)$ be FFNs and
r_j and s_i be the the weight vectors for the parameters e_j's and experts x_i's
respectively satisfying $r_j > 0$, $s_i > 0$ and $\sum_{j=1}^m r_j = 1$, $\sum_{i=1}^n s_i = 1$, then the
operator $FFSWA : \Lambda^n \to \Lambda$ is defined as*
$$FFSWA(F_{e_{11}}, F_{e_{12}}, F_{e_{13}}, ..., F_{e_{nm}}) = \boxplus_{j=1}^m r_j (\boxplus_{i=1}^n s_i F_{e_{ij}}).$$

Theorem 1. *Let $F_{e_{ij}} = (\alpha_{ij}, \beta_{ij})(i = 1, 2, ...n; j = 1, 2, ...m)$ be FFNs, the
aggregated value using the FFSWA operator is again a FFSN obtained by the
following expression*
$$FFSWA(F_{e_{11}}, F_{e_{12}}, F_{e_{13}}, ..., F_{e_{nm}}) = \left(\sqrt[3]{1 - \prod_{j=1}^m (\prod_{i=1}^n (1 - \alpha_{ij}^3)^{s_i})^{r_j}}, \prod_{j=1}^m \right.$$
$$\left. (\prod_{i=1}^n \beta_{ij}^{s_i})^{r_j} \right).$$

Proof: For $n = 1$, we get $s_1 = 1$
$$FFSWA(F_{e_{11}}, F_{e_{12}}, F_{e_{13}}, ..., F_{e_{1m}}) = \boxplus_{j=1}^m r_j F_{e_{1j}}$$
$$\boxplus_{j=1}^m r_j F_{e_{ij}} = r_1 F_{e_{11}} \boxplus r_2 F_{e_{12}} \boxplus ...... \boxplus r_m F_{e_{1m}}$$
$$r_1 F_{e_{11}} \boxplus r_2 F_{e_{12}} = (\sqrt[3]{1 - [(1 - \alpha_{11}^3)^{r_1} (1 - \alpha_{12}^3)^{r_2}]}, \beta_{11}^{r_1} \beta_{12}^{r_2})$$
$$r_1 F_{e_{11}} \boxplus r_2 F_{e_{12}} \boxplus r_3 F_{e_{13}} = (\sqrt[3]{1 - [(1 - \alpha_{11}^3)^{r_1} (1 - \alpha_{12}^3)^{r_2} (1 - \alpha_{13}^3)^{r_3}]}, \beta_{11}^{r_1} \beta_{12}^{r_2} \beta_{13}^{r_3})$$
Hence $\boxplus_{j=1}^m r_j F_{e_{1j}} = (\sqrt[3]{1 - \prod_{j=1}^m (1 - \alpha_{1j}^3)^{r_j}}, \prod_{j=1}^m \beta_{1j}^{r_j}) =$
$$(\sqrt[3]{1 - \prod_{j=1}^m (\prod_{i=1}^1 (1 - \alpha_{ij}^3)^{s_i})^{r_j}}, \prod_{j=1}^m (\prod_{i=1}^1 \beta_{ij}^{s_i})^{r_j}).$$
For $m = 1$, we get $r_1 = 1$ we have
$$FFSWA(F_{e_{11}}, F_{e_{21}}, F_{e_{31}}, ..., F_{e_{n1}}) = \boxplus_{i=1}^n s_i F_{e_{i1}} = (\sqrt[3]{1 - \prod_{i=1}^n (1 - \alpha_{i1}^3)^{s_i}},$$
$$\prod_{i=1}^n \beta_{i1}^{s_i}) = (\sqrt[3]{1 - \prod_{j=1}^1 (\prod_{i=1}^n (1 - \alpha_{ij}^3)^{s_i})^{r_j}}, \prod_{j=1}^1 (\prod_{i=1}^n \beta_{ij}^{s_i})^{r_j}).$$
Thus the expression is true for $n = 1$, $m = 1$.

Assume the expression is true for $m = t_1 + 1$, $n = t_2$ and $m = t_1$, $n = t_2 + 1$ that is

$$\boxplus_{j=1}^{t_1+1} r_j (\boxplus_{i=1}^{t_2} s_i F_{e_{ij}}) = (\sqrt[3]{1 - \prod_{j=1}^{t_1+1}(\prod_{i=1}^{t_2}(1 - \alpha_{ij}^3)^{s_i})^{r_j}}, \prod_{j=1}^{t_1+1}(\prod_{i=1}^{t_2} \beta_{ij}^{s_i})^{r_j})$$

$$\boxplus_{j=1}^{t_1} r_j (\boxplus_{i=1}^{t_2+1} s_i F_{e_{ij}}) = (\sqrt[3]{1 - \prod_{j=1}^{t_1}(\prod_{i=1}^{t_2+1}(1 - \alpha_{ij}^3)^{s_i})^{r_j}}, \prod_{j=1}^{t_1}(\prod_{i=1}^{t_2+1} \beta_{ij}^{s_i})^{r_j}).$$

Now for $m = t_1 + 1, n = t_2 + 1$ we get

$$\boxplus_{j=1}^{t_1+1} r_j \boxplus_{i=1}^{t_2+1} s_i F_{e_{ij}} = \boxplus_{j=1}^{t_1+1} r_j (\boxplus_{i=1}^{t_2} s_i F_{e_{ij}} \boxplus s_{t_2+1} F_{e_{(t_2+1)j}}) = \boxplus_{j=1}^{t_1+1} r_j \boxplus_{i=1}^{t_2} s_i F_{e_{ij}}$$

$$\boxplus \boxplus_{j=1}^{t_1+1} r_j s_{t_2+1} F_{e_{(t_2+1)j}} = (\sqrt[3]{1 - \prod_{j=1}^{t_1+1}(\prod_{i=1}^{t_2}(1 - \alpha_{ij}^3)^{s_i})^{r_j}}, \prod_{j=1}^{t_1+1}(\prod_{i=1}^{t_2} \beta_{ij}^{s_i})^{r_j})$$

$$\boxplus (\sqrt[3]{1 - \prod_{j=1}^{t_1+1}((1 - \alpha_{(t_2+1)j}^3)^{s_{t_2+1}})^{r_j}}, \prod_{j=1}^{t_1+1}(\beta_{(t_2+1)j}^{s_{t_2+1}})^{r_j}) =$$

$$(\sqrt[3]{1 - \prod_{j=1}^{t_1+1}(\prod_{i=1}^{t_2+1}(1 - \alpha_{ij}^3)^{s_i})^{r_j}}, \prod_{j=1}^{t_1+1}(\prod_{i=1}^{t_2+1} \beta_{ij}^{s_i})^{r_j}).$$

Hence it is true for $m = t_1 + 1, n = t_2 + 1$ and thus it is true by induction for all $n, m \geq 1$.

We have $0 \leq \alpha_{ij} \leq 1 \implies 0 \leq \alpha_{ij}^3 \leq 1 \implies 0 \leq 1 - \alpha_{ij}^3 \leq 1 \implies$
$0 \leq \prod_{j=1}^{m}(\prod_{i=1}^{n}(1 - \alpha_{ij}^3)^{s_i})^{r_j} \leq 1 \implies 0 \leq 1 - \prod_{j=1}^{m}(\prod_{i=1}^{n}(1 - \alpha_{ij}^3)^{s_i})^{r_j} \leq 1$
$0 \leq \beta_{ij} \leq 1 \implies 0 \leq \beta_{ij}^3 \leq 1 \implies 0 \leq \prod_{j=1}^{m}(\prod_{i=1}^{n} \beta_{ij}^{s_i})^{r_j} \leq 1.$

Consider $[\sqrt[3]{1 - \prod_{j=1}^{m}(\prod_{i=1}^{n}(1 - \alpha_{ij}^3)^{s_i})^{r_j}}]^3 + [\prod_{j=1}^{m}(\prod_{i=1}^{n} \beta_{ij}^{s_i})^{r_j}]^3 = 1 - \prod_{j=1}^{m}(\prod_{i=1}^{n}(1-\alpha_{ij}^3)^{s_i})^{r_j} + \prod_{j=1}^{m}(\prod_{i=1}^{n}(\beta_{ij}^3)^{s_i})^{r_j} \leq 1 - \prod_{j=1}^{m}(\prod_{i=1}^{n}(1-\alpha_{ij}^3)^{s_i})^{r_j} + \prod_{j=1}^{m}(\prod_{i=1}^{n}(1 - \alpha_{ij}^3)^{s_i})^{r_j} = 1.$

Thus we obtain a FFSN using the FFWA operator.

Definition 17. *Let $F_{e_{ij}} = (\alpha_{ij}, \beta_{ij})(i = 1, 2, ...n; j = 1, 2, ...m)$ be FFNs and r_j and s_i be the the weight vectors for the parameters e_j's and experts x_i's respectively satisfying $r_j > 0$, $s_i > 0$ and $\sum_{j=1}^{m} r_j = 1$, $\sum_{i=1}^{n} s_i = 1$, then the operator $FFSWG : \Lambda^n \to \Lambda$ is defined as*
$FFSWG(F_{e_{11}}, F_{e_{12}}, F_{e_{13}}, ..., F_{e_{nm}}) = \boxtimes_{j=1}^{m}(\boxtimes_{i=1}^{n} F_{e_{ij}}^{s_i})^{r_j}.$

Theorem 2. *Let $F_{e_{ij}} = (\alpha_{ij}, \beta_{ij})(i = 1, 2, ...n; j = 1, 2, ...m)$ be FFNs, the aggregated value using the FFSWG operator is again a FFSN obtained by the following expression*
$FFSWG(F_{e_{11}}, F_{e_{12}}, F_{e_{13}}, ..., F_{e_{nm}}) = (\prod_{j=1}^{m}(\prod_{i=1}^{n} \alpha_{ij}^{s_i})^{r_j},$
$\sqrt[3]{1 - \prod_{j=1}^{m}(\prod_{i=1}^{n}(1 - \beta_{ij}^3)^{s_i})^{r_j}}).$

Proof: For $n = 1$, we get $s_1 = 1$
$FFSWA(F_{e_{11}}, F_{e_{12}}, F_{e_{13}}, ..., F_{e_{1n}}) = \boxtimes_{j=1}^{m} F_{e_{1j}}^{r_j} = (\prod_{j=1}^{m} \alpha_{ij}^{r_j},$
$\sqrt[3]{1 - \prod_{j=1}^{m}(1 - \beta_{ij}^3)^{r_j}}) = (\prod_{j=1}^{m}(\prod_{i=1}^{1} \alpha_{ij}^{s_i})^{r_j}, \sqrt[3]{1 - \prod_{j=1}^{m}(\prod_{i=1}^{1}(1 - \beta_{ij}^3)^{s_i})^{r_j}})$
For $m = 1$, we get $r_1 = 1$ we have
$FFSWA(F_{e_{11}}, F_{e_{21}}, F_{e_{31}}, ..., F_{e_{n1}}) = \boxtimes_{i=1}^{n} F_{e_{i1}}^{s_i} = (\prod_{i=1}^{n} \alpha_{ij}^{s_i},$
$\sqrt[3]{1 - \prod_{i=1}^{n}(1 - \beta_{ij}^3)^{s_i}}) = (\prod_{j=1}^{1}(\prod_{i=1}^{n} \alpha_{ij}^{s_i})^{r_j}, \sqrt[3]{1 - \prod_{j=1}^{1}(\prod_{i=1}^{n}(1 - \beta_{ij}^3)^{s_i})^{r_j}})$
Thus the expression is true for $n = 1, m = 1$.

Assume the expression is true for $m = t_1 + 1, n = t_2$ and $m = t_1, n = t_2 + 1$ that is

$$\boxtimes_{j=1}^{t_1+1}(\boxtimes_{i=1}^{t_2} F_{e_{ij}}^{s_i})^{r_j} = (\prod_{j=1}^{t_1+1}(\prod_{i=1}^{t_2} \alpha_{ij}^{s_i})^{r_j}, \sqrt[3]{1 - \prod_{j=1}^{t_1+1}(\prod_{i=1}^{t_2}(1 - \beta_{ij}^3)^{s_i})^{r_j}})$$ and

$$\boxtimes_{j=1}^{t_1}(\boxtimes_{i=1}^{t_2+1} F_{e_{ij}}^{s_i})^{r_j} = (\prod_{j=1}^{t_2}(\prod_{i=1}^{t_2+1} \alpha_{ij}^{s_i})^{r_j}, \sqrt[3]{1 - \prod_{j=1}^{t_1}(\prod_{i=1}^{t_2+1}(1 - \beta_{ij}^3)^{s_i})^{r_j}})$$

Now for $m = t_1 + 1, n = t_2 + 1$ we get

$$\boxtimes_{j=1}^{t_1+1}(\boxtimes_{i=1}^{t_2+1} F_{e_{ij}}^{s_i})^{r_j} = \boxtimes_{j=1}^{t_1+1}(\boxtimes_{i=1}^{t_2} F_{e_{ij}}^{s_i} \boxtimes F_{e_{(t_2+1)j}}^{s_{t_2+1}})^{r_j} = \boxtimes_{j=1}^{t_1+1}(\boxtimes_{i=1}^{t_2} F_{e_{ij}}^{s_i})^{r_j} \boxtimes$$

$$\boxtimes_{j=1}^{t_1+1}(F_{e_{(t_2+1)j}}^{s_{t_2+1}})^{r_j} = (\prod_{j=1}^{t_1+1}(\prod_{i=1}^{t_2} \alpha_{ij}^{s_i})^{r_j}, \sqrt[3]{1 - \prod_{j=1}^{t_1+1}(\prod_{i=1}^{t_2}(1 - \beta_{ij}^3)^{s_i})^{r_j}}) \boxtimes$$

$$(\prod_{j=1}^{t_1+1}(\alpha_{(t_2+1)j}^{s_{t_2+1}})^{r_j}, \sqrt[3]{1 - \prod_{j=1}^{t_1+1}(1 - \beta_{ij}^3)^{s_{t_2+1}})^{r_j}})$$

$$= (\prod_{j=1}^{t_1+1}(\prod_{i=1}^{t_2+1} \alpha_{ij}^{s_i})^{r_j}, \sqrt[3]{1 - \prod_{j=1}^{t_1+1}(\prod_{i=1}^{t_2+1}(1 - \beta_{ij}^3)^{s_i})^{r_j}})$$

Hence it is true for $m = t_1 + 1, n = t_2 + 1$ and thus by induction it is true for all $n, m \geq 1$.

$$0 \leq \alpha_{ij} \leq 1 \implies 0 \leq \alpha_{ij}^{s_i} \leq 1 \implies 0 \leq \prod_{j=1}^{m}(\prod_{i=1}^{n} \alpha_{ij}^{s_i})^{r_j} \leq 1$$

$$0 \leq \beta_{ij} \leq 1 \implies 0 \leq \prod_{i=1}^{n}(1 - \beta_{ij})^{s_i} \leq 1 \implies 0 \leq \prod_{j=1}^{m}(\prod_{i=1}^{n}(1 - \beta_{ij})^{s_i})^{r_j} \leq$$

$$1 \implies 0 \leq \sqrt[3]{1 - \prod_{j=1}^{m}(\prod_{i=1}^{n}(1 - \beta_{ij})^{s_i})^{r_j}} \leq 1.$$

$$[\prod_{j=1}^{t_1+1}(\prod_{i=1}^{t_2+1} \alpha_{ij}^{s_i})^{r_j}]^3 + [\sqrt[3]{1 - \prod_{j=1}^{m}(\prod_{i=1}^{n}(1 - \beta_{ij}^3)^{s_i})^{r_j}}]^3 = \prod_{j=1}^{t_1+1}(\prod_{i=1}^{t_2+1}$$

$$(\alpha_{ij}^3)^{s_i})^{r_j} + 1 - \prod_{j=1}^{m}(\prod_{i=1}^{n}(1 - \beta_{ij}^3)^{s_i})^{r_j}) \leq \prod_{j=1}^{t_1+1}(\prod_{i=1}^{t_2+1}(1 - \beta_{ij}^3)^{s_i})^{r_j} +$$

$$1 - \prod_{j=1}^{m}(\prod_{i=1}^{n}(1 - \beta_{ij}^3)^{s_i})^{r_j}) = 1.$$ Hence the aggregate using FFSWG operator is a FFSN.

3.2 Decision-Making Using the Suggested Operators

A MCDM problem involves the evaluation of a set of alternatives by considering multiple criteria.

Let $Y = \{y_1, y_2, y_3, ...y_t\}$ be the set of alternatives, which are valuated by a set of n experts, $x_1, x_2, x_3, ...x_n$, with weight vector $s = (s_1, s_2, ...s_n)^T$, $s_i > 0$, $\sum_{i=1}^{n} s_i = 1$ under the parameters $E = \{e_1, e_2, ...e_m\}$ with weight vector $r = (r_1, r_2, ...r_m)^T$, $r_j > 0$, $\sum_{j=1}^{m} r_j = 1$. These experts give their assessment for the alternatives in terms of FFSNs, $F_{e_{ij}} = (\alpha_{ij}, \beta_{ij})$ satisfying $\alpha_{ij}^3 + \beta_{ij}^3 \leq 1$.

Step 1: Formulate the decision matrix $D = (F_{e_{ij}})_{n \times m}$ where $F_{e_{ij}}$ is the Fermatean fuzzy number $(\alpha_{ij}, \beta_{ij})$.

$$D_{n \times m} = \begin{bmatrix} (\alpha_{11}, \beta_{11}) & .... & (\alpha_{1m}, \beta_{1m}) \\ (\alpha_{21}, \beta_{21}) & .... & (\alpha_{2m}, \beta_{2m}) \\ . & .... & . \\ . & .... & . \\ . & .... & . \\ (\alpha_{n1}, \beta_{n1}) & .... & (\alpha_{nm}, \beta_{nm}) \end{bmatrix}$$

Step 2: Normalize the decision matrix by converting the assessment values of cost type parameters to benefit type parameters by applying the formula

$$\gamma_{ij} = \begin{cases} F^c_{e_{ij}} & \text{for cost type parameters} \\ F_{e_{ij}} & \text{for benefit type parameters} \end{cases}$$

Step 3: Aggregate the FFSNs $F_{e_{ij}}$ using FFSWA (or FFSWG) operator to get a collective value v_k for each alternative $y_k (k = 1, 2, \ldots\ldots t)$.

Step 4: Obtain the value of score function of v_k for each alternative.

Step 5: Rank the alternatives according to the values of score function and accuracy function of v_k and choose the best alternative(s).

Example: A manufacturing company wants to purchase raw material for production. Four suppliers S_1, S_2, S_3, S_4 are ready to provide the raw material. The company wants to choose a supplier from these four suppliers. The parameters influencing this selection includes

1. Net price
2. quality
3. financial position
4. delivery

with weight vector $(0.3, 0.3, 0.15, 0.25)^T$. A group of five experts c_1, c_2, c_3, c_4, c_5 is constituted to assess these four suppliers and make the decision. The weight vector of these experts is $(0.3, 0.1, 0.2, 0.15, 0.25)^T$.

Following are steps used in finding a suitable supplier. Let

1. e_1: Net price
2. e_2: quality
3. e_3: financial position
4. e_4: delivery

Here e_1 is a cost type parameter and e_2, e_3, e_4 are benefit type parameters.

Step 1: Experts evaluate the suppliers and provide their evaluations as FFSNs. The decision matrix corresponding to each supplier is given in Tables 1, 2, 3, 4.

Table 1. Fermatean fuzzy soft matrix of Supplier 1.

	Net Price	Quality	Financial Position	Delivery
c_1	$(0.90, 0.60)$	$(0.60, 0.70)$	$(0.79, 0.62)$	$(0.50, 0.50)$
c_2	$(0.75, 0.70)$	$(0.52, 0.95)$	$(0.65, 0.85)$	$(0.85, 0.58)$
c_3	$(0.56, 0.89)$	$(0.20, 0.55)$	$(0.10, 0.35)$	$(0.59, 0.84)$
c_4	$(0.70, 0.30)$	$(0.92, 0.60)$	$(0.90, 0.45)$	$(0.40, 0.12)$
c_5	$(0.10, 0.80)$	$(0.70, 0.15)$	$(0.25, 0.25)$	$(0.65, 0.45)$

Table 2. Fermatean fuzzy soft matrix of Supplier 2.

	Net Price	Quality	Financial Position	Delivery
c_1	$(0.49, 0.91)$	$(0.40, 0.40)$	$(0.91, 0.48)$	$(0.60, 0.22)$
c_2	$(0.98, 0.36)$	$(0.67, 0.74)$	$(0.86, 0.63)$	$(0.77, 0.39)$
c_3	$(0.24, 0.80)$	$(0.19, 0.48)$	$(0.96, 0.39)$	$(0.51, 0.10)$
c_4	$(0.80, 0.32)$	$(0.41, 0.93)$	$(0.25, 0.73)$	$(0.75, 0.75)$
c_5	$(0.50, 0.73)$	$(0.58, 0.66)$	$(0.79, 0.75)$	$(0.49, 0.20)$

Table 3. Fermatean fuzzy soft matrix of Supplier 3.

	Net Price	Quality	Financial Position	Delivery
c_1	$(0.10, 0.93)$	$(0.87, 0.39)$	$(0.52, 0.95)$	$(0.80, 0.78)$
c_2	$(0.35, 0.68)$	$(0.17, 0.23)$	$(0.64, 0.90)$	$(0.20, 0.80)$
c_3	$(0.52, 0.92)$	$(0.64, 0.79)$	$(0.88, 0.62)$	$(0.23, 0.30)$
c_4	$(0.44, 0.99)$	$(0.35, 0.97)$	$(0.59, 0.86)$	$(0.69, 0.74)$
c_5	$(0.81, 0.40)$	$(0.98, 0.21)$	$(0.40, 0.45)$	$(0.30, 0.96)$

Table 4. Fermatean fuzzy soft matrix of Supplier 4.

	Net Price	Quality	Financial Position	Delivery
c_1	$(0.89, 0.43)$	$(0.16, 0.65)$	$(0.92, 0.58)$	$(0.37, 0.93)$
c_2	$(0.50, 0.95)$	$(0.47, 0.24)$	$(0.71, 0.85)$	$(0.46, 0.91)$
c_3	$(0.78, 0.78)$	$(0.32, 0.98)$	$(0.40, 0.60)$	$(0.92, 0.56)$
c_4	$(0.38, 0.58)$	$(0.83, 0.10)$	$(0.79, 0.70)$	$(0.99, 0.14)$
c_5	$(0.90, 0.10)$	$(0.66, 0.77)$	$(0.425, 0.51)$	$(0.82, 0.20)$

Step 2: Normalize the decision matrices (Tables 5, 6, 7, 8).

Table 5. Normalized Fermatean fuzzy soft matrix of Supplier 1.

	Net Price	Quality	Financial Position	Delivery
c_1	$(0.60, 0.90)$	$(0.60, 0.70)$	$(0.79, 0.62)$	$(0.50, 0.50)$
c_2	$(0.70, 0.75)$	$(0.52, 0.95)$	$(0.65, 0.85)$	$(0.85, 0.58)$
c_3	$(0.89, 0.56)$	$(0.20, 0.55)$	$(0.10, 0.35)$	$(0.59, 0.84)$
c_4	$(0.30, 0.70)$	$(0.92, 0.60)$	$(0.90, 0.45)$	$(0.40, 0.12)$
c_5	$(0.80, 0.10)$	$(0.70, 0.15)$	$(0.25, 0.25)$	$(0.65, 0.45)$

Table 6. Normalized Fermatean fuzzy soft matrix of Supplier 2.

	Net Price	Quality	Financial Position	Delivery
c_1	$(0.91, 0.49)$	$(0.40, 0.40)$	$(0.91, 0.48)$	$(0.60, 0.22)$
c_2	$(0.98, 0.36)$	$(0.67, 0.74)$	$(0.86, 0.63)$	$(0.77, 0.39)$
c_3	$(0.80, 0.24)$	$(0.19, 0.48)$	$(0.96, 0.39)$	$(0.51, 0.10)$
c_4	$(0.32, 0.80)$	$(0.41, 0.93)$	$(0.25, 0.73)$	$(0.75, 0.75)$
c_5	$(0.73, 0.50)$	$(0.58, 0.66)$	$(0.79, 0.75)$	$(0.49, 0.20)$

Table 7. Normalized Fermatean fuzzy soft matrix of Supplier 3.

	Net Price	Quality	Financial Position	Delivery
c_1	$(0.93, 0.10)$	$(0.87, 0.39)$	$(0.52, 0.95)$	$(0.80, 0.78)$
c_2	$(0.68, 0.35)$	$(0.17, 0.23)$	$(0.64, 0.90)$	$(0.20, 0.80)$
c_3	$(0.92, 0.52)$	$(0.64, 0.79)$	$(0.88, 0.62)$	$(0.23, 0.30)$
c_4	$(0.99, 0.44)$	$(0.35, 0.97)$	$(0.59, 0.86)$	$(0.69, 0.74)$
c_5	$(0.40, 0.81)$	$(0.98, 0.21)$	$(0.40, 0.45)$	$(0.30, 0.96)$

Table 8. Normalized Fermatean fuzzy soft matrix of Supplier 4.

	Net Price	Quality	Financial Position	Delivery
c_1	$(0.43, 0.89)$	$(0.16, 0.65)$	$(0.92, 0.58)$	$(0.37, 0.93)$
c_2	$(0.95, 0.50)$	$(0.47, 0.24)$	$(0.71, 0.85)$	$(0.46, 0.91$
c_3	$(0.78, 0.78)$	$(0.32, 0.98)$	$(0.40, 0.60)$	$(0.92, 0.56)$
c_4	$(0.58, 0.38)$	$(0.83, 0.10)$	$(0.79, 0.70)$	$(0.99, 0.14)$
c_5	$(0.10, 0.90)$	$(0.66, 0.77)$	$(0.425, 0.51)$	$(0.82, 0.20)$

Step 3: Aggregate the opinions of experts using FFSWA operator to obtain the following:

$$v_1 = (0.7010, 0.4468) \qquad v_2 = (0.7564, 0.4223)$$
$$v_3 = (0.8337, 0.4762) \qquad v_4 = (0.7480, 0.5542)$$

Step 4: Calculate the score function for each v_k

$$S(v_1) = 0.2553 \qquad S(v_2) = 0.3575$$
$$S(v_3) = 0.4715 \qquad S(v_4) = 0.2483$$

Step 5: Rank the suppliers using the above values of score function.

Here $v_3 > v_2 > v_1 > v_4$, thus Supplier 3 is the suitable supplier.

If we employ FFSWG operator for this practical example then the aggregated opinion of each expert is

$$v_1 = (0.5426, 0.6812) \qquad v_2 = (0.5746, 0.6046)$$

$v_3 = (0.5831, 0.7645) \qquad v_4 = (0.4631, 0.7986)$

The corresponding score values are:

$S(v_1) = -0.1564 \quad S(v_2) = -0.0313 \quad S(v_3) = -0.2486 \quad S(v_4) = -0.4099$

Here $v_2 > v_1 > v_3 > v_4$. Thus the Supplier 2 is the suitable supplier.

4 Conclusions

In this work we have presented the definition of Fermatean fuzzy soft set and discussed certain basic operations on it. Two aggregation operators to compile the information represented by Fermatean fuzzy soft sets is proposed. In the end a decision-making problem by making use of these two operators is illustrated.

Further many other aggregation operations can be obtained for this new extension of soft set which could have potential applications in decision making.

References

1. Atanassov, K.T.: Intuitionistic fuzzy sets. Fuzzy Sets Syst. **20**(1), 87–96 (1986)
2. Atanassov, K.T.: On Intuitionistic Fuzzy Sets Theory. Springer, Heidelberg (2012)
3. Xu, Z.: Intuitionistic fuzzy aggregation operators. IEEE Trans. Fuzzy Syst. **15**(6), 1179–1187 (2007)
4. Garg, H.: Generalized intuitionistic fuzzy interactive geometric interaction operators using Einstein t-norm and t-conorm and their application to decision making. Comput. Ind. Eng. **101**, 53–69 (2016)
5. Garg, H.: Novel intuitionistic fuzzy decision making method based on an improved operation laws and its application. Eng. Appl. Artif. Intell. **60**, 164–174 (2017)
6. Huang, J.Y.: Intuitionistic fuzzy hamacher aggregation operators and their application to multiple attribute decision making. J. Intell. Fuzzy Syst. **27**(1), 505–513 (2014)
7. Garg, H.: Intuitionistic fuzzy hamacher aggregation operators with entropy weight and their applications to multi-criteria decision-making problems. Iranian J. Sci. Technol. Trans. Electr. Eng. **43**(3), 597–613 (2019)
8. Chen, S.M., Chang, C.H.: Fuzzy multiattribute decision making based on transformation techniques of intuitionistic fuzzy values and intuitionistic fuzzy geometric averaging operators. Inf. Sci. **352**, 133–149 (2016)
9. Yager, R.R.: Pythagorean fuzzy subsets. In: 2013 joint IFSA world congress and NAIFS annual meeting (IFSA/NAFIPS), pp. 57–61. IEEE (2013)
10. Yager, R.R.: Pythagorean membership grades in multicriteria decision making. IEEE Trans. Fuzzy Syst. **22**(4), 958–965 (2013)
11. Yager, R.R., Abbasov, A.M.: Pythagorean membership grades, complex numbers, and decision making. Int. J. Intell. Syst. **28**(5), 436–452 (2013)
12. Peng, X., Yuan, H.: Fundamental properties of Pythagorean fuzzy aggregation operators. Fund. Inform. **147**(4), 415–446 (2016)
13. Garg, H.: A new generalized Pythagorean fuzzy information aggregation using Einstein operations and its application to decision making. Int. J. Intell. Syst. **31**(9), 886–920 (2016)

14. Zeng, S., Mu, Z., Baležentis, T.: A novel aggregation method for Pythagorean fuzzy multiple attribute group decision making. Int. J. Intell. Syst. **33**(3), 573–585 (2018)
15. Garg, H.: Confidence levels based Pythagorean fuzzy aggregation operators and its application to decision-making process. Comput. Math. Organ. Theory **23**(4), 546–571 (2017). https://doi.org/10.1007/s10588-017-9242-8
16. Yager, R.R.: Generalized orthopair fuzzy sets. IEEE Trans. Fuzzy Syst. **25**(5), 1222–1230 (2016)
17. Liu, P., Wang, P.: Some q-rung orthopair fuzzy aggregation operators and their applications to multiple-attribute decision making. Int. J. Intell. Syst. **33**(2), 259–280 (2018)
18. Senapati, T., Yager, R.R.: Fermatean fuzzy sets. J. Ambient. Intell. Humaniz. Comput. **11**(2), 663–674 (2019). https://doi.org/10.1007/s12652-019-01377-0
19. Senapati, T., Yager, R.R.: Fermatean fuzzy weighted averaging/geometric operators and its application in multi-criteria decision-making methods. Eng. Appl. Artif. Intell. **85**, 112–121 (2019)
20. Molodtsov, D.: Soft set theory-first results. Comput. Math. Appl. **37**(4–5), 19–31 (1999)
21. Maji, P.K., Biswas, R., Roy, A.: Soft set theory. Comput. Math. Appl. **45**(4–5), 555–562 (2003)
22. Maji, P.K., Biswas, R., Roy, A.: Fuzzy soft sets. J. Fuzzy Math. **9**(3), 589–602 (2001)
23. Maji, P.K., Biswas, R., Roy, A.: Intuitionistic fuzzy soft sets. J. Fuzzy Math. **9**(3), 677–692 (2001)
24. Peng, X.D., Yang, Y., Song, J.P., Jiang, Y.: Pythagorean fuzzy soft set and its application. Comput. Eng. **41**(7), 224–229 (2015)
25. Arora, R., Garg, H.: Robust aggregation operators for multi-criteria decision-making with intuitionistic fuzzy soft set environment. Scientia Iranica. Trans. E Ind. Eng. **25**(2), 931–942 (2018)
26. Guleria, A., Bajaj, R.K.: On Pythagorean fuzzy soft matrices, operations and their applications in decision making and medical diagnosis. Soft. Comput. **23**(17), 7889–7900 (2018). https://doi.org/10.1007/s00500-018-3419-z
27. Naeem, K., Riaz, M., Peng, X., Afzal, D.: Pythagorean fuzzy soft MCGDM methods based on TOPSIS, VIKOR and aggregation operators. J. Intell. Fuzzy Syst. **37**(5), 6937–6957 (2019)

A Fast Computing Model for Despeckling Ultrasound Images

Febin Iyyath Pareedpillai(✉) [ID] and Jidesh Padikkal [ID]

Department of Mathematical and Computational Sciences, National Institute of Technology, Karnataka Surathkal, Srinivasanagar, Mangalore 575025, India
{febin.177ma501,jidesh}nitk.edu.in

Abstract. Ultrasound imaging is a highly preferred diagnostic method due to its non- invasive nature; however, the presence of speckle noise degrades the quality of the images captured by this modality. Among the plentiful researches that happened in the field of despeckling, the non-local total variation methods have demonstrated promising results by maintaining relevant details and edges present in images. Nevertheless, the model is computationally expensive as it has to deal with large size matrices in computing the results, which in turn restricts its applicability in real-time environments. This study contributes a fast and numerically stable Non-local Total Variation Model for despeckling Ultrasound images. Multi-core GPU processors are employed for computing the parallelized algorithms developed using a fast converging Split-Bregman iterative scheme. A comprehensive evaluation is performed on the basis of execution time to demonstrate the efficiency of the model.

Keywords: Non-local total variation · Split-bregman iteration · GPU Acceleration

1 Introduction

Ultrasound imaging has enormous applications in the medical domain as it lacks any ionizing radiations. It also acts as a cost-effective diagnostic technique in numerous medical procedures. Nowadays, real-time applications are prevalent, and many researchers are striving to replace the existing diagnostic systems with their automated counterparts. Real-time ultrasound images have a wide range of usage, which includes monitoring the fetal heart function, measuring blood flow, etc. It is also used for guided surgeries or robotic surgeries and needle positioning in biopsies. The ultrasound beams that enter a human body hits on different tissue boundaries and get reflected, scattered, or absorbed. The scattering of this signal occurs when the size of the object on which it hits is smaller than the wavelength. In some cases, the scattering of the signal leads to a combination of

Supported by Science and Engineering Research Board, India under the Project Grant No. ECR/2017/000230.

© Springer Nature Singapore Pte Ltd. 2021
A. Awasthi et al. (Eds.): CSMCS 2020, CCIS 1345, pp. 217–228, 2021.
https://doi.org/10.1007/978-981-16-4772-7_17

constructive and destructive interferences known as speckle. This speckle-noise degrades the visibility of the image and which in turn limits its application. Hence, speckle reduction is an essential preprocessing step for all ultrasound image-based diagnostic systems. As speckles interfere with the details in the data, their analysis becomes difficult. Moreover, the speckles are observed to be data-correlated interventions making their removal apparently tedious [1]. Many existing restoration methods assume the noise as data-uncorrelated and additive., such as Gaussian [2,3]. However, a hand full of methods are introduced for data-correlated noise such as multiplicative Gamma or Rayleigh [4,5]. The speckles being multiplicative they are data-dependent and as evident from many works in the literature they tend to follow a Gamma law in their distribution. The initial works in this direction such as Lee et al. [6], Kuan [7] and Frost [8] are based on the coefficient of variation (CoV). Further modification were proposed by incorporating the CoV in anisotropic diffusion models (inspired by the Perona-Malik model [9]) see SRAD [10] and its variants OSRAD [11], DPAD [12] etc. Variational models captured the attention of many researchers due to its elegant performance in terms of restoring the data, see ROF [2] model and its variants [13]. Variational models derived from the Bayesian framework were introduced for various data distributions see [1,4,14] for Gamma, Rayleigh and Poisson distributions respectively. Non-local variational framework inspired by the non-local means was first introduced to image processing by Gilboa et al. in [15]. The non-local models works on the principle of similarity between the patches selected globally therefore, the restoration appears more natural and the details are preserved more carefully compared to the local pixel-wise averaging filters [15–17]. However, their computations are more expensive, one has to deal with large matrices when computing the non-local identities. Therefore, the usage non-local models in real-time applications is limited. This has paved the way to introduce fast computing methods that run parallelly under multi-processing environments. As the cost of multi-core processing units have become manageable, GPU based computing environments have become ubiquitous. The GPU based implementations of Non-local means (NLM) filters are given in[18–20]. Similarly, GPU bases accelerations of many popular restoration algorithms are introduced in the recent years, see [21,22]. In this study, we try extent the GPU computing facility to non-local variational framework (by parallelizing the algorithms) implemented using the fast converging Split-Bregman scheme to improve the computational efficiency of the model.

2   Methodology

In the despeckling method designed here, we perform a log transformation on the data to address the multiplicative nature of the noise. The multiplicative noise becomes additive in the log domain. So the multiplicative Gamma in the original domain can be approximated with additive Gaussian in the log domain [23]. The formulation of the model is done using the Bayesian framework. The maximum a posteriori (MAP) estimate for the Gaussian distribution has already been studied in [2], where authors have used TV norm as the regularization prior. The TV norm has the drawback of losing textures and relevant details. To improve this method, we use the non-local total variation, which is efficient in preserving local gradients and textures in the restored image. Gilboa and Osher [15] proposed the Non-Local gradient of a function I, for a pair of points or pixels (x, y) as below:

$$\nabla_{NL} I(x, y) = (I(y) - I(x)) \sqrt{w(x, y)}, \tag{1}$$

where the weight $w(x, y)$ depends on similarity between the pixels x and y. It is defined based on the similarity of intensity vectors $v(N_x)$ and $v(N_y)$, where N_x and N_y denotes a square neighbourhood of fixed size centered around pixels x and y respectively. The non-local weight calculation as given below:

$$w(x, y) = \frac{1}{C(x)} e^{-\frac{(G_\sigma * |v(N_x) - v(N_y)|^2)}{h^2}}, \tag{2}$$

where G_σ denotes the Gaussian kernel with standard deviation σ, $C(x)$ is a normalizing constant, and h is a filtering parameter. A classical algorithm for non-local weight calculation is given in Algorithm 1. The non-local regularization functional for Gaussian denoising is as below (see [15,16]):

$$\min_{I} \left\{ \int_{\Omega} \left(|\nabla_{NL} I| + \lambda \|I - I_0\|_2^2 \right) dx dy \right\}. \tag{3}$$

Using Gradient descent method it can be solved as

$$I^{k+1} = I^k + \nabla t \left(\nabla \cdot \left(\frac{\nabla_{NL} I}{|\nabla_{NL} I|} \right) + \lambda (I - I_0) \right), \tag{4}$$

where I and I_0 denotes the restored and noisy images respectively and ∇t represents the time step. However the conventional Gradient descent optimization is slow in convergence and depends on the chosen time step. To address this issue, we are using the Split-Bregman (SB) based iterative scheme. The Split-Bregman iteration introduces a constrain $d = \nabla I$ and an auxiliary variable b into the traditional non-local TV model. As a result, the optimization problem can split into the following I, d and b subproblems.

$$I^{k+1} = \min_{I} \left\{ \frac{\lambda}{2} \|I - I_0\|_2^2 + \frac{\beta}{2} \|d - \nabla_{NL} I - b^k\|_2^2 \right\}, \tag{5}$$

$$d^{k+1} = \min_{d} \left\{ \|d\| + \frac{\beta}{2} \|d - \nabla_{NL} I - b^k\|_2^2 \right\}, \tag{6}$$

$$b^{k+1} = b^k + \nabla_{NL} I - d^{k+1}. \tag{7}$$

After applying Euler-Lagrange equation the minimization problem in Eq. (5) takes the following form

$$I^{k+1} = \frac{\lambda I_0 + \beta \nabla \cdot (d^k - b^k)}{\lambda - \beta \Delta_{NL}}, \tag{8}$$

where I_d represents identity matrix. The problem in Eq. (6) is solved using shrinkage operator as follows

$$d^{k+1} = shrink(\nabla I + b^k, \frac{1}{\beta}) = \frac{\nabla I + b^k}{|\nabla I + b^k|} \max(|\nabla I + b^k| - \frac{1}{\beta}, 0). \qquad (9)$$

1 **Input** $I \leftarrow$ Noisy Digital image of size $M \times N$
2 **Output** Non-local Weight W
3 **begin**
4   Initialize $SW = 7 \times 7, PW = 5 \times 5$
5   **for** *each pixel of I* **do**
6   Initialize W=0
7   **for** *each pixel of SW* **do**
8   **for** *each pixel of PW* **do**
9   Compute Euclidean distance d
10   end
11   Compute the weight of the pixel within the SW as $W = e^{(-d/h^2)}$
12   end
13   Normalize the weight
14   end
15 end

Algorithm 1: Classical non-local Weight Algorithm

In order to make this restoration method fast enough to work with real-time data, we introduce a parallel implementation based on GPU. Here we use the NVIDIA's parallel computation API called Compute Unified Devise Architecture (CUDA). The CUDA includes the concept of host and device systems where the host system is the CPU responsible for serial computations, and the device system is the GPU that performs the parallel computations. A CUDA platform provides support for allocating and deallocating device memory. Copying of data from the host to device and vise versa. The parallel functions that execute on GPU are known as kernels. A typical CUDA program is heterogeneous with CPU and GPU operations. The operations like memory allocation, data initialization, thread allocation, and kernel calls come under the CPU operations. The GPU operations are defined as separate kernels for parallel processing. In a CUDA program, threads are arranged in the form of several grids. One grid has many thread blocks and each of which will contain several threads. In CUDA programming, each thread can be uniquely identified using its blockID and threadID. Threads in a block can communicate with each other through shared memory. A grid has its own global memory, which is shared by all the threads. In addition to this, each thread also has its storage registers.

For a Non-local TV algorithm, the gradient is calculated as in Eq. (1), where the computation of non-local weight takes most of the execution time. In a given image I, for each pixel (x, y) a Search Window (SW) and Patch Window (PW) is defined around it. Afterward, the PW around referring pixel $I(x, y)$ is compared with all such PWs possible inside the defined SW region. Instead of comparing each pixel with its immediate 4 or 8 neighbours, this method compares PW around pixel $I(x, y)$ with PWs around all other pixels in a fixed SW. The classical derivative operation works locally; hence they are not robust enough to handle noisy images. In contrast, non-local gradients make use of the redundancy in data and help to preserve textures. The classical implementation of non-local weight is given in Algorithm 1, where the complexity is $O(N * M(SW^2)(PW^2))$ i.e., it depends on image size (N*M), SW size, and PW size.

The devised Fast Non-local TV method includes three kernels as given in Algorithm 2; these are *nonlocalWeightGradient, nonlocalDivergence,* and *gradientDescent*. The number of threads in a block is fixed based on the empirical study to get better result. Here we use the SIMD (Single Instruction Multiple Data) parallelism by making all the tasks pixel-wise independent. For each pixel, corresponding non-local weight and non-local gradient are calculated concurrently using Algorithm 3. This makes the complexity independent of size of image. Initially, as in Algorithm 2, a Gradient descent optimization method is used for creating the parallel Non-local TV, and it is implemented with C++ and CUDA.

---

1 **Input** $I^0 \leftarrow$ Noisy Digital image of size $M \times N$
2 **Output** Restored Digital image I of size $M \times N$
3 **begin**
4   Initialize $\epsilon = 0.0001$
5   **while** $\|I^{k+1} - I^k\|/\|I^{k+1}\| < \epsilon$ **do**
6   Call *nonlocalWeightGradient* kernel to calculate weight W and non-local gradient $\nabla_{NL} I^k$.
7   Call *nonlocalDivergence* kernel to find divergence of $\frac{\nabla_{NL} I^k}{|\nabla_{NL} I^k|}$.
8   Call *gradientDescent* kernel for getting updated I^{k+1}
9   **end**
10   update I as I^{k+1}
11 **end**

Algorithm 2: GPU based non-local TV

---

1 **Input** $I \leftarrow$ Noisy Digital image of size $M \times N$
2 **Output** Gradient of the image ∇I and weight W
3 **begin**
4   Initialize $SW = 7 \times 7, PW = 5 \times 5$
5   Idx=ThreadId.x+blockIdx.x*blockDim.x
6   Extract pixel location (i_x, i_y) from Idx
7   **for** *each pixel of Search Window around* (i_x, i_y) **do**
8   Extract neighbour location (j_x, j_y)
9   **for** *each pixel of Patch Window around* (i_x, i_y) *and* (j_x, j_y) **do**
10   Compute Euclidean distance d
11   **end**
12   Compute the weight of the pixel as $W = e^{(-d/h^2)}$
13   Calculate $\nabla I = (I(i_x, i_y) - I(j_x, j_y)) * \sqrt{W}$
14   **end**
15 **end**

Algorithm 3: *nonlocalWeightGradient* Algorithm

We have initiated 320 threads per block, and the number of blocks is calculated as $\frac{N*M}{Thredsper Block}$, where $M*N$ represents the number of pixels in the image. Generally, threads can be initiated in three dimensions, but we are using only one dimension, and all of the operations are done in a row major fashion. The conventional Gradient descent optimization takes more iterations to converge to a final restored result, and as these iterations are executed serially. More the iterations are, so is the execution time. To further improve the execution time, we suggest the parallel implementation of the Split-Bregman (SB) iterative scheme. It works using the Eqs. (8), (9), and (7). We have implemented the SB based non-local TV as six kernels that work independently for each pixel. The kernel calls are made in the order, as in Algorithm 4.

The parallel kernels used in this method are *calcWeight*, *calcNLgrad*, *calcDiff*, *calcDiv*, *splitBreg*, and *findDB*. The kernel for non-local weight calculation (*calcWeight*) is given in Algorithm 5 where each thread is allocated to a pixel and corresponding SW around that, and different PWs inside it are selected parallelly in all the initiated threads. Hence the weight for each pixel is calculated concurrently. The non-local weight estimated needs to be used in the next kernel call for non-local gradient calculation (*calcNLgrad*), and consequently, a synchronization barrier is added in between to avoid possible race conditions. In *calcNLgrad* kernel ∇I is estimated based on Eq. (1); see Algorithm 6 for details. Similarly, the kernels *calcDiff* and *calcDiv* are executed to evaluate $\nabla.(d^k - b^k)$. The kernel *findDB* is explained in Algorithm 7, where auxiliary variables d^{k+1} and b^{k+1} are updated accordingly. Finally, I^{k+1} is refreshed in each iteration as in Eq. (8) using the kernel *splitBreg*.

---

1 **Input** $I^0 \leftarrow$ Noisy Digital image of size $M \times N$
2 **Output** Restored Digital image I of size $M \times N$
3 **begin**
4 | Initialize $\epsilon = 0.0001, d = 0, b = 0$
5 | **while** $\|I^{k+1} - I^k\|/\|I^{k+1}\| < \epsilon$ **do**
6 | | Call *calcWeight* kernel to calculate weight W.
7 | | Call *calcNLgrad* kernel to calculate non-local gradient $\nabla_{NL}I^k$.
8 | | Call kernels *calcDiff* and *calcDiv* to find $\nabla.(d^k - b^k)$.
9 | | Call *splitBreg* kernel for getting updated I^{k+1}
10 | | Update the value of d and b by calling kernel *findDB*
11 | **end**
12 | update I as I^{k+1}
13 **end**

Algorithm 4: GPU based non-local TV with SB

3   Results and Discussion

We have compared the GPU based parallel algorithm with the sequential implementation of the same. The different GPUs used for the comparison are Nvidia's

```
1 Input I ← Noisy Digital image of size M × N
2 Output weight W
3 begin
4 |   Initialize SW = 7 × 7, PW = 5 × 5
5 |   Idx=ThreadId.x+blockIdx.x*blockDim.x
6 |   Extract pixel location (i_x, i_y) from Idx
7 |   for each pixel of Search Window around (i_x, i_y) do
8 |   |   Extract neighbour location (j_x, j_y)
9 |   |   for each pixel of Patch Window around (i_x, i_y) and (j_x, j_y) do
10 |  |   |   Compute Euclidean distance d
11 |  |   end
12 |  |   Compute the weight of the pixel as W = e^{(-d/h^2)}
13 |  end
14 end
```

Algorithm 5: *calcWeight* Algorithm

```
1 Input I ← Noisy Digital image of size M × N, W ← weight
2 Output Non-local gradient of the image ∇I
3 begin
4 |   Initialize SW = 7 × 7, PW = 5 × 5
5 |   Idx=ThreadId.x+blockIdx.x*blockDim.x
6 |   Extract pixel location (i_x, i_y) from Idx
7 |   for each pixel of Search Window around (i_x, i_y) do
8 |   |   Extract neighbour location (j_x, j_y) Calculate
   |   |   ∇I = (I(i_x, i_y) − I(j_x, j_y)) * √W
9 |   end
10 end
```

Algorithm 6: *calcNLgrad* Algorithm

```
1 Input ∇I ← Non-local gradient of the image
2 Output variables d^{k+1} and b^{k+1}
3 begin
4 |   Idx=ThreadId.x+blockIdx.x*blockDim.x
5 |   Based on current thread ID Idx extract corresponding pixel I(i_x, i_y)
6 |   for each pixel of Search Window around (i_x, i_y) do
7 |   |   Compute b^{k+1} = b^k + ∇I − d^{k+1} (Refer equation (7))
8 |   |   Calculate d^{k+1} = (∇I+b^k)/|∇I+b^k| max(|∇I + b^k| − 1/β, 0) (Refer equation (9))
9 |   end
10 end
```

Algorithm 7: *findDB* Algorithm

Tesla K40 (Kepler Architecture) and V100 (Volta Architecture). The Nvidia Tesla K40 GPU has compute ability 3.5, 15 Streaming Multiprocessors(SM), and 192 cores per SM, running at 745 MHz, which means it has 2880 CUDA cores in Kepler architecture. It has 12 GB global memory, and 48 KB shared memory. The maximum number of threads per multiprocessor is 2048, and the maximum threads per block are 1024. Threads can be allocated in three dimensions; the maximum dimension size of the thread block is (1024,1024,64). The comparison based on the size of the image is given in Fig. 1 and Table 1. From this, we can infer that the execution time needed for sequential implementation depends heavily on input image size, which is not the case in parallel implementation.

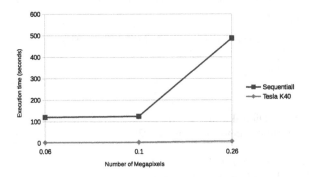

Fig. 1. Performance comparison of Tesla K40 with Sequential

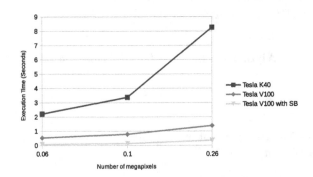

Fig. 2. Performance comparison of Tesla K40, V100 and V100 with SB

The Tesla V100 with compute ability 7.0 has also been used for comparison. It has 5120 CUDA cores with 16 GB memory. In addition to the conventional Gradient descent optimization, for fast convergence, we have also compared the GPU based non-local TV with Split-Bregman (SB), see Fig. 2 for details. From the visual comparison, we can infer that the clarity obtained by using both conventional non-local TV and fast non-local TV methods is comparable, see

Figs. 3, 4 and 5. The computational time needed for the traditional algorithm is based on the input image size. For an input image of size $N \times N$ the complexity for weight calculation is $O(N^2 \times SW^2 \times PW^2)$. As we do pixel-wise operations in parallel, the GPU algorithms proposed here only depend on the search window dimension. We have conducted a study on the basis of window size and details of which can see in Fig. 6 and Table 2. As we can see from these results, the small window size helps in fast computations. For our studies, we have fixed the window size to 7×7. From the comparison shown in Table 1, we can infer that the Tesla K40 is giving a 60 times speedup over the corresponding serial code. From Fig. 2, we can see that the computational advancement obtained by Tesla V100 is comparatively high with an average speedup of 356 times. However, the GPU algorithm with split-Bregman optimization has shown a significant advancement that supports the use of this algorithm in real-time restoration tasks. For images of size 512×512 this method is giving an average speedup of 1354 times.

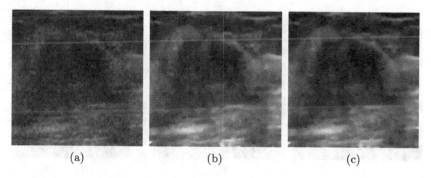

(a)   (b)   (c)

Fig. 3. Restored results (a) Original noisy image [24] (b) restored result using conventional Non-local TV (c) restored result using Fast Non-local TV

Table 1. Execution time taken by sequential and parallel implementation

Image Size	Sequential	TeslaK40	TeslaV100	Tesla V100 with SB
256×256	120.06	2.21	0.53	0.08
320×320	124.75	3.37	0.78	0.13
512×512	487.48	8.26	1.38	0.36

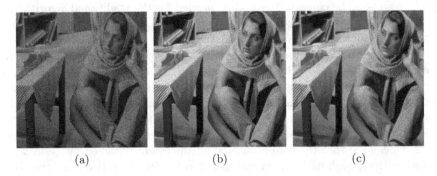

(a)   (b)   (c)

Fig. 4. Restored results for test image 1 with added gamma noise of variance 0.15 (a) Noisy image (b) restored result using conventional Non-local TV (c) restored result using Fast Non-local TV

(a)   (b)   (c)

Fig. 5. Restored results for test image 2 with added gamma noise of variance 0.1 (a) Noisy image (b) restored result using conventional Non-local TV (c) restored result using Fast Non-local TV

Table 2. Execution time taken on the basis of window size

Image Size	Window size		
	5×5	7×7	11×11
256×256	0.32	0.53	0.95
320×320	0.44	0.78	1.45
512×512	0.64	1.38	3.36

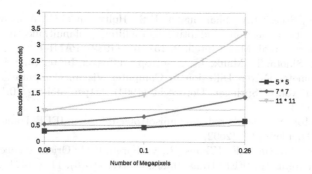

Fig. 6. Performance comparison on the basis of window size

4 Conclusion

We have developed a computationally efficient and fast converging non-local TV model for ultrasound image restoration. The devised methods are tested with different Nvidia GPUs in the CUDA platform. The experimental results obtained show that the GPU based non-local TV algorithms are useful in providing quick restored results. Due to the fast converging nature, the Split-Bregman based GPU algorithm has outperformed gradient descent based GPU method with an average speedup of 1354 times. This advancement achieved in execution time assures the efficient restoration of ultrasound images in real-time scenarios.

Acknowledgments. The authors would like to thank Science and Engineering Research Board, India for providing financial support under the Project Grant No. ECR/2017/000230.

References

1. Jidesh, P., Holla, S.: Non-local total variation regularization models for image restoration. Comput. Electr. Eng. **67**, 114–133 (2018)
2. Rudin, L.I., Osher, S., Emad, F.: Nonlinear total variation based noise removal algorithms. Physica D Nonlinear Phenomena **60**(1), 259–268 (1992)
3. Buades, A., Coll, B., Morel, J.M. (ed.): A non-local algorithm for image denoising, vol. 2. IEEE (2005)
4. Aubert, G., Aujol, J.: A variational approach to removing multiplicative noise. SIAM J. Appl. Math. **68**(4), 925–946 (2008)
5. Mäkitalo, M., Foi, A., Fevralev, D., Lukin, V.: Denoising of single-look SAR images based on variance stabilization and nonlocal filters. In: Proceedings of the 13th International Conference on Mathematical Methods in Electromagnetic Theory (2010)
6. Lee, J.S.: Digital image enhancement and noise filtering by use of local statistics. IEEE Trans. Pattern Anal. Mach. Intell. **2**, 165–168 (1980). PAMI-2
7. Kuan, D.T., Sawchuk, A.A., Strand, T.C., Chavel, P.: Adaptive noise smoothing filter for images with signal-dependent noise. IEEE Trans. Pattern Anal. Mach. Intell. **2**, 165–177 (1985). PAMI-7

8. Frost, V.S., Stiles, J.A., Shanmugan, K.S., Holtzman, J.C.: A model for radar images and its application to adaptive digital filtering of multiplicative noise. IEEE Trans. Pattern Anal. Mach. Intell. **2**, 157–166 (1982). PAMI-4
9. Perona, P., Shiota, T., Malik, J.: Anisotropic diffusion. In: ter Haar Romeny B.M. (eds.) Geometry-Driven Diffusion in Computer Vision. Computational Imaging and Vision, vol. 1. Springer, Dordrecht (1994). https://doi.org/10.1007/978-94-017-1699-4_3
10. Yu, Y., Acton, S.: Speckle reducing anisotropic diffusion. IEEE Trans. Image Process. **11**(11), 1260–1270 (2002)
11. Krissian, K., Westin, C.F., Kikinis, R., Vosburgh, K.G.: Oriented speckle reducing anisotropic diffusion. IEEE Trans. Image Process. **16**(5), 1412–1424 (2007)
12. Aja-Fernández, S., Alberola-López, C.: On the estimation of the coefficient of variation for anisotropic diffusion speckle filtering. IEEE Trans. Image Process. **15**(9), 2694–2701 (2006)
13. Vese, L.A., Osher, S.J.: Image denoising and decomposition with total variation minimization and oscillatory functions. J. Math. Imaging Vis. **20**, 7–18 (2004)
14. Le, T., Chartrand, R., Asaki, T.J.: A variational approach to reconstructing images corrupted by poisson noise. J. Math. Imaging Vis. **27**(3), 257–263 (2007)
15. Gilboa, G., Osher, S.: Nonlocal operators with applications to image processing. SIAM J. Multiscale Model. Sim. **7**(3), 1005–1028 (2008)
16. Liu, X., Huang, L.: A new nonlocal total variation regularization algorithm for image denoising. Math. Comput. Simul. **97**, 224–233 (2014)
17. Jidesh, P., Banothu, B.: A new nonlocal total variation regularization algorithm for image denoising. Comput. Electr. Eng. **70**, 631–646 (2018)
18. Márques, A., Pardo, A.: Implementation of non local means filter in GPUs. In: Ruiz-Shulcloper, J., Sanniti di Baja, G. (eds.) CIARP 2013. LNCS, vol. 8258, pp. 407–414. Springer, Heidelberg (2013). https://doi.org/10.1007/978-3-642-41822-8_51
19. De Michele, P., Cuomo, S., Piccialli, F.: 3D data denoising via nonlocal means filter by using parallel GPU strategies. Comput. Math. Methods Med. (2014). https://doi.org/10.1155/2014/523862
20. Honzatko, D., Krulis, M.: Accelerating block-matching and 3d filtering method for image denoising on GPUs. J. Real-Time Image Proc. **16**(6), 2273–2287 (2019)
21. Davy, A., Ehret, T.: GPU acceleration of NL-means, BM3D and VBM3D. J. Real-Time Image Proc. **18**(1), 57–74 (2020). https://doi.org/10.1007/s11554-020-00945-4
22. Pfeger, S.G., Plentz, P.D.M., Rocha, R.C.O., et al.: Real-time video denoising on multicores and GPUs with Kalman-based and bilateral filters fusion. J. Real-Time Image Proc. **16**(5), 1629–1642 (2017)
23. Yahya, N., Kamel, N.S., Malik, A.S.: Subspace-based technique for speckle noise reduction in ultrasound images. Biomed. Eng. Online **13**(1), 154 (2014)
24. Rodrigues, S.R.P.: Breast ultrasound image (2017)
25. Gulo Carlos, A.S.J., et al.: Efficient parallelization on GPU of an image smoothing method based on a variational model. J. Real-Time Image Proc. **16**, 1249–1261 (2019)

General Computing

Non-linear Convection in Couple Stress Fluid with Non-classical Heat Conduction Under Magnetic Field Modulation

Maria Thomas[1]($\boxtimes$), K. Sangeetha George[2], and S. Pranesh[1]

[1] Department of Mathematics, CHRIST (Deemed to be University),
Bengaluru, Karnataka, India
maria.thomas@res.christuniversity.in
[2] Department of Science, Christ Academy Institute for Advanced Studies,
Bengaluru, Karnataka, India

Abstract. A theoretical examination of thermal convection for a couple stress fluid which is electrically conducting and possessing significant thermal relaxation time is explored under time dependent magnetic field. Fourier's law fails for a diverse area of applications such as fluids subjected to rapid heating, strongly confined fluid and nano-devices and hence a non-classical heat conduction law is employed. The heat transport in the system is examined and quantified employing the Lorenz model. The Nusselt number is deduced to quantitate the transfer of heat.

Keywords: Magnetic field modulation · Maxwell-Cattaneo law · Couple stress fluid

1 Introduction

Heat transfer problems are in general studied by considering heat conduction law by Fourier. Although the law explains the phenomena of transfer of heat in numerous situations and real life scenarios, it violates the principle of causality. This is because Fourier's law along with the energy equation gives a parabolic profile for the temperature field vaticinating the heat propagation speed as infinite. This shortcoming was pointed out by Maxwell [9]. Cattaneo [2] addressed this drawback by appending a transient term involving the relaxation time to Fourier's law which resulted in a hyperbolic energy equation or wave equation. Straughan and Franchi [19] was the first to study the propagation of thermal waves for Bénard convection. Straughan [17,18] investigated convection in Newtonian fluid and Darcy porous material, respectively, with the rate change in heat flux as proposed by Christov [4] and found that the implications of the thermal relaxation time on both the systems are significant. Stranges et al. [15] considered fluids with noticeable relaxation time for studying thermal convection. The possibility of a bistable mode was seen as a consequence of the occurrence of stationary as well as oscillatory mode of convection. An analysis of the steady

© Springer Nature Singapore Pte Ltd. 2021
A. Awasthi et al. (Eds.): CSMCS 2020, CCIS 1345, pp. 231–242, 2021.
https://doi.org/10.1007/978-981-16-4772-7_18

convection in non-Fourier fluids was carried out using a dynamical approach with low-order by Stranges et al. [16].

One of the non-Newtonian fluid theories called couple stress fluids proposed by Stokes [14] can be employed to understand the dynamics of fluids with couple stresses along with the classical Cauchy stress, such as in the case of rheologically complex fluids like animal and human blood , polymeric suspensions,liquids containing long-chain molecules, lubrication and liquid crystals. Experimental studies on the flow of blood (Cockelet [5], Goldsmith and Skalak [6]) have shown that flow of blood under certain conditions deviates from the characteristics seen in the flow of Newtonian fluid. Pranesh and George [10] considered an electrically conducting fluid with couple stresses and examined the stability of magneto-convection of the fluid under modulated temperature walls. Ramesh [12] investigated an asymmetric channel with homogeneous porous medium and couple stress fluid. Kumar et al. [8] used Ginzburg-Landau equation to understand the non-linearity of two component convection in fluids with suspensions under gravity modulation.

The discovery of the presence of magnetic fields in sunspots and the finding that their interaction with convection leads to the relative coolness and hence the darkness in them drew the attention of researchers to magneto-convection. Circulation and convection in small passages can be activated or improved with magnetic forces. Magneto-convection arises on account of the interference of the applied magnetic field with the flow of a fluid which is electrically conducting. Bhadauria and Kiran [1] analysed the non-linearity employing the Ginzburg-Landau equation of a system with sinusoidal and time dependent external magnetic field applied to an electrically conducting fluid layer. Kiran et al. [7] investigated nonlinear time dependant magneto-convection and have discussed the consequence of various parameters governing the system. They concluded that oscillatory mode gives better results in comparison to stationary mode for modulated magnetic field.

This paper attempts to determine the impacts of non-Fourier law and time-periodic magnetic field on Bénard convection in a fluid with suspended particles. The study deduces the convection amplitude in terms of parameters governing the system. An estimation of the transfer of heat is also carried out by finding the Nusselt number.

2   Mathematical Formulation

A layer of fluid with couple stresses is enclosed by two parallel and horizontal plates. The problem under consideration is schematically graphed in Fig. 1. The system is studied in Cartesian co-ordinates. The origin is placed on the lower plate and the z-axis is in the vertical upward direction.

The equations governing the system, assuming the Boussinesq approximation, are given by (Siddheshwar and Pranesh [13] and Stranges et al. [16])

Continuity equation

$$\nabla.\vec{q} = 0, \tag{1}$$

Conservation of momentum

$$\rho_0\left[\frac{\partial \vec{q}}{\partial t} + (\vec{q}.\nabla)\vec{q}\right] = -\nabla P + \rho \vec{g}(t) + \mu \nabla^2 \vec{q} - \mu'\nabla^4 \vec{q} + \mu_m(\vec{H}.\nabla)\vec{H}, \qquad (2)$$

Conservation of energy

$$\frac{\partial T}{\partial t} + (\vec{q}.\nabla)T = -\nabla.\vec{Q}, \qquad (3)$$

Maxwell-Cattaneo heat flux law

$$\tau\left[\frac{\partial \vec{Q}}{\partial t} + \vec{q}.\nabla\vec{Q} - \vec{Q}.\nabla\vec{q} - \vec{Q}\nabla.\vec{q}\right] = -\vec{Q} - \kappa\nabla T, \qquad (4)$$

Equation of state

$$\rho = \rho_0\left[1 - \gamma(T - T_0)\right], \qquad (5)$$

Magnetic induction equations

$$\nabla.\vec{H} = 0, \qquad (6)$$

$$\frac{\partial \vec{H}}{\partial t} + (\vec{q}.\nabla)\vec{H} = (\vec{H}.\nabla)\vec{q} + \nu_m\nabla^2\vec{H}. \qquad (7)$$

where $\vec{q}$ defines the velocity, $\vec{g}$: the acceleration due to gravity, T: the temperature, T_0: the reference temperature, ΔT: the temperature difference between the plates, P: the hydromagnetic pressure, ρ: the density, ρ_0: the density at $T = T_0$, τ: the relaxation time, $\vec{Q}$: the heat flux vector, μ: the dynamic viscosity, γ: the coefficient of thermal expansion, μ': the couple stress viscosity, κ: the thermal conductivity, $\vec{H}$: the intensity of the magnetic field, ν_m: the magnetic viscosity, μ_m: the magnetic permeability.

A sinusoidal magnetic field varying with respect to time with ω and ϵ as the modulation frequency and the amplitude of magnetic modulation respectively, is imposed vertically on the system which is mathematically represented by

$$\vec{H} = H_0\left[1 + \epsilon\cos(\omega t)\right]\hat{k}. \qquad (8)$$

2.1   Basic State

The layer of fluid is stationary in the basic state and is represented by

$$\vec{q_b}(z) = \vec{0},\; T = T_b(z),\; P = P_b(z),\; \rho = \rho_b(z),\; \vec{Q} = \vec{Q_b}(z),\; \vec{H_b} = H_0\hat{k}. \qquad (9)$$

Substitution of Eq. (9) in Eqs. (1)–(7), the equations for the basic state are deduced as follows:

$$\frac{\partial p_b}{\partial z} = -\rho_b g, \qquad (10)$$

$$\vec{Q_b} = -\kappa\frac{\partial T_b}{\partial z}, \qquad (11)$$

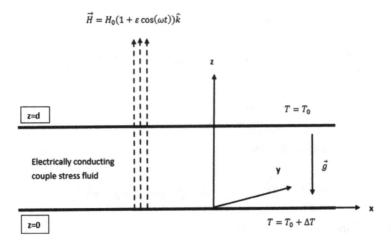

Fig. 1. Physical configuration of the problem.

$$\frac{d^2 T_b}{dz^2} = 0,$$ (12)

$$\rho_b = \rho_0[1 - \gamma(T_b - T_0)].$$ (13)

A perturbation to the basic state is applied as follows:

$$\vec{q} = \vec{q_b} + \vec{q}', \ \vec{Q} = \vec{Q_b} + \vec{Q}', \ T = T_b + T', \ P = P_b + P', \ \rho = \rho_b + \rho', \ \vec{H}' = \vec{H_b} + \vec{H}',$$ (14)

where $\vec{q}'$, ρ', P', T', $\vec{Q}'$ and $\vec{H}'$ represents the small perturbed quantities.

Substitution of Eq. (14) in Eqs. (1)–(7) and using Eqs. (10)–(13) gives

$$\nabla.\vec{q}' = 0,$$ (15)

$$\rho_0\left(\frac{\partial \vec{q}'}{\partial t} + (\vec{q}'.\nabla)\vec{q}'\right) = -\nabla P' + \mu\nabla^2\vec{q}' - \mu'\nabla^4\vec{q}' - \rho' g\hat{k}$$

$$+ \mu_m H_0\left[1 + \epsilon\cos(\omega t)\right]\frac{\partial \vec{H}'}{\partial z} + \mu_m(\vec{H}'.\nabla)\vec{H}',$$ (16)

$$\frac{\partial T'}{\partial t} + (\vec{q}'.\nabla)T' - w'\frac{\partial T_b}{\partial z} = -\nabla.\vec{Q}',$$ (17)

$$\tau\left[\frac{\partial \vec{Q}'}{\partial t} - \frac{\kappa}{d}\frac{\Delta T}{} \frac{\partial W'}{\partial z} + \vec{q}.\nabla\vec{Q}' - \vec{Q}'.\nabla\vec{q}\right] = -\vec{Q}' - \kappa\nabla T',$$ (18)

$$\rho' = -\rho_0\gamma T',$$ (19)

$$\nabla\vec{H}' = 0,$$ (20)

$$\frac{\partial \vec{H}'}{\partial t} + (\vec{q}'.\nabla)\vec{H}' = (\vec{H}'.\nabla)\vec{q}' + H_0\left[1 + \epsilon\cos(\omega t)\right]\frac{\partial w'}{\partial z} + \nu_m\nabla^2\vec{H}'.$$ (21)

Equations (15)–(21) represents the expressions for perturbed state.

Since the study explores the two-dimensional disturbances, only the xz-plane is considered. Hence Ψ, the stream function and Φ, the magnetic potential are taken as

$$(u', w') = \left(\frac{\partial \Psi}{\partial z}, -\frac{\partial \Psi}{\partial x}\right), \ (H'_x, H'_z) = \left(\frac{\partial \Phi}{\partial z}, -\frac{\partial \Phi}{\partial x}\right). \tag{22}$$

The x and z component of Eq. (16) are evaluated. The resulting equations are cross differentiated to eliminate the pressure, P. $\vec{Q}'$ is eliminated between Eqs. (17) and (18) by operating divergence on Eq. (18). The equations thus obtained is non-dimensionalized using

$$(x^*, y^*) = \left(\frac{x}{d}, \frac{z}{d}\right), \ \omega^* = \frac{\omega}{\left(\frac{\kappa}{d^2}\right)}, \ t^* = \frac{t}{\left(\frac{d^2}{\kappa}\right)}, \ \Psi^* = \frac{\Psi}{\kappa}, \ T^* = \frac{T}{\Delta T}, \ \Phi^* = \frac{\Phi}{dH_0}.$$

$$\tag{23}$$

Ignoring the asterisk, the non-dimensionalized equations which govern the systems are given by:

$$\frac{1}{Pr}\frac{\partial}{\partial t}(\nabla^2 \Psi) - \frac{1}{Pr}J(\nabla^2 \Psi, \Psi) = \nabla^4 \Psi - C\nabla^6 \Psi - RT_x$$

$$+ \frac{QPr}{Pm}[1 + \epsilon\cos(\omega t)]\frac{\partial}{\partial z}(\nabla^2 \Phi) - \frac{QPr}{Pm}J(\nabla^2 \Phi, \Phi), \tag{24}$$

$$M\left[\frac{\partial^2 T}{\partial t^2} - 2\,J(\Psi, T_t) - J(\Psi_t, T) + \Psi_{xt} - J(\Psi, \Psi_x) - \Psi_z J(\Psi_x, T) + \Psi_x J(\Psi_z, T)\right.$$

$$\left. - \Psi_z J(\Psi, T_x) + \Psi_x J(\Psi, T_z)\right] + T_t + \Psi_x - J(\Psi, T) = \nabla^2 T, \tag{25}$$

$$\Phi_t - J(\Psi, \Phi) = \left[1 + \epsilon\cos(\omega t)\right]\Psi_z + \frac{Pr}{Pm}\nabla^2 \Phi. \tag{26}$$

where

$M = \dfrac{\tau\kappa}{d^2}$: defines the Cattaneo number, $C = \dfrac{\mu'}{\mu d^2}$: the couple stress parameter, $R = \dfrac{\gamma g \Delta T d^3}{\mu\kappa}$: the Rayleigh number, $Pr = \dfrac{\mu}{\rho_0\kappa}$: the Prandtl number, $Q = \dfrac{\mu_m H_0^2 d^2}{\mu\nu_m}$: the Chandrasekhar number, $Pm = \dfrac{\mu_m}{\rho_0\nu_m}$: the magnetic Prandtl number.

Equations (24), (25) and (26) are considered for boundaries that are isothermal and stress-free with vanishing couple stresses. Thus the equations are solved for

$$\Psi = D^2\Psi = 0, \ D\Phi = 0, \ T = 0 \text{ at } z = 0 \text{ and } z = 1. \tag{27}$$

2.2 Linear Stability Analysis

The procedure followed by Venezian [20] is applied to derive the critical value of the Rayleigh number and the corresponding wave number for the case without modulations. It is given by

$$R = \frac{k^6(1 + Ck^2)}{\alpha^2(1 + Mk^2)} + \frac{Q\pi^2k^2}{\alpha^2(1 + Mk^2)}, \qquad (28)$$

where α is the wave number and $k^2 = \pi^2 + \alpha^2$.

Limiting cases:

$C = 0$, $M = 0$ and $Q = 0$ in Eq. (28) gives $R = \dfrac{k^6}{\alpha^2}$ which is the critical Rayleigh number for a Newtonian fluid (Chandrasekhar [3]).

$M = 0$ and $Q = 0$ in Eq. (28) gives $R = \dfrac{k^6(1 + Ck^2)}{\alpha^2}$ which is the critical Rayleigh number obtained by Siddheshwar and Pranesh [13].

$M = 0$ in Eq. (28) gives $R = \dfrac{k^6(1 + Ck^2)}{\alpha^2} + \dfrac{Q\pi^2k^2}{\alpha^2}$, which is the critical Rayleigh number obtained by Pranesh and George [10].

$C = 0$ in Eq. (27) gives $R = \dfrac{k^6}{\alpha^2(1 + Mk^2)} + \dfrac{Q\pi^2k^2}{\alpha^2(1 + Mk^2)}$ which is the critical Rayleigh number obtained by Pranesh and Kiran [11].

2.3 Non-linear Analysis

A stability analysis involving non-linearity is carried out to quantize the heat transport, and to examine the non-Fourier characteristics and the impact of modulated magnetic field and suspended particles on the heat transport. This task is performed using truncated Fourier series (Siddheshwar and Pranesh [13]). The temperature field is disturbed by the intersection of Φ and T, and Ψ and T. A change in the temperature profile produces an alteration in the horizontal mean, i.e., it leads to the formation of a factor in terms of $\sin(2\pi z)$, and the zonal velocity field results in the formation of a factor in terms of $\sin(2\pi\alpha x)$.

The Fourier series representation characterizing the finite amplitude free convection in a minimal form is taken to be

$$\Psi = a(t)\sin(\pi\alpha x)\sin(\pi z) \qquad (29)$$

$$\Phi = f(t)\sin(\pi\alpha x)\cos(\pi z) + g(t)\sin(2\pi\alpha x) \qquad (30)$$

$$T = h(t)\cos(\pi\alpha x)\sin(\pi z) + e(t)\sin(2\pi z) \qquad (31)$$

where the dynamics of the system estimate $a(t)$, $h(t)$, $e(t)$, $g(t)$ and $f(t)$ which are the amplitudes.

Substitution of Eqs. (29), (30) and (31) in Eqs. (24), (25), (26) yields

$$\dot{a}(t) = -Pr\eta k_1^2 a(t) - \frac{R\,Pr\pi\alpha}{k_1^2}h(t) - \pi[1 + \epsilon\cos(\omega t)]\frac{QPr^2}{Pm}f(t) \qquad (32)$$

$$\dot{h}(t) = x(t) \tag{33}$$

$$\dot{x}(t) = \pi\alpha\eta Prk_1^2 a(t) + \frac{\pi^2\alpha^2 R\ Pr}{k_1^2}h(t) + \frac{\pi^2\alpha QPr^2}{Pm}[1 + \epsilon\cos(\omega t)]f(t)$$
$$- \frac{1}{2M}x(t) - \frac{\pi\alpha}{2M}a(t) - \frac{k_1^2}{2M}h(t) \tag{34}$$

$$\dot{e}(t) = y(t) \tag{35}$$

$$\dot{y}(t) = 2\pi^2\alpha a(t)x(t) - Pr\pi^2\alpha k_1^2\eta a(t)h(t) - \frac{\pi^3\alpha^2 RPr}{k_1^2}h(t)^2 - \frac{\pi^3\alpha QPr^2}{Pm}f(t)h(t)$$
$$+ \pi^3\alpha^2 a(t)^2 - \frac{1}{M}y(t) + \frac{\pi^2\alpha}{2M}a(t)h(t) - \frac{4\pi^4}{M}e(t) \tag{36}$$

$$\dot{f}(t) = -2\pi^2\alpha a(t)g(t) + [1 + \epsilon\cos(\omega t)]\pi a(t) - \frac{Pr}{Pm}k_1^2 f(t) \tag{37}$$

$$\dot{g}(t) = \frac{-\pi^2\alpha}{2}a(t)f(t) - 4\pi^2\alpha^2\frac{Pr}{Pm}g(t) \tag{38}$$

where $k_1^2 = \pi^2(\alpha^2 + 1)$ and $\eta = 1 + Ck_1^2$.

It follows that

$$\frac{\partial\dot{a}}{\partial a} + \frac{\partial\dot{x}}{\partial x} + \frac{\partial\dot{y}}{\partial y} + \frac{\partial\dot{f}}{\partial f} + \frac{\partial\dot{g}}{\partial g} = -\left(Prk_1^2\eta + \frac{1}{2M} + \frac{1}{M} + k_1^2\frac{Pr}{Pm} + 4\pi^2\alpha^2\frac{Pr}{Pm}\right). \tag{39}$$

Equation (39) is always negative and hence can concluded that the system is bounded and dissipative.

2.4   Heat Transport

Analyzing the characteristics of heat transport is significant in the study of convection problems with couple stress as an enhancement in the critical Rayleigh number is detected more effectively by the characteristics shown in the transport of heat. It should be noted here that the transport of heat in the basic state takes place only through conduction.

If the heat transport per unit area is given by J, then

$$J = -\kappa\left\langle\frac{\partial T_{total}}{\partial z}\right\rangle_{z=0}, \tag{40}$$

where

$$T_{total} = T_0 - \Delta T\frac{z}{d} + T(x, z, t). \tag{41}$$

and the angular brackets represents the horizontal average. Substitution of Eq. (31) in Eq. (41) and the resulting expression in Eq. (40) gives

$$J = \frac{\kappa\Delta T}{d} - \frac{\kappa\Delta T}{d}2\pi e(t) \tag{42}$$

The Nusselt number is determined by

$$Nu = \frac{J}{\left(\frac{\kappa\Delta T}{d}\right)} = 1 - 2\pi e(t). \tag{43}$$

3   Results and Discussion

The non-Fourier effects of a couple stress fluid layer with modulated magnetic field is analyzed in this paper using a stability analysis which is weakly nonlinear. The characteristics of the system and transfer of heat in it are explored using the amplitude equations. The implications of system governing parameters on heat transport are studied using the graph of Nusselt number.

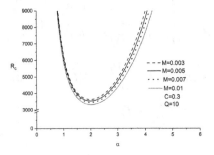

Fig. 2. Graph of R_c vs. α for a range of values of Cattaneo number.

Fig. 3. Graph of R_c vs. α for a range of values of couple stress parameter.

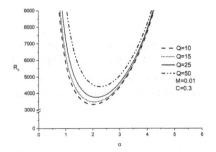

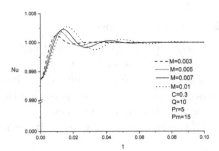

Fig. 4. Graph of R_c vs. α for a range of values of Chandrasekhar number.

Fig. 5. Variation in Nu with regard to time for distinct Cattaneo number.

In Fig. 2 it can be observed that as Cattaneo number, M, increases, the critical Rayleigh number, R_c decreases. M represents the non-Fourier effects. An increase in M results in the contraction of the convective cells and reduction in the wave number as M varies inversely with square of the characteristic length, d, which advances the setting in of convection.

The influence of couple stress parameter, C, on the critical Rayleigh number is shown in Figs. 3. As C increases, R_c increases. C characterizes the amount of suspended particles in the fluid layer. Hence as C increases, more amount of energy is required for the onset of convection resulting in an enhancement of the stability of the fluid layer.

Figure 4 depicts that an increase in Chandrasekhar number, Q, leads to an increase in R_c. Q characterizes the influence of magneto-convection. When a fluid layer is subjected to an external magnetic field, a current is induced in the system. The combination of this current and the magnetic field results in Lorenz force. Lorenz force acts against the velocity, in the opposite direction, increasing the viscosity of the fluid layer. Hence Q results in a delay in the outset of convection, thereby stabilizing the system.

Figure 5 shows the features on heat transfer because of the Cattaneo number. A decrease in the temperature propagation with an increment in M is observed for a very small time period. M is proportional to the relaxation time. Hence the more the relaxation time is, lesser the heat transfer in the system. However this effect reverses and the transfer of heat is enhanced as M increases due to the decrease in the critical Rayleigh number as given in the explanation for Fig. 2.

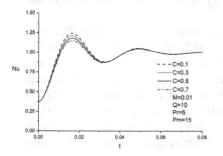 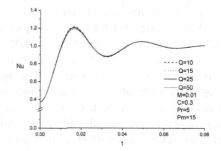

Fig. 6. Variation in Nu with regard to time for distinct couple stress parameter.   **Fig. 7.** Variation in Nu with regard to time for distinct Chandrasekhar number.

Figure 6 explores the influence of couple stress parameter on heat transfer. It is observed that the amplitude of convection decrease with a rise in the value of C as the R_c increases with an increase in C as seen in Fig. 3.

In Fig. 7, the impact of Chandrasekhar number on transfer of heat is shown. As Q increases, there is a reduction in the heat transfer. As discussed in Fig. 4, R_c enhances with an increment in Q resulting in reduced heat transfer.

Figure 8 depicts the influence of Prandtl number, Pr, on heat transfer. It is seen that Pr diminishes the transport of heat. Couple stress fluids are more viscous compared to clean fluids and hence Prandtl number greater than 1 is considered. Since Prandtl number varies inversely with the thermal diffusivity, $Pr > 1$ implies less thermal diffusivity. This results in diminishing thermal boundary layer thickness and the temperature profile which consequently results in reduced heat transfer.

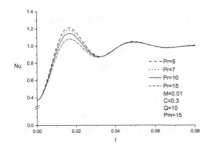
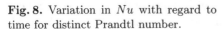
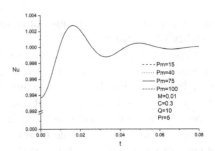

Fig. 8. Variation in Nu with regard to time for distinct Prandtl number.

Fig. 9. Variation in Nu with regard to time for distinct magnetic Prandtl number.

Figure 9 shows the influence of magnetic Prandtl number, Pm, and the graph shows that the effect is insignificant. Pm reflects the rate of viscous diffusion to that of magnetic diffusion. Since the viscosity of couple stress fluid is high, higher values of Pm are considered for the study.

Figures $10, 11, 12, 13, 14, 15$ depict streamlines for $t = 0.03, 0.05, 0.07, 0.1,$ $0.3, 0.5$ respectively where $M = 0.01$, $C = 0.3$, $Q = 10$, $Pr = 5$, $Pm = 15$, $\omega = 10$ and $\varepsilon = 0.1$. In Fig. 10(a) it can be seen that the magnitude of the streamlines is small for small t. As the time t progress, there is an increase in the magnitude of streamlines which indicate that convection is taking place. Convection is seen to become faster on further increasing of time t. As there is no change in the magnitude of streamlines after $t = 0.3$, the system is observed to have achieved steady state beyond $t = 0.3$. Streamlines attain their maximum size beyond $t = 0.3$.

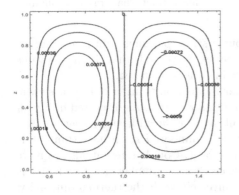
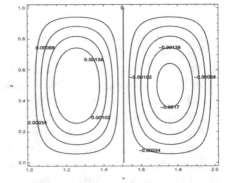

Fig. 10. Streamlines at $t = 0.03$.

Fig. 11. Streamlines at $t = 0.05$.

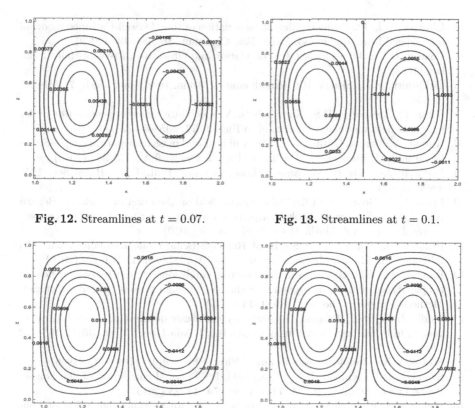

Fig. 12. Streamlines at $t = 0.07$.   **Fig. 13.** Streamlines at $t = 0.1$.

Fig. 14. Streamlines at $t = 0.3$.   **Fig. 15.** Streamlines at $t = 0.5$.

4   Conclusion

This paper investigates non-Fourier effects in a fluid system with couple stresses and externally applied modulated magnetic field. Cattaneo number inhibits the heat transport for a short period of time. This influence is seen to reverse resulting in a raise in heat transfer as time progresses. Couple stress parameter, Chandrasekhar number and Prandtl number inhibit the heat transport. The influence of magnetic Prandtl number on heat transport is seen to be insignificant.

References

1. Bhadauria, B.S., Kiran, P.: Weak nonlinear analysis of magneto-convection under magnetic field modulation. Phys. Scr. **89**, 095209 (2014)
2. Cattaneo, C.: Sulla condizione del calore. Atti Del Semin. Matem. E Fis. Della Univ. Modena. **3**, 83–101 (1948)
3. Chandrasekhar, S.: Hydrodynamic and Hydromagnetic Stability. Clarendon Press, Oxford (1961)

4. Christov, C.I.: On frame indifferent formulation of the Maxwell-Cattaneo model of finite-speed heat conduction. Mech. Res. Commun. **36**, 481–486 (2009)
5. Cokelet, G.R.: Biomechanics, Its Foundation and Objectives. Prentice-Hall, Hoboken (1963)
6. Goldsmith, H.L., Skalak, R.: Hemodynamics. Annu. Rev. Fluid Mech. **7**, 231–247 (1975)
7. Kiran, P., Bhadauria, B.S., Narasimhulu, Y.: Oscillatory magneto-convection under magnetic field modulation. Alexandria Eng. J. **57**, 445–453 (2018)
8. Kumar, A., Vanita, Gupta, V.K.: Study of heat and mass transport in couple-Stress liquid under g-jitter effect. Ain Shams Eng. J. Online **9**(4), 973–984 (2016)
9. Maxwell, J.C.: On the dynamical theory of gases. Phil. Trans. Royal Soc. **157**, 49–88 (1867)
10. Pranesh, S., George, S.: Effect of magnetic field on the onset of Rayleigh- Bénard convection in Boussinesq-Stokes Suspensions with time periodic boundary temperatures. Int. J. Appl. Math. Mech. **6**(16), 38–55 (2010)
11. Pranesh, S., Kiran, R.V.: Study of Rayleigh-Bénard magneto convection in a micropolar fluid with Maxwell-Cattaneo law. Appl. Math. **1**, 470–480 (2010)
12. Ramesh, K.: Effects of slip and convective conditions on the peristaltic flow of couple stress fluid in an asymmetric channel through porous medium. Comput. Methods Programs Biomed. **135**, 1–14 (2016)
13. Siddheshwar, P.G., Pranesh, S.: An analytical study of linear and non-linear convection in Boussinesq-Stokes suspensions. Int. J. Non-Linear Mech. **39**(1), 165–172 (2004)
14. Stokes, V.K.: Couple stresses in fluids. Phys. Fluids **9**(9), 1709–1715 (1966)
15. Stranges, D.F., Khayat, R.E., Albaalbaki, B.: Thermal convection of non-Fourier fluids. Linear Stability. Int. J. Therm. Sci. **74**, 14–23 (2013)
16. Stranges, D.F., Khayat, R.E., Debruyn, J.: Finite thermal convection of non-Fourier fluids. Int. J. Therm. Sci. **104**, 437–447 (2016)
17. Straughan, B.: Porous convection with Cattaneo heat flux. Int. J. Heat Mass Transf. **53**, 2808–2812 (2010)
18. Straughan, B.: Thermal convection with the Cattaneo-Christov model. Int. J. Heat Mass Transf. **53**, 95–98 (2010)
19. Straughan, B., Franchi, F.: Bénard convection and the Cattaneo law of heat conduction. Proc. Royal Soc. Edinburgh **96A**, 175–178 (1984)
20. Venezian, G.: Effect of modulation on the onset of thermal convection. J. Fluid Mech. **35**(2), 243–254 (1969)

A Characterization for V_4-Vertex Magicness of Trees with Diameter 5

Muhammed Sabeel Kollaran[iD], Appattu Vallapil Prajeesh[iD],
and Krishnan Paramasivam[✉][iD]

Department of Mathematics, National Institute of Technology Calicut,
673601 Kozhikode, India
{sabeel_p160078ma,prajeesh_p150078ma,sivam}@nitc.ac.in

Abstract. Let G be an undirected simple graph with vertex set $V(G)$ and the edge set $E(G)$ and $\mathcal{A}$ be an additive Abelian group with the identity element 0. A function $l : V(G) \to \mathcal{A} \setminus \{0\}$ is said to be a $\mathcal{A}$-vertex magic labeling of G if there exists an element μ of $\mathcal{A}$ such that $w(v) = \sum_{u \in N(v)} l(u) = \mu$ for any vertex v of G. A graph G having $\mathcal{A}$-vertex magic labeling is called a $\mathcal{A}$-vertex magic graph. If G is $\mathcal{A}$-vertex magic graph for every non-trivial additive Abelian group $\mathcal{A}$, then G is called a group vertex magic graph. In this article, a characterization for the $\mathcal{A}$-vertex magicness of any tree T with diameter 5, is given, when $\mathcal{A} \cong \mathbb{Z}_2 \oplus \mathbb{Z}_2$.

Keywords: Group vertex magic graph · Tree · Diameter · Interior neighborhood

2010 AMS Subject Classification: 05C78 · 05C25.

1 Introduction

In this article, we deal with undirected and simple finite graphs. We employ $V(G)$ for the set of vertices and $E(G)$ for the set of edges of G. The neighborhood $N_G(v)$ of a vertex v of G is the set of all adjacent vertices of v, and the degree $deg_G(v)$ of v is the number of vertices in $N_G(v)$ and if v is a vertex with $deg_G(v) = 1$, then v is called a pendant. Note that for an arbitrary vertex v of G, $N_G(v) = N_{int}(v) \cup N_{ext}(v)$, where $N_{int}(v) = \{u \in V(G) : uv \in E(G) \ and \ deg_G(v) > 1\}$ and $N_{ext}(v) = \{u \in V(G) : uv \in E(G) \ and \ deg_G(v) = 1\}$ and any vertex in $N_{int}(v)$ is an interior vertex. A tree is a connected graph containing no cycles. For a graph G, the vertices of minimum eccentricity are called central vertices of G. A tree with odd diameter $k > 1$, has two central vertices, and we denote them by v_{c_1} and v_{c_2}. The bi-star graph $B_{r,s}$ is a graph, which is obtained by connecting two copies of stars $K_{1,r}$ and $K_{1,s}$ by adding an edge between central vertices of two stars. For two graphs G_1 and G_2, $G_1 \odot G_2$ is the graph obtained by picking a copy of G_1 and $|G_2|$ number copies of G_2 and then connecting all vertices of i-th copy of G_2 to i-th vertex of G_1 by new edges and is called the

A. Awasthi et al. (Eds.): CSMCS 2020, CCIS 1345, pp. 243–249, 2021.
https://doi.org/10.1007/978-981-16-4772-7_19

corona product of G_1 and G_2. We consider only additive Abelian groups $\mathcal{A}$ with the identity element 0. The Klein's-4 group $V_4 = \{a, b : 2a = 2b = 2(a + b) = 0\}$, which also can be represented as $\mathbb{Z}_2 \oplus \mathbb{Z}_2 = \{(0, 0), (1, 0), (0, 1), (1, 1)\}$ under component-wise addition modulo 2 with $(0, 0)$ as identity element. For more, notation and terminology in graph theory and group theory, we refer [1] and [2], respectively.

The motivation for magic labeling arises from magic squares. Magic squares are $(n \times n)$-arrays consisting the numbers $1, 2, \cdots, n^2$, each appearing once, along with the condition that the sum of numbers in each row, column, the main diagonal, and the main backward diagonal are equal and is $n(n^2 + 1)/2$. For more details related to magic square, refer [3].

Lee et al. [4] in 2001 introduced the notion of group-magic graphs. For an arbitrary Abelian group $\mathcal{A}$, a graph G is called $\mathcal{A}$-magic graph, if there exists a function $l : E(G) \rightarrow A \setminus \{0\}$ such that for any vertex v of G, the induced function $l^+ : V(G) \rightarrow A$ given by $l^+(v) = \displaystyle\sum_{uv \in E(G)} l(uv)$, is a constant function.

Further studies on group-magic graphs, can be found in [6–10] including specialized studies on V_4-magicness of trees and group magicness of certain product graphs.

In 2019, Kamatchi et al. [5], introduced the notion of group vertex magic graphs.

Definition 1. [5] *$\mathcal{A}$ be an additive Abelian group with the identity element 0. A function $l : V(G) \rightarrow A \setminus \{0\}$ is a $\mathcal{A}$-vertex magic labeling of a graph G if there exists a group element μ of $\mathcal{A}$ such that for any vertex v of G, the weight of v, $w(v) = \displaystyle\sum_{x \in N(v)} l(x)$ is μ. A graph G that receives such a labeling is called an $\mathcal{A}$-vertex magic graph and such weight μ is called magic constant. For any non-trivial Abelian group $\mathcal{A}$, if G is $\mathcal{A}$-vertex magic, then G is a group vertex magic graph.*

Observation 1. [5] *If a graph G is regular, then G is group vertex magic. One can label all vertices of G by g, where g is an element of any Abelian group $\mathcal{A}$ with $g \neq 0$.*

In [5], several results related to V_4-vertex magic graphs are proved and a characterization of all V_4-vertex magic trees up to diameter 4, is obtained.

Lemma 1. [5] $n \geq 2$. *Then any element $g \in V_4$ can be expressed as $g = g_1 + \cdots + g_n$, where $g_i \in V_4$ and $g_i \neq 0$ for all $i \in \{1, \cdots, n\}$.*

Theorem 1. [5] *If T is a tree with n vertices and if T is of diameter 2, then*

(i) T is V_4-vertex magic.
(ii) T is group vertex magic, when n is even.

Theorem 2. [5] *If* $T = B_{m,n}$ *is a tree with diameter* 3, *then* T *is* V_4-*vertex magic if and only if* $m, n \geq 2$.

Theorem 3. [5] *A tree of diameter* 4 *is* V_4-*vertex magic if and only if at least one of the following conditions hold.*

(i)   *All internal vertices have at least two pendant vertices as neighbors.*

(ii) *The central vertex is of odd degree and have no pendant vertex as its neighbor.*

(iii) *The central vertex is odd degree with exactly one pendant vertex as its neighbor and all other internal vertices have at least two pendant vertices as its neighbors.*

Note that the following observations are useful while proving the V_4-vertex magicness of a graph G having pendants.

Observation 2. *Let* G *be a graph with at least one pendant. Suppose that* G *is* V_4-*vertex magic graph with labeling* l *such that* $l(u) = g$, *for some vertex* u *having a pendant. Then the magic constant* $\mu = g$, *with respect to* l. *Thus the identity element* 0 *cannot be the magic constant.*

Thus from the above observation, we see that the magic constant of any V_4-vertex magic graph having a vertex v with $|N_{ent}(v)| = 1$ is always non-identity element. On the other hand, the magic constant of any V_4-vertex magic graph having all vertices v with $|N_{ext}(v)| > 1$, may or may not be identity element.

Observation 3. *Given that* $g \in V_4$. *Suppose* v *is a vertex of a graph* G *such that the labels of all the vertices in* $N_{int}(v)$ *are known. If* $|N_{ext}(v)| > 1$, *then by Lemma 1, it is possible to label all vertices of* $N_{ext}(v)$ *such that* $w(v) = g$. *On the other hand, if* $N_{ext}(v)$ *is* $\{u\}$, *then it is possible to label* u *such that* $w(v) = g$, *provided that* $g \neq \sum_{x \in N_{int}(v)} l(x)$.

The above observation guarantees that for a given group element in V_4, if the labels of all interior neighbor's of a vertex are known, then it is possible to predict the labels of all exterior neighbor's of the same vertex in such a way that the weight of the vertex is the given group element.

2   Main Results

In this section, certain structural properties of V_4-vertex magic graphs are discussed and finally, a characterization for any tree of diameter 5 to be V_4-vertex magic, is given.

Theorem 4. *Suppose* G *is a graph such that all internal vertices have more than one pendant. Then* G *is* V_4-*vertex magic.*

Proof. Let $a \in V_4 \setminus \{0\}$. Now label any vertex v of G by a, whenever $\deg_G(v) > 1$. Consequently, all the pendants of G have weight a. Now, we need to label the remaining vertices in such a way that every internal vertex of G, must have same weight a. Suppose v is an internal vertex of G. Then the weight of v, $w(v) = \sum_{x \in N_{int}(v)} l(x) + \sum_{y \in N_{ext}(v)} l(y) = |N_{int}(v)|a + \sum_{y \in N_{ext}(v)} l(y)$. Now, if v is the vertex with $|N_{int}(v)|$ is odd, then by Lemma 1, label all the vertices of $N_{ext}(v)$ such that $\sum_{y \in N_{ext}(v)} l(y) = 0$. If $|N_{int}(v)|$ is even, then again by Lemma 1, label all the vertices of $N_{ext}(v)$ such that $\sum_{y \in N_{ext}(v)} l(y) = a$. Therefore, in either case $w(v) = a$. This completes the proof.

Corollary 1. *For any graph G, $G \odot \overline{K}_n$ is a V_4-vertex magic graph, where $n > 1$.*

The above corollary guarantees that any graph can be an induced subgraph of some V_4-vertex magic graph. Therefore, V_4-vertex magic graphs do not have any forbidden structures.

In the next theorem, we construct a larger V_4-vertex magic graph from the existing one. Note that K_n is V_4-vertex magic for all $n \geq 1$. Let $K_n^\dagger$ be the graph derived from K_n by augmenting a new vertex $u^\dagger$ to exactly one vertex of K_n by an edge.

Theorem 5. *Let $n > 1$. Then $K_n^\dagger$ is V_4-vertex magic if and only if n is even.*

Proof. Consider the graph $K_n^\dagger$ with the vertex set $\{v_0, v_1, \cdots, v_{n-1}\} \cup \{u^\dagger\}$, where v_i's are the vertices of K_n and v_0 is adjacent to $u^\dagger$. Suppose l is a V_4-vertex magic labeling of $K_n^\dagger$. Then by Observation 2, the magic constant $\mu = l(v_0)$. Now, we have,

$$w(v_0) = \sum_{i=1}^{n-1} l(v_i) + l(u^\dagger), \qquad (1)$$

and for $j \neq 0$,

$$w(v_j) = \sum_{i=0, \, i \neq j}^{n-1} l(v_i). \qquad (2)$$

Adding Eq. (1) and (2), we get, $l(v_j) = l(u^\dagger) + l(v_0)$ for $j \neq 0$.

Now, if n is odd, then $w(v_1) = l(v_0) + (n - 2)(l(u^\dagger) + l(v_0)) = l(u^\dagger)$. But $\mu = l(v_0)$ and hence $l(v_0) = l(u^\dagger)$, which gives $l(v_1) = 0$, a contradiction. Hence, $K_n^\dagger$ cannot be V_4-vertex magic graph for any odd integer n.

If n is even, let $l(v_0) = a$, $l(u^\dagger) = b$ and $l(v_j) = a+b$ for all $j \in \{1, \cdots, n-1\}$, which yields a V_4-vertex magic labeling with $\mu = a$.

Now, we give a typical representation of a tree of diameter 5 with two central vertices v_{c_1} and v_{c_2} and their internal and external neighborhood's in Fig. 1.

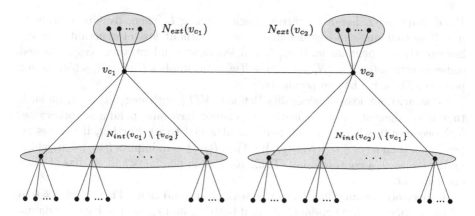

Fig. 1. A typical diameter 5 tree

The following two theorems give a characterization for the V_4-vertex magicness of diameter 5 trees, when either both of the central vertices have an odd number of internal vertices or both of the central vertices have even number of internal vertices.

Theorem 6. *Let T be a tree of diameter 5 with central vertices v_{c_1} and v_{c_2} such that both $|N_{int}(v_{c_1})|$ and $|N_{int}(v_{c_2})|$ are even. Then T is V_4-vertex magic if and only if none of the vertices are of degree 2 and both central vertices have at least one pendant.*

Proof. Suppose T has a V_4-vertex magic labeling l. By Observation 2, the magic constant $\mu \neq 0$ and we assume $\mu = a \neq 0$. Then, $|N_{ext}(v_{c_i})| > 1$ for $i = 1, 2$ (otherwise one of central vertex will have label 0, a contradiction). Therefore, $l(v_{c_i}) = a$, $i = 1, 2$, and hence no vertex has degree 2 (for if $N(x) = \{u_1, u_2\}$ for some x, where u_1 is one of the central vertex, then $l(u_2) = 0$, a contradiction again.)

Conversely, assume that T is a tree with given conditions. Label all internal vertices of T with $a \neq 0$ and thus, all the pendants of T have weight a, and for $i = 1, 2$, label the vertices of $N_{ext}(v_{c_i})$ in such a way that $\sum_{u \in N_{ext}(v_{c_i})} l(u) = a$. Thus, $w(v_{c_i}) = |N_{int}(v_{c_i})|a + a = a$, for both $i = 1, 2$. Now, by Observation 2, it is straight forward to label all remaining pendants of T in such a way that T is V_4-vertex magic.

Theorem 7. *Let T be a tree of diameter 5 with central vertices v_{c_1} and v_{c_2} such that $|N_{int}(v_{c_1})|$ is odd and $|N_{int}(v_{c_2})|$ is odd. Then T is V_4-vertex magic if and only if both of the following conditions hold:*

(i) whenever one of the central vertex has exactly one pendant, the other central vertex has no pendants.
(ii) whenever a non-central vertex is of degree 2, then it's neighboring central vertex has no pendants and the other central vertex has at least one pendant.

Proof. Suppose T has a V_4-vertex magic labeling l. Then, by Observation 2, $\mu \neq 0$ so that one can assume that $\mu = a$. Without loss of generality, let v_{c_1} has exactly one pendant w. If v_{c_2} has at least one pendant, then $l(v_{c_2}) = a$ and consequently, $w(v_{c_1}) = |N_{int}(v_{c_1})|a + l(w) = a$, implies $l(w) = 0$, which is not possible. Thus, v_{c_2} has no pendants.

Now, without loss of generality, let $u \in V(T)$ with $\deg_T(u) = 2$, be such that u is adjacent to v_{c_1}. Clearly, v_{c_1} cannot have any pendants (otherwise, $l(x)$ becomes 0, where x is the pendant attached to u). Further, if v_{c_2} has no pendants, then $w(v_{c_2}) = (|N_{int}(v_{c_2})| - 1)a + l(v_{c_1}) = a$, implies $l(v_{c_1}) = a$, which makes $l(x) = 0$, a contradiction. Thus, v_{c_1} has no pendants and v_{c_2} has at least one pendant.

Conversely, assume that T is a tree with given conditions. Then label $a \neq 0$ to all the vertices having pendants. Also, if both v_{c_1} and v_{c_2} do not have pendants, label both by a and hence $w(v_{c_i}) = |N_{int}(v_{c_i})|a = a$ for $i = 1, 2$. Now, if only one of the central vertex has no pendant, label it by $b \neq 0$, and by Lemma 1, label the pendants of other central vertex in such a way that the label-sum is $a + b$. Hence, we have $w(v_{c_1}) = a = w(v_{c_2})$. By Observation 2, in either case, it is not hard to label the rest of the pendants of T in such a way that T is V_4-vertex magic with $\mu = a$.

The following theorem discusses the V_4-vertex magicness of diameter 5 trees, when one central vertex has an odd number of internal vertices and another central vertex has an even number of internal vertices.

Theorem 8. *Let T be a tree of diameter 5 with two central vertices v_{c_1} and v_{c_2} such that $|N_{int}(v_{c_i})|$ is odd and $|N_{int}(v_{c_j})|$ is even, for $i, j \in \{1, 2\}$ and $i \neq j$. Then T is V_4-vertex magic if and only if all of the following conditions hold:*

(i) v_{c_j} has atleast one pendant and v_{c_j} is not adjacent to a two degree vertex.
(ii) if v_{c_i} has a pendant, then v_{c_i} is not adjacent to a two degree vertex.
(iii) v_{c_i} cannot have exactly one pendant.

Proof. Suppose T has a V_4-vertex magic labeling l. Then by Observation 2 the magic constant is a, where $a \neq 0$.

Proof for (i). On contrary, suppose that v_{c_j} has no pendants. Then $w(v_{c_j}) = (|N_{int}(v_{c_j})| - 1)a + l(v_{c_i}) = a + l(v_{c_i}) = a$, contradicts $l(v_{c_i}) \neq 0$. Therefore, v_{c_j} has a pendant and $l(v_{c_j}) = a$. Now, if possible v_{c_j} is adjacent to a vertex u of degree 2 and y is the adjacent vertex of u other than v_{c_j}, then because of the fact $w(u) = a$, we have $l(y)$ is 0, a contradiction.

Proof for (ii). If v_{c_i} has a pendant, using the same arguments as above, v_{c_i} cannot have a degree 2 neighbor.

Proof for (iii). Suppose, v_{c_i} has exactly one pendant (say) x. Since $l(v_{c_j}) = a$, then $l(x)$ is 0, a contradiction.

Conversely, assume T is a tree with given conditions and let $a, b \in V_4 \setminus \{0\}$. Then label all the vertices having pendants with a. If v_{c_i} has pendants, label $N_{ext}(v_{c_i})$ such that its label sum is 0 and label $N_{ext}(v_{c_j})$ in such a way that sum of the labels is a, hence both v_{c_i} and v_{c_j} will have weight a. If v_{c_i} has no pendants, then label v_{c_i} with b and $N_{ext}(v_{c_j})$ such that its label sum is b, then both v_{c_i}

and v_{c_j} will have weight a. In both cases, with the help of 2, label all the rest of the vertices of T in such a way that T becomes V_4-vertex magic with magic constant a.

Thus, the above three Theorems have completely characterized the existence of V_4-vertex magicness of all trees T of diameter 5.

3   Conclusion

In this article, we characterized all the V_4-vertex magic trees of diameter 5 and proved V_4-vertex magicness of certain classes of graphs. Note that for any tree with diameter ≤ 5, all the internal vertices are always adjacent to one of the central vertex(s). However, for trees of diameter k, where $k > 5$, the central vertex(s) need not be adjacent to all the internal vertices and that crucial factor increases the number of cases to be considered and analyzed. Thus, we give the following problem.

Problem 1. Characterize all V_4-vertex magic trees T with diameter > 5.

References

1. Bondy, J.A., Murty, U.S.R.: Graph theory with applications. Elsevier Science Publishing Co., Inc., New York (1976)
2. Herstein, I. N.: Topics in algebra, John Wiley & Sons, New York (2006)
3. Colbourn, C., Dinitz, J. (eds.): The CRC Handbook of Combinatorial Designs, Chapman and Hall/CRC, Boca Raton, FL (2007)
4. Lee, S.M., Sun, H., Wen, I.: On group-magic graphs. J. Comb. Math. Comb. Comput. **38**, 197–207 (2001)
5. Kamatchi, N., Paramasivam, K., Prajeesh, A.V., Muhammed Sabeel, K., Arumugam, S.: On group vertex magic graphs. AKCE Int. J. Graphs Comb. **17**(1), 461–465 (2020)
6. Lee, S.M., Saba, F., Salehi, E., Sun, H.: On the V_4-magic graphs. Congr. Numer. **156**, 59–67 (2002)
7. Low, R.M., Lee, S.M.: On the products of group-magic graphs. Australas. J. Comb. **34**, 41–48 (2006)
8. Shiu, W.C., Lam, P.C.B., Sun, P.K.: Construction of group-magic graphs and some a-magic graphs with a of even order. In: Proceedings of the Thirty-Fifth Southeastern International Conference on Combinatorics, Graph Theory and Computing. Congressus Numerantium, vol. 167, pp. 97–107 (2004)
9. Low, R.M., Lee, S.M.: On group-magic Eulerian graphs. J. Comb. Math. Comb. Comput. **50**, 141–148 (2004)
10. Shiu, W.C., Low, R.M.: Group-magic labelings of graphs with deleted edges. Australas. J. Comb. **57**, 3–19 (2013)

Modelling

Data-Driven Regression-Based Compartmental Model to Identify the Dynamical Behavior of Dengue Incidences in Urban Colombo

K. K. W. H. Erandi$^{(\boxtimes)}$, S. S. N. Perera , and A. C. Mahasinghe

Research and Development Centre for Mathematical Modelling,
Department of Mathematics, University of Colombo, Colombo, Sri Lanka

Abstract. Dengue is a mosquito–borne viral disease that has rapidly spread in tropical and subtropical regions. Understanding the transmission dynamics of dengue incidences is vital in developing appropriate strategies to control potential outbreaks of the diseases. In literature, the $SIR - UV$ model is one of the mostly–used mathematical model to describe the dynamics of the disease, of which the solvability of the relevant system of differential equations seems to be uncertain. However, this system can be reduced down into a two–dimensional system by considering quasi–equilibrium for infected vector population. Once the reduced system is considered, its solution requires the estimation of per–capita vector density, which becomes an infeasible task due to lack of resources as well as complexity of external impacts. Since the per–capita vector density heavily depends upon the climate factor, we propose a generalized linear regression formula to estimate this by capturing the seasonal pattern of dengue dynamics according to the effect of climate factors in urban Colombo. Further, we compare the dynamical behaviour of the infected host population in the simplified model with reported dengue incidences. Our simulation shows that the dynamical behaviour of the infected host population captures the seasonality of the reported dengue incidences.

Keywords: Dengue transmission · SIR–UV model · Quasi–equilibrium · Regression formula

1 Introduction

Dengue is a mosquito–borne viral disease that has rapidly spread in tropical and subtropical regions during the past few decades. There are currently four billion people living in high–risk areas of dengue fever transmission, with 390 million cases reported annually [1]. Identified in Sri Lanka for the first time

This work was supported by grant RPHS/2016/D/05 of the National Science Foundation, Sri Lanka.

© Springer Nature Singapore Pte Ltd. 2021
A. Awasthi et al. (Eds.): CSMCS 2020, CCIS 1345, pp. 253–259, 2021.
https://doi.org/10.1007/978-981-16-4772-7_20

in 1962, it has now gained the status of an endemic disease. Statistics of the epidemiology unit of the Ministry of Health, Sri Lanka indicate that 185,969 dengue cases were reported in 2017. Every year, the government and the private sector spend a significant amount of funds on individual healthcare and anti-dengue campaigns. This is the context in which examining the dynamics of dengue transmission becomes vital.

In order to capture the transmission of the dengue virus, one might naturally look into well–known mathematical models of disease transmission based on classical compartments. In compartment models, the population is divided into categories for which the interactions and dynamics are formulated mathematically. Being one such model, the $SIR - UV$ classification divides the host population into three compartments: susceptible (S), infected (I) and recovered (R) and vectors population into two compartments: susceptible (U) and infected (V) [2]. Thus, the $SIR-UV$ model consists of a system of differential equations which describes the dynamical behavior of the disease in terms of interactions between these compartments. Since dengue is a vector–borne disease, the spreading trend of the disease depends on the infected vector density. Accordingly, the number of adult vectors in an area is a major factor which affects the dynamical behavior of the disease. However, calculating this vector density is practically an infeasible task. Thus, the classical $SIR - UV$ model has been modified by improving its predictability power. First, we considered the fact that vector dynamics occur in a faster time scale compares to host dynamics and derived a two dimensional compartmental model with aid of classical $SIR - UV$ models and a function for per–capita vector density has been introduced. Further, per-capita vector density is a highly sensitive parameter that heavily depends upon climatic factors such as rainfall and temperature. According to [5,7], reported dengue incidences show a strong correlation with rainfall in Colombo with different time lags. Motivated from all these, we examine how to use climate factors in order to yield a formula for per–capita vector density.

Generalized regression formula is particularly interesting when there is time–dependence between variables and possible to model a temporal structure with more reliable predictions. In this study, we examine how to use multiple climate factors and reported dengue data to develop a general regression formula to capture the per–capita vector population. Once the generalized linear regression formula is developed, the two dimensional IR model has been moderated with the developed regression formula to capture the periodic pattern of dengue according to effect of climate factors in urban Colombo.

2   Method

2.1   Two Dimensional IR Model for Dengue Transmission

A modified version of the $SIR - UV$ model for dengue transmission has been developed in [2], which represented interactions between human and vectors in terms of a normalized system of non–linear ordinary differential equations. In this study, we simplify these differential equations, making justifiable assumptions for a real–world situation as in [3]. First, we convert the model in [2] into a dimensionless form. Let S, I, R, U and V denote the susceptible human, infected human, recovered human, susceptible vector and infected vector densities respectively. Let μ_h, η, μ_v and γ denote the birth and death rates for human, recruitment rate of vectors, death rates for vectors and recovery rate for human. Let the adequate contact rate of vectors to human and human to vectors be denoted by β_h and β_v. Also let N_v denotes the vector population and N_h the total human population. Accordingly, Eqs. (1a) to (1e) represent a normalized $SIR - UV$ model.

$$\frac{dS}{dt} = \mu_h(1 - S) - \beta_h \frac{N_v}{N_h} VS, \tag{1a}$$

$$\frac{dI}{dt} = \beta_h \frac{N_v}{N_h} VS - (\mu_h + \gamma_h)I, \tag{1b}$$

$$\frac{dR}{dt} = \gamma_h I - \mu_h R, \tag{1c}$$

$$\frac{dU}{dt} = \eta - \mu_v U - \beta_v UI, \tag{1d}$$

$$\frac{dV}{dt} = \beta_v UI - \mu_v V. \tag{1e}$$

This is reducible further, as the sum of susceptible, infected and recovered human density is one. Thus, we reduce Eqs. (1a), (1b) and (1c) into two dimensional system and replace S by $1 - I - R$. A similar reduction is applicable to Eqs. (1d) and (1e).

$$\frac{dI}{dt} = \beta_h nV(1 - I - R) - (\mu_h + \gamma_h)I, \tag{2a}$$

$$\frac{dR}{dt} = \gamma_h I - \mu_h R, \tag{2b}$$

$$\frac{dV}{dt} = \beta_v(1 - V)I - \mu_v V. \tag{2c}$$

Since, there is no method to estimate the infected vector density in practical manner and simulation results of three dimensional system in Eq. (2) provides infected vector density with certain assumptions which cannot be validated, we consider the idea of quasi–equilibrium for vector density to overcome this problem.

Since the average life expectancy for female mosquito is 6 weeks and the average life expectancy for human is 72 years, vector dynamics achieve the equilibrium faster than host. Therefore, we take into consideration that vector density is in its quasi–equilibrium [6]. Let $\beta = \frac{\beta_v}{\mu_v}$. Then U^* and V^* denote the quasi–equilibrium values for vector density as follows.

$$U^* = \frac{1}{\beta I + 1}, \quad V^* = \frac{\beta I}{\beta I + 1}. \tag{3}$$

Now substitute the quasi–equilibrium value V^* for the infected vector density in Eq. (2a). Then we obtain the reduced two–dimensional quasi–equilibrium IR model as in Eq. (4).

$$\frac{dI}{dt} = \beta_h n \frac{\beta I}{\beta I + 1}(1 - I - R) - (\mu_h + \gamma_h)I, \tag{4a}$$

$$\frac{dR}{dt} = \gamma_h I - \mu_h R. \tag{4b}$$

Since n is a time–dependent term as it is proportional to time–dependent external force on dengue spreading such as climate change. Our next task is to define a function for n to capture the seasonal effect on dengue dynamics.

2.2   Linear Regression Formula for Per–capita Vector Density

Generalized linear regression formula measures expected value of dependent parameter through a linear combination of known parameters [4]. The theory of the generalized linear regression model requires data that generally follow a multivariate normal distribution with the same covariance. Let D_t, R_t, MT_t and mT_t denote number of dengue incidences, rainfall, maximum temperature and minimum temperature at time t. In our study t corresponds to the time in weeks. We use $glmfit$ package in MATLAB to estimate parameters b_0, b_1, b_2, b_3 and b_4 based on distribution fitting. With the assumption of linearity $n(t)$ is expressed as follows.

$$n(t) = b_0 + b_1 D_t + b_2 R_t + b_3 MT_t + b_4 mT_t. \tag{5}$$

Recall that per–capita vector density has been changed with climate factors and dengue incidences. Moreover, dengue incidences have correlated with climate factors with time lags. Therefore, it is possible to obtain Eq. 6 by including time lag into rainfall, maximum temperature and minimum temperature variables in Eq. 5. We include time lag l_D, l_R, l_{MT} and l_{mT} to reported dengue incidences, rainfall, maximum temperature and minimum temperature variables respectively.

$$n(t) = b_0 + b_1 D_{t-l_D} + b_2 R_{t-l_R} + b_3 MT_{t-l_{MT}} + b_4 mT_{t-l_{mT}} \tag{6}$$

Now the simplified IR model can be restated as,

$$\frac{dI}{dt} = \beta_h n(t)\frac{\beta I}{\beta I + 1}(1 - I - R) - (\mu_h + \gamma_h)I, \tag{7a}$$

$$\frac{dR}{dt} = \gamma_h I - \mu_h R, \tag{7b}$$

with

$$n(t) = b_0 + b_1 D_{t-l_D} + b_2 R_{t-l_R} + b_3 MT_{t-l_{MT}} + b_4 mT_{t-l_{mT}}. \tag{7c}$$

3 Results and Discussion

For this study it has been considered the weekly reported dengue incidence gained from Epidemiology unit, Department of Health, Sri Lanka, weekly rainfall, maximum temperature and minimum temperature data from Meteorological Department, Sri Lanka from 2006 to 2016 for CMC area. To calculate the time delay between dengue incidences and climate factors, the Pearson correlation formula has been used and plotted the correlation measure with time lags from 0 to 20 weeks (refer Fig. 1). Results illustrated by Fig. 1 indicated that the highest correlation occurs between dengue incidence and rainfall data with 10—weeks time delay, maximum temperature data with 16—weeks time delay, minimum temperature data with 13—weeks time delay. Furthermore, reported

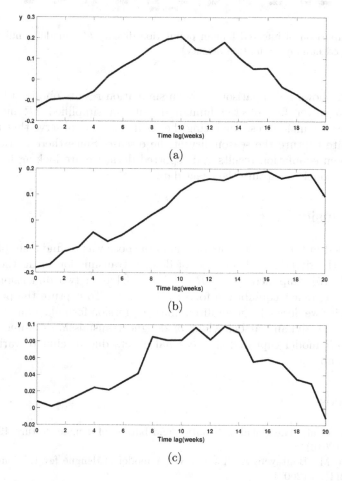

Fig. 1. Cross–correlation between (a) reported dengue data and rainfall data (b) reported dengue data and maximum temperature and (c) reported dengue data and minimum temperature

dengue data with 4–weeks time delay has been used to define a function for per—capita vector density. Then *glmfit* package in MATLAB has been used to estimate parameters b_0, b_1, b_2, b_3 and b_4, based on distribution fitting. Moreover, to assess the distribution *histfit* function in MATLAB has been used.

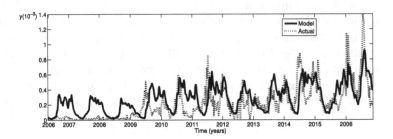

Fig. 2. Comparison of infected human population density (IR model simulation and actual infected human population density)

Figure 2 shows a comparison between simulation results obtained by MAT-LAB ode45 solver for infected human density in simplified IR model and reported dengue incidences. Analyzing the results, we can observe that modeled infection rate capture the seasonality of the disease. Since there is a disagreement between simulation results and reported data, we are looking forward to extend the model to capture the exact data.

4   Conclusion

As mentioned in the introduction, our main purpose was to find a simple model to capture the dynamical behaviour of dengue transmission using the $SIR - UV$ model. We simplified the $SIR - UV$ model to a two dimensional form by considering quasi–equilibrium for vector density. To capture the per–capita vector density we derived a generalized linear regression formula. Our simulation shows that the dynamical behaviour of dengue transmission obtained by the simplified IR model captured the seasonal effects due to climate variation in urban Colombo.

References

1. Bhatt, S., et al.: The global distribution and burden of dengue. Nature **496**(7446), 504–507 (2013)
2. Derouich, M., Boutayeb, A., Twizell, E.: A model of dengue fever. Biomed. Eng. Online **2**(1), 4 (2003)
3. Erandi, K.K.W.H., Perera, S.S.N., Mahasinghe, A.C.: Analysis and forecast of dengue incidence in urban Colombo, Sri Lanka. Theor. Biol. Med. Model. **18**(1), 1–19 (2021)

4. McCullagh, P.: Generalized linear models. Routledge (2019)
5. Pathirana, S., Kawabata, M., Goonatilake, R.: Study of potential risk of dengue disease outbreak in Sri Lanka using GIS and statistical modelling. J. Rural Tr. Public Health **8**, 8 (2009)
6. Rashkov, P., Venturino, E., Aguiar, M., Stollenwerk, N., Kooi, B.W.: On the role of vector modeling in a minimalistic epidemic model. Math. Biosci. Eng. **16**(5), 4314–4338 (2019)
7. Wickramaarachchi, W.P.T.M., Perera, S.S.N.: A mathematical model with control to analyse the dynamics of dengue disease transmission in urban Colombo. J. Nat. Sci. Found. Sri Lanka **46**(1), 41–49 (2018)

Ball Convergence of Multipoint Methods for Non-linear Systems

Ioannis K. Argyros[1], Santhosh George[2], and Shobha M. Erappa[3(✉)]

[1] Department of Mathematical Sciences, Cameron University,
Lawton, OK 73505, USA
`iargyros@cameron.edu`
[2] Department of Mathematical and Computational Sciences, National Institute
of Technology Karnataka, Mangalore 575 025, India
`sgeorge@nitk.ac.in`
[3] Department of Mathematics, Manipal Institute of Technology,
Manipal, Karnataka 576104, India
`shobha.me@manipal.edu`

Abstract. We study Multipoint methods using only the first derivative. Earlier studies use higher than three order derivatives not on the methods. Moreover Lipschitz constants are used to find error estimates not presented in earlier papers. Numerical examples complete this paper.

Keywords: Ball convergence · Lipschitz constant · Radius of convergence.

1 Introduction

In this study, we seek a $x_* \in D$ solving

$$F(x) = 0. \tag{1.1}$$

Here $F : D \subseteq M \to M (M = \mathbb{R}^m; M = \mathbb{C}^m)$.

Chebyshev's, Halley's, Euler's, Super Halley's, [2,5,7,11,18] use the second derivative F'' making them expensive in nature. In this paper, we consider the 4th, 6th and 8th order methods, respectively,

$$
\begin{aligned}
y_n &= x_n - A_n^{-1} F(x_n) \\
x_{n+1} &= y_n - B_n A_n^{-1} F(y_n),
\end{aligned} \tag{1.2}
$$

$$
\begin{aligned}
y_n &= x_n - A_n^{-1} F(x_n) \\
z_n &= y_n - B_n A_n^{-1} F(y_n) \\
x_{n+1} &= z_n - B_n A_n^{-1} F(z_n)
\end{aligned} \tag{1.3}
$$

and

$$
\begin{aligned}
y_n &= x_n - A_n^{-1} F(x_n) \\
z_n &= y_n - B_n A_n^{-1} F(y_n) \\
w_n &= z_n - B_n A_n^{-1} F(z_n) \\
x_{n+1} &= w_n - B_n A_n^{-1} F(w_n),
\end{aligned} \tag{1.4}
$$

© Springer Nature Singapore Pte Ltd. 2021
A. Awasthi et al. (Eds.): CSMCS 2020, CCIS 1345, pp. 260–269, 2021.
https://doi.org/10.1007/978-981-16-4772-7_21

$A_n = DF(x_n) - F'(x_n)$, $B_n = -\frac{3}{2}I + \frac{1}{2}A_n^{-1}F'(y_n)$ and $DF(x_n)$ is a matrix such that the entries of the vector $F(x_n)$ are on its diagonal and the other elements of the matrix are zero. These methods were introduced and studied in [2,4–6,13,17]. The motivation and efficiency of these methods were also discussed in these references. But the existence of fifth and ninth derivative, is needed for convergence, respectively restricting the applicability.

Define function ϕ on $D = [-\frac{1}{2}, \frac{3}{2}]$ as

$$\phi(t) = \begin{cases} 0, & t = 0 \\ t^3 \ln t^2 + t^5 - t^4, & t \neq 0 \end{cases}$$

so

$$\phi'(t) = 3t^2 \ln t^2 + 5t^4 - 4t^3 + 2t^2, \quad \phi'(1) = 3,$$
$$\phi''(t) = 6t \ln t^2 + 20t^3 - 12t^2 + 10t,$$
$$\phi'''(t) = 6 \ln t^2 + 60t^2 - 24t + 22,$$

so function ϕ''' is not bounded. There are numerous methods [1–19]. But these local results do not provide a computable radius of convergence.

We deal with these methods in Sect. 2. We use Computational Order of Convergence(COC) or Approximate Computational Order of Convergence (ACOC) that do not need high order derivatives (see Remark 1). We can do the same to other methods [1–19].

Next, the convergence is given in Sect. 2 with examples in Sect. 3.

2 Analysis

Consider $d \in [0,1)$, L_0, $L > 0$ and $K \geq 1$. Let functions p_0 and p on $[0, +\infty)$ as $p_0(t) = d + L_0 t$, $p(t) = d + 2L_0 t$ and parameters r_{p_0}, r_p and r_A by $r_{p_0} = \frac{1-d}{L_0}$, $r_p = \frac{1-d}{2L_0}$ and $r_A = \frac{2}{2L_0+L}$. Notice that $0 \leq r_p < r_{p_0} < \frac{1}{L_0}$ and $0 < r_A < \frac{1}{L_0}$. Then, consider g_1 and h_1 on $[0, r_{p_0})$ as

$$g_1(t) = \frac{1}{2(1 - L_0 t)}[Lt + \frac{2Kp(t)}{1 - p_0(t)}],$$

$$h_1(t) = g_1(t) - 1.$$

Suppose that

$$h_1(0) = g_1(0) - 1 = \frac{Kd}{1-d} - 1 < 0. \tag{2.5}$$

We have that $h_1(0) < 0$ and $h_1(t) \to \infty$ as $t \to r_{p_0}$. By the IVT, function h_1 has roots in $(0, r_{p_0})$. Call by r_1 the least such root. Then, consider g_2 and h_2 on $[0, r_1]$ as

$$g_2(t) = [1 + \frac{(4K + 3d)K}{2(1 - p_0(t))^2}]g_1(t)$$

and

$$h_2(t) = g_2(t) - 1.$$

Suppose that

$$h_2(0) = g_2(0) - 1 = \frac{Kd}{1-d}[1 + \frac{(4K+3d)K}{2(1-d)^2}] - 1 < 0. \tag{2.6}$$

Then, we obtain that $h_2(0) < 0$ and $h_2(r_1) = \frac{(4K+3d)K}{2[1-p_0(r_1)]^2} > 0$. Let r_2 be the smallest root of function h_2 on $(0, r_1)$. Koreover, consider functions g_3 and h_3 on $[0, r_2]$ as

$$g_3(t) = [1 + \frac{(4K+3d)K}{2(1-p_0(t))^2}]g_2(t)$$

and

$$h_2(t) = g_3(t) - 1.$$

Assume

$$h_3(0) = g_3(0) - 1 = \frac{Kd}{1-d}[1 + \frac{(4K+3d)K}{2(1-d)^2}]^2 - 1 < 0. \tag{2.7}$$

Then, we get that $h_3(0) < 0$ and $h_3(r_2) = \frac{(4K+3d)K}{2(1-p_0(r_2))^2} > 0$. Let r_3 be the smallest root of function h_3 on $(0, r_2)$. Furthermore, consider functions g_4 and h_4 on $[0, r_3]$ as

$$g_4(t) = [1 + \frac{(4K+3d)K}{2(1-p_0(t))^2}]g_3(t)$$

and

$$h_4(t) = g_4(t) - 1.$$

Assume

$$h_4(0) = g_4(0) - 1 = \frac{Kd}{1-d}[1 + \frac{(4K+3d)K}{2(1-d)^2}]^3 - 1 < 0. \tag{2.8}$$

Then, we get that $h_4(0) < 0$ and

$$h_4(r_3) = \frac{(4K+3d)K}{2(1-p_0(r_3))^2} > 0. \tag{2.9}$$

Let r_4 be the smallest root of function h_4 on $(0, r_3)$. Then,

$$0 \le g_1(t) < 1, \tag{2.10}$$

$$0 \le p_0(t) < 1, \tag{2.11}$$

$$0 < p(t), \tag{2.12}$$

$$0 \le g_2(t) < 1, \tag{2.13}$$

$$0 \le g_3(t) < 1 \tag{2.14}$$

and

$$0 \le g_4(t) < 1, \tag{2.15}$$

for each $t \in [0, r_4)$. Notice also that $(2.8) \Rightarrow (2.7) \Rightarrow (2.6) \Rightarrow (2.5)$. Define polynomials H_{i+1} on $[0,1]$, $i = 0, 1, 2, 3$ by

$$H_{i+1}(t) = Kt(2(1 - t)^2 + (4K + 3t)K)^i - 2^i(1 - t)^{i+1}. \tag{2.16}$$

We have that $H_{i+1}(0) = -2^i < 0$ and $H_{i+1}(1) = K^{i+1}(4K + 3)^i > 0$. Let d_{i+1} be the smallest root of polynomial H_{i+1} on $(0,1)$. Then, conditions (2.5)–(2.8) hold, since

$$\frac{Kd_1}{1 - d_1} < 1, \tag{2.17}$$

$$\frac{Kd_2}{1 - d_2}\left[1 + \frac{(4K + 3d_2)K}{(1 - d_2)^2}\right] < 1, \tag{2.18}$$

$$\frac{Kd_3}{1 - d_3}\left[1 + \frac{(4K + 3d_3)K}{(1 - d_3)^2}\right] < 1 \tag{2.19}$$

and

$$\frac{Kd_4}{1 - d_4}\left[1 + \frac{(4K + 3d_4)K}{(1 - d_4)^2}\right] < 1 \tag{2.20}$$

respectively. Consider conditions $(\mathcal{C}^4)$:
There exist a simple $x_* \in D$, $L_0, L > 0$, $K \geq 1$ with

$(\mathcal{C}_1)$ $F : D \subseteq R^m \to R^m$ is a differentiable mapping with $F(x_*) = 0$.
$(\mathcal{C}_2)$
$$\|F'(x_*)^{-1}(F'(x) - F'(x_*))\| \leq L_0\|x - x_*\|, \ x \in D. \tag{2.21}$$

Set $U_0 = D \cap U(x_*, \frac{1}{L_0})$.
$(\mathcal{C}_3)$
$$\|F'(x_*)^{-1}(F'(x) - F'(y))\| \leq L\|x - y\|, \ x, y \in U_0. \tag{2.22}$$

$(\mathcal{C}_4)$ $\|F'(x_*)^{-1}F'(x)\| \leq K$, $x \in U_0$.
$(\mathcal{C}_5)$ $\|F'(x_*)^{-1}DF(x)\| \leq d^*$, $x \in U_0$.
$(\mathcal{C}_6)$ $\bar{U}(x_*, r^*) \subseteq D$, where

$$d^* = \begin{cases} d_2 \ for \ method \ (1.2) \\ d_3 \ for \ method \ (1.3) \\ d_4 \ for \ method \ (1.4) \end{cases}$$

and

$$r^* = \begin{cases} r_2 \ for \ method \ (1.2) \\ r_3 \ for \ method \ (1.3) \\ r_4 \ for \ method \ (1.4). \end{cases}$$

and
$(\mathcal{C}_7)$ condition (2.12) holds.
$(\mathcal{C}_8)$ There exists $\bar{r} \geq r_4$ such that $L_0\bar{r} < z$. Let $U_1 = D \cap U(x^*, r)$.

Theorem 21. *Suppose conditions* $(\mathcal{C}^4)$ *hold. Then,* $\{x_n\} \subset U(x_*, r_4)$, $\lim\limits_{n \to \infty} \{x_n\} = x_*$, *and for* $x_0 \in U(x_*, r_4) - \{x_*\}$, $e_n = \|x_n - x_*\|$,

$$\|y_n - x_*\| \leq g_1(e_n)e_n \leq e_n < r_4, \tag{2.23}$$

$$\|z_n - x_*\| \leq g_2(e_n)e_n \leq e_n, \tag{2.24}$$

$$\|w_n - x_*\| \leq g_3(e_n)e_n \leq e_n \tag{2.25}$$

and

$$\|x_{n+1} - x_*\| \leq g_4(e_n)e_n \leq e_n, \tag{2.26}$$

where "g" functions and the convergence radius are defined previously.

Moreover, x_* *is unique in* U_1 *as a solution of equation* $F(x) = 0$

Proof. In view of $x_0 \in U(x_*, r_4) - \{x_*\}$, condition (c_2)

$$\|F'(x_*)^{-1}(F'(x_0) - F'(x_*))\| \leq L_0 e_0 < L_0 r_4 < 1 \tag{2.27}$$

Using (2.27) and the Banach Perturbation Lemma for mappings [2,8,10,16], $F'(x_0)$ is invertible, with

$$\|F'(x_0)^{-1}F'(x_*)\| \leq \frac{1}{1 - L_0 e_0} < \frac{1}{1 - L_0 r_4}. \tag{2.28}$$

We can write by condition $(\mathcal{C}_1)$ that

$$F(x_0) = F(x_0) - F(x_*) = \int_0^1 F'(x_* + \theta(x_0 - x_*))(x_0 - x_*)d\theta. \tag{2.29}$$

we also have $\|x_* + \theta(x_0 - x_*) - x_*\| = \theta e_0 < r_4$. So, $x_* + \theta(x_0 - x_*) \in U(x_*, r_4)$. Then using condition $(\mathcal{C}_4)$ and (2.29) we get that

$$\|F'(x_*)^{-1}F(x_0)\| = \|\int_0^1 F'(x_*)^{-1}(F'(x_* + \theta(x_0 - x_*))(x_0 - x_*)d\theta\|$$
$$\leq K e_0 \tag{2.30}$$

Next, we show that $A_0^{-1} \in L(\mathbb{R}^m, \mathbb{R}^m)$. Using conditions $(\mathcal{C}_2)$, $(\mathcal{C}_3)$, $(\mathcal{C}_5)$ we get in turn that

$$\|(-F'(x_*))^{-1}(A_0 - F'(x_*))\| = \|(-F'(x_*))^{-1}(DF(x_0) - F'(x_0) + F'(x_*))\|$$
$$\leq \|F'(x_*)^{-1}DF(x_0)\| + \|F'(x_*)^{-1}(F'(x_0) - F'(x_*))\|$$
$$\leq d_4 + L_0\|x_0 - x^*\|$$
$$= p_0(e_0) < 1. \tag{2.31}$$

It follows from (2.31) that $A_0^{-1} \in L(\mathbb{R}^m, \mathbb{R}^m)$ and

$$\|A_0^{-1}F'(x_*)\| \leq \frac{1}{1 - p_0(e_0)}. \tag{2.32}$$

Hence y_0, z_0, w_0 and x_1 exist by method (1.4) for $n = 0$. We also have by (2.31) the estimate,

$$
\begin{aligned}
\|F'(x_*)^{-1}(A_0 - F'(x_0))\| &\leq \|F'(x_*)^{-1}(A_0 - F'(x_*))\| \\
&\quad + \|F'(x_*)^{-1}(F'(x_0) - F'(x_*))\| \\
&\leq p_0(e_0) + L_0\|x_0 - x^*\| \\
&= p(e_0).
\end{aligned}
\tag{2.33}
$$

By (1.4) for $n = 0$, (2.9)–(2.12), conditions $(\mathcal{C}_1)$–$(\mathcal{C}_4)$ (2.28) and (2.30):

$$
\begin{aligned}
\|y_0 - x_*\| &= \|x_0 - x_* - F'(x_0)^{-1}F(x_0)\| \\
&\quad + \|(F'(x_0)^{-1} - A_0^{-1})F(x_0)\| \\
&\leq \|F'(x_0)^{-1}F'(x_*)\|\left\|\int_0^1 F'(x_*)^{-1}(F'(x_* + \theta(x_0 - x_*)) - F'(x_0))(x_0 - x_*)d\theta\right\| \\
&\quad + \|F'(x_0)^{-1}F'(x_*)\|\|F'(x_*)^{-1}(A_0 - F'(x_0))\|\|F'(x_*)^{-1}F(x_0)\| \\
&\leq \frac{Le_0^2}{2(1 - L_0e_0)} \\
&\quad + \frac{Kp(e_0)e_0}{(1 - L_0e_0)(1 - p_0(e_0))} \\
&= g_1(e_0)e_0 < e_0 < r_4,
\end{aligned}
\tag{2.34}
$$

so (2.23) holds for $n = 0$ and $y_0 \in U(x_*, r_4)$. From (2.30) (for $y_0 = x_0$), we have

$$
\|F'(x_*)^{-1}F(y_0)\| \leq K\|y_0 - x_*\|.
\tag{2.35}
$$

Next, we need an estimate on $\|B_0\|$. Using the definition of B_0 we can write in turn that

$$
\begin{aligned}
B_0 &= -I + \frac{1}{2}A_0^{-1}(F'(y_0) + F'(x_0) - DF(x_0)) \\
&= A_0^{-1}[F'(x_0) - DF(x_0) + 1/2(F'(y_0) + F'(x_0) - DF(x_0))] \\
&= \frac{1}{2}A_0^{-1}[3(F'(x_0) - DF(x_0)) + F'(y_0)].
\end{aligned}
\tag{2.36}
$$

Using conditions $(\mathcal{C}_4)$, $(\mathcal{C}_5)$, (2.32) and (2.36) we have that

$$
\begin{aligned}
\|B_0\| &\leq \frac{1}{2}\|A_0^{-1}F'(x_*)\|[3\|F'(x_*)^{-1}F'(x_0)\| \\
&\quad + \|F'(x_*)^{-1}F'(y_0)\| + 3\|F'(x_*)^{-1}DF(x_0)\|] \\
&\leq \frac{4K + 3d_4}{2(1 - p_0(e_0))}.
\end{aligned}
\tag{2.37}
$$

In view of (1.4) for $n = 0$, (2.9), (2.13) and (2.35)–(2.37), we have that

$$
\begin{aligned}
\|z_0 - x_*\| &\leq \|y_0 - x_*\| + \|B_0\|\|A_0^{-1}F'(x_*)\|\|F'(x_*)^{-1}F(y_0)\| \\
&\leq \|y_0 - x_*\| + \frac{K(4K + 3d)}{2(1 - p_0(e_0))^2}\|y_0 - x_*\| \\
&\leq [1 + \frac{K(4K + 3d)}{2(1 - p_0(e_0))^2}]g_1(e_0)e_0 \\
&= g_2(e_0)e_0 < e_0 < r_4,
\end{aligned}
\tag{2.38}
$$

showing $z_0 \in U(x_*, r_4)$ if $n = 0$. Then, we get by (2.30) (for $z_0 = x_0$) that

$$\|F'(x_*)^{-1}F(z_0)\| \leq K\|z_0 - x_*\|. \tag{2.39}$$

Moreover, by (2.9), (2.14), (2.38) and (2.39), we have that

$$\|w_0 - x_*\| \leq \|z_0 - x_*\| + \frac{K(4K + 3d)\|z_0 - x_*\|}{2(1 - p_0(e_0))^2}$$

$$\leq [1 + \frac{K(4K + 3d)}{2(1 - p_0(e_0))^2}]g_2(e_0)e_0$$

$$= g_3(e_0)e_0 < e_0 < r_4, \tag{2.40}$$

which shows (2.25), for $n = 0$ and $w_0 \in U(x_*, r_4)$. Then, we have by (2.30) (for $w_0 = x_0$) that

$$\|F'(x_*)^{-1}F(w_0)\| \leq K\|w_0 - x_*\|. \tag{2.41}$$

In view of (2.9), (2.15), (2.37), (2.39) and (2.40), we get that

$$\|x_1 - x_*\| \leq \|w_0 - x_*\| + \frac{K(4K + 3d)\|w_0 - x_*\|}{2(1 - p_0(e_0))^2}$$

$$\leq [1 + \frac{K(4K + 3d)}{2(1 - p_0(e_0))^2}]g_3(e_0)e_0$$

$$= g_4(e_0)e_0 < e_0 < r_4, \tag{2.42}$$

so (2.26) holds, for $n = 0$ and $x_1 \in U(x_*, r_4)$. Substitute x_0, y_0, z_0, w_0, x_1 by x_k, y_k, z_k, w_k, x_{k+1}, respectively above we terminate the induction By $\|x_{k+1} - x_*\| < \|x_k - x_*\| < r_4$, we deduce that $x_{k+1} \in U(x_*, r_4)$ and $\lim_{k \to \infty} x_k = x_*$. Consider, $Q = \int_0^1 F'(æa + \theta(x_* - æa)d\theta$ for some $æa \in U_1$ with $F(æa) = 0$. By (2.10), $(\mathcal{C}_8)$ and we obtain that

$$\|F'(x_*)^{-1}(Q - F'(x_*))\| \leq \int_0^1 L_0\|æa + \theta(x_* - æa) - x_*\|d\theta$$

$$\leq L_0 \int_0^1 (1 - \theta)\|x_* - æa\|d\theta \leq \frac{L_0}{2}T < 1, \tag{2.43}$$

and we get $x_* = æa$ from Q is invertible, $0 = F(x_*) - F(æa) = Q(x_* - æa)$.

Next, we present the corresponding results for method (1.3) and method (1.2), respectively are given by specializing Theorem 21.

Corollory 22. *Assume conditions* $(\mathcal{C}_3)$ *hold. Then,* $\{x_n\} \subset U(x^*, r_3)$, $\lim_{n \to \infty} \{x_n\} = x^*$, *and for* $x_0 \in U(x_*, r_3) - \{x_*\}$,

$$\|y_n - x_*\| \leq g_1(e_n)e_n < e_n < r_3,$$

$$\|z_n - x_*\| \leq g_2(e_n)e_n < e_n,$$

$$\|x_{n+1} - x_*\| \leq g_3(e_n)e_n,$$

Moreover, $x_* \in U_1$ *is unique as a solution of* $F(x) = 0$.

Corollory 23. *Assume conditions* (C_2) *hold. Then,* $\{x_n\} \subset U(x_*, r_2)$, $\lim\limits_{n \to \infty} \{x_n\} = x_*$, *and for* $x_0 \in U(x_*, r_2) - \{x_*\}$,

$$\|y_n - x_*\| \le g_1(e_n)e_n < e_n < r_2,$$

$$\|x_{n+1} - x_*\| \le g_2(e_n)e_n < e_n,$$

Moreover, the limit point x_* *is the only solution of equation* $F(x) = 0$ *in* U_1.

Remark 24. *1. By* (C_2) *and*

$$\|F'(x_*)^{-1}F'(x)\| = \|F'(x_*)^{-1}(F'(x) - F'(x_*)) + I\|$$
$$\le 1 + \|F'(x_*)^{-1}(F'(x) - F'(x_*))\| \le 1 + L_0\|x - x_*\|,$$

(C_4) *is not needed if* K *is chosen for each* $t \in [0, \frac{1}{L_0})$, *as*

$$K(t) = 1 + L_0 t$$

or as

$$K = K(t) = 2.$$

2. Not using higher than one derivatives, COC and ACOC are given by

$$\mu = \ln\left(\frac{\|x_{n+1} - x_*\|}{e_n}\right) / \ln\left(\frac{\|x_n - x_*\|}{\|x_{n-1} - x_*\|}\right)$$

or

$$\mu_1 = \ln\left(\frac{\|x_{n+1} - x_n\|}{\|x_n - x_{n-1}\|}\right) / \ln\left(\frac{\|x_n - x_{n-1}\|}{\|x_{n-1} - x_{n-2}\|}\right),$$

respectively. This way we obtain in practice the order of convergence in a way that avoids the bounds involving estimates using estimates higher than the first Fréchet derivative of operator F *(Table 1).*

3   Numerical Illustrations

The results are tested using three examples

Example 31. *Considering again example of the introduction one obtains,* $L = L_0 = 146.6629073$, *and* $K = 2$. *So, we obtain* $d = 0$, $r_{p_0} = 0.0068$, $r_p = 0.0034$, $d_1 = 0.3333$, $d_2 = 0.0151$, $d_3 = 0.0061$, $d_4 = 0.0014$.
Notice that the conditions in [5, 6] are not satisfied.

Table 1. Radius of convergence of Example 31

	r_1	r_2	r_3	r_4	r_A
$L = L_0 = 146.6629073, K = 2.$	0.0011	0.0002	0.0003	0.0003	0.0045

Example 32. *Set $Y = X = \mathbb{R}^3, x_* = (0,0,0)^T$. Introduce F on D by*

$$F(u) = (e^{v_1} - 1, \frac{e-1}{2}v_2^2 + v_2, v_3)^T,$$

for $u = (v_1, v_2, v_3)^T$. Then,

$$F'(u) = \begin{bmatrix} e^{v_1} & 0 & 0 \\ 0 & (e-1)v_2 + 1 & 0 \\ 0 & 0 & 1, \end{bmatrix}.$$

So, $L_0 = e - 1, K = L = e^{\frac{1}{L_0}}$, if $D = \bar{U}(0,1)$ leading to $d = 0$, $r_{p_0} = 0.5820$, $r_p = 0.2910$, $d_1 = 0.3585$, $d_2 = 0.0210$, $d_3 = 0.0099$, $d_4 = 0.0026$.

The parameters with $L_0 = L = K = e$ are $d = 0$, $r_{p_0} = 0.3679$, $r_p = 0.1839$, $d_1 = 0.2689$, $d_2 = 0.0061$, $d_3 = 0.0015$, $d_4 = 0.0002$. Hence, by Table 2, $r_R = 0.3725 \leq r_A$.

Table 2. Radius of convergence of Example 32

	r_1	r_2	r_3	r_4	r_A
$L_0 = e - 1, K = L = e^{\frac{1}{L_0}}$,	0.0997	0.0172	0.0297	0.0299	0.3827
$L_0 = L = K = e$	0.0475	0.0038	0.0071	0.0071	0.2453

Example 33. *Let $X = Y = C[0,1]$, $F''(x) = B(x)$, where*

$$F(\psi)(x) = \psi(x) - 5 \int_0^1 x\tau\psi(\tau)^3 d\tau. \qquad (3.44)$$

hence, we obtain

$$F'(\psi(\lambda))(x) = \lambda(x) - 15 \int_0^1 x\tau\psi(\theta)^2\lambda(\tau)d\tau, \quad \text{for each } \lambda \in D.$$

Then, we have for $x_* = 0$, $D = \bar{U}(0,1)$ that $L_0 = 15$, $L = 30$, so $d = 0$, $r_{p_0} = 0.0667$, $r_p = 0.0333$, $d_1 = 0.3333$, $d_2 = 0.0151$, $d_3 = 0.0061$, $d_4 = 0.0014$ and with $L_0 = L = 30, K = 2$ are $d = 0$, $r_{p_0} = 0.0333$, $r_p = 0.0167$, $d_1 = 0.3333$, $d_2 = 0.0151$, $d_3 = 0.0061$, $d_4 = 0.0014$. Notice that in Table 3 the radius of convergence is larger under our approach. Moreover, $r_R = 0.0222 \leq r_A$.

Table 3. Radius of convergence of Example 33

	r_1	r_2	r_3	r_4	r_A
$L_0 = 15, L = 30,$	0.0099	0.0014	0.0025	0.0025	0.0333
$L_0 = L = 30, K = 2$	0.0053	0.0008	0.0013	0.0013	0.0222

References

1. Argyros, I.K.: Computational Theory of Iterative Solvers Series: Studies in Computational Mathematics. In: Chui, C.K., Wuytack, L., et al. (eds.) Elsevier Publ. Co., New York, U.S.A (2007)
2. Argyros, I.K., Magréñan, A.A.: A Contemporary Study of Iterative Methods, Elsevier. Academic Press), New York (2018)
3. Argyros, I.K., Magréñan, A.A.: Iterative Methods and their Dynamics with Applications. CRC Press, New York, USA (2017)
4. Argyros, I.K., George, S.: Mathematical Modeling for the Solution of Equations and Systems of Equations with Applications. Nova Publishers, New York, Volume-III (2019)
5. Cordero, A., Torregrosa, J.R., Vassileva, M.P.: Increasing the order of convergence of iterative scheme for solving non-linear systems. J. Comput. Appl. Math. **2012**,(2012). https://doi.org/10.1016/j.cam.2012.11.024
6. Cordero, A., Hueso, J.L., Martinez, E., Torregrosa, J.: A modified Newton-Jarratta's composition. Numer. Algor. **55**, 87–99 (2010)
7. Ezquerro, J.A., Hernández, M.A.: On the R-order of the Halley method. J. Math. Anal. Appl. **303**, 591–601 (2005)
8. Kantorovich, L.V., Akilov, G.P.: Functional Analysis. Pergamon Press, Oxford (1982)
9. Parhi, S.K., Gupta, D.K.: Recurrence relations for a Newton-like method in Banach spaces. J. Comput. Appl. Math. **206**(2), 873–887 (2007)
10. Potra, F.A., Ptak, V.: Nondiscrete induction and iterative processes. Pitman Advanced Publishing Program **103** (1984)
11. Ren, H., Wu, Q., Bi, W.: New variants of Jarratt method with sixth-order convergence. Numer. Algor. **52**(4), 585–603 (2009)
12. Rheinboldt, W.C.: An adaptive continuation process for solving systems of nonlinear equations, Mathematical models and numerical methods. In: Tikhonov, A.N., et al. (eds.) 3(19), Banach Center, Warsaw Poland, pp. 129–142 (1978)
13. Rostamy, D., Bakhtiari, P.: New efficient multipoint iterative method for solving nonlinear systems. Appl. Math. Comput. **266**, 350–356 (2015)
14. Sharma, J.R., Arora, H.: Efficient Jarratt-like methods for solving systems of nonlinear equations. Calcolo **51**(1), 193–210 (2013). https://doi.org/10.1007/s10092-013-0097-1
15. Sharma, J.R., Guha, R.K., Sharma, R.: An efficient fourth order weighted-Newton method for systems of nonlinear equations. Numer. Algor. **62**, 307–323 (2013)
16. Sharma, J.R., Arora, H.: Improved Newton-like solvers for solving systems of nonlinear equations. SEMA J. **74**(2), 147–163 (2017)
17. Soleymani, F., Lotfi, T., Bakhtiari, P.: A multi-step class of iterative methods for nonlinear systems. Optimization Letters **8**(3), 1001–1015 (2013). https://doi.org/10.1007/s11590-013-0617-6
18. Traub, J.F.: Iterative Methods for the Solution of equations. Prentice hall, New York (1964)
19. Wang, X., Kou, J.: Semilocal convergence of a modified multi-point Jarratt method in Banach spaces under general continuity conditions. Numer. Algor. **60**, 369–390 (2012)

Author Index

Printed in the United States
by Baker & Taylor Publisher Services

Printed in the United States
by Baker & Taylor Publisher Services